The Philosophy of Science

A Collection of Essays

Series Editor

Lawrence Sklar
University of Michigan

A GARLAND SERIES
READINGS IN PHILOSOPHY

ROBERT NOZICK, *ADVISOR*
HARVARD UNIVERSITY

Series Contents

1. Explanation, Law and Cause
2. The Nature of Scientific Theory
3. Theory Reduction and Theory Change
4. Probability and Confirmation
5. Bayesian and Non-Inductive Methods
6. The Philosophy of Physics

Theory Reduction and Theory Change

Edited with an introduction by

Lawrence Sklar

University of Michigan

Routledge
Taylor & Francis Group

NEW YORK AND LONDON

First published by Garland Publishing, Inc

This edition published 2013 by Routledge

711 Third Avenue   2 Park Square, Milton Park
New York   Abingdon
NY 10017   Oxon, OX14 4RN

Library of Congress Cataloging-in-Publication Data

Philosophy of science : a collection of essays / edited with an
   introduction by Lawrence Sklar.
        p.   cm.
        "A Garland series"—Ser. t.p.
        Includes bibliographical references.
        Contents: 1. Explanation, law, and cause — 2. The nature of
scientific theory — 3. Theory reduction and theory change —
4. Probability and confirmation — 5. Bayesian and non-inductive
methods — 6. The philosophy of physics.
        ISBN 0-8153-2700-5 (v.1 : alk. paper) — ISBN 0-8153-2701-3
(v. 2 : alk. paper) — ISBN 0-8153-2702-1 (v. 3 : alk. paper) —
ISBN 0-8153-2703-X (v. 4 : alk. paper) — ISBN 0-8153-3492-3
(v. 5 : alk. paper) — ISBN 0-8153-3493-1 (v. 6 : alk. paper)
        1. Science—Philosophy. 2. Science—Methodology. 3. Physics—
Philosophy. I. Sklar, Lawrence.

Q175.P51227 1999
        501—dc21   99-40012
                                                            CIP

SET ISBN   9780815326991
POD ISBN   9780415870719
    Vol1   9780815327004
    Vol2   9780815327011
    Vol3   9780815327028
    Vol4   9780815327035
    Vol5   9780815334927
    Vol6   9780815334934

Contents

vii   Introduction

A. Theory Reduction

1   Unity of Science as a Working Hypothesis
Paul Oppenheim and Hilary Putnam

35   Types of Inter-Theoretic Reduction
Lawrence Sklar

51   Special Sciences
(Or: The Disunity of Science as a Working Hypothesis)
J.A. Fodor

B. Theory Change

70   Explanation, Reduction, and Empiricism
P.K. Feyerabend

140   The Nature and Necessity of Scientific Revolutions
Thomas S. Kuhn

159   Revolutions as Changes of World View
Thomas S. Kuhn

185   The Road Since Structure
Thomas S. Kuhn

197   Explanation and Reference
Hilary Putnam

221   A Confutation of Convergent Realism
Larry Laudan

253   Realism, Approximate Truth, and Philosophical Method
Richard Boyd

290   Varieties of Progress
Philip Kitcher

327   Acknowledgments

Introduction

Science is not static. Hypotheses and theories are constructed, and their suitability for describing and explaining the world is tested by means of observation and experiment. But hypotheses and theories rarely, if ever, occupy a permanent and unchanging place in our scheme of accepted best scientific conclusions.

Usually hypotheses and theories that have been accepted for shorter, or longer, periods are later found wanting in the light of more refined experimental evidence or in the light of new theoretical insights. In such cases these previously accepted bodies of scientific thought lose their preferred status. They become "once accepted" instead of "now accepted" hypotheses. But even if a theory has been displaced in scientific esteem by some successor theory, that need not mean that it will be considered entirely discarded. Often it is claimed that the older theory, although now replaced by some newer account of the world, still has a place to play in science. Sometimes it is said that the older theory "reduces" to the newer account, even when accepting the newer account entails rejecting aspects of the older theory.

In other cases the replacement of older by newer theory is thought of as entailing something other than a rejection of the older theory. Rather, it is often claimed, the older theory remains an accepted component of scientific belief. It is just that, so it is said, the older theory now survives as a less fundamental, or, perhaps, more specialized, theory, a theory whose status is now better understood in the light of the newer theory that offers a "deeper," more fundamental, or, perhaps, more general account of the world. Sometimes it is said that in the light of the newer theory the older requires some modifications, but still retains its "essential correctness."

But in all of these cases there may still be claims to the effect that one ought to speak of the older theory as "reducing" to the newer theory.

In some cases scientists are inclined to think not just of individual theories, but of whole scientific disciplines, as reducing, the one to the other. It is often suggested, for example, that eventually biology may be reduced to chemistry and chemistry to physics.

In each such claim of one theory or, one discipline, reducing to another, the methodological philosopher of science would like to know just exactly what the reductive relationship consists in Is there one sense of one theory reducing to another, or are there many, quite distinct, relationships that theories can bear to one another that have been called, at one time or another, inter-theoretical reductions? If there are many senses of reduction, how are they related to one another?

Many accounts of theory reduction demand the existence of some important connection between the concepts of the reduced theory and the concepts of the reducing theory. If the newer theory is to somehow absorb or encompass the older, and if the older theory is not simply framed in a fragment of the newer theory's concepts, some connection must be drawn between how the older theory characterized the world and how the world is conceptually framed in the newer account.

Sometimes it is argued that there must be "bridging laws" connecting concepts of the older and the newer theory. Others have argued that true reductions require the specification of generalized identifications of the entities described by the older theory with the members of some class of entities described by the newer theory. An example is the identifying of light waves with a species of electromagnetic weave in the reduction of physical optics to general electromagnetic theory.

It is also important to notice that in many cases the details of the reductive relationship may be highly dependent on the specific nature of the theories involved in the reductions. In these cases many deep special scientific problems may arise in trying to say just what some specific reduction really amounts to.

Consider, for example, the alleged reduction of pre-relativistic theories of space and time to the later relativistic accounts. Subtle limiting notions must be invoked to explain how the concepts of the older theory "fit in" to an account of the world framed in the newer theory's terms. An even more problematic case is the alleged reduction of pre-quantum physics to the quantum description of the world. Here the project of saying just how the theories are to be conceptually related to one another turns out to be a major task for the theoretical science itself.

Whereas some methodologists emphasize the role played in science by the reductive relationship among theories, other philosophers have pointed to the ways predecessor and successor theories may be such that we cannot speak of the reduction of the former to the latter. Here attention is being focused on those cases where reduction has not taken place, but not because the earlier theory has been fully discarded by the later science as being simply wrong-headed in its picture of the world. Rather, it is the fact that the ways in which the two theories deal conceptually with the world may so differ that any simple "derivation" of the one theory from the other, even a derivation assisted by some general bridging principles, is impossible.

Yet it may still be possible to maintain, despite the existence of this conceptual "misfit" between the older and the newer theory, that it is because the world is as described by the deeper or more general theory that the world described by the less "foundational" or more restricted theory is (more or less) correctly describable by that more restrictive theory. Sometimes it is said that even with the inability to find a natural connection between the concepts of the older theory and those of the newer theory, we ought to think of all the features of the world described by the older theory as having their causal explanation in the newer theory's terms.

For a simple example, consider the fact that we are very unlikely to find any concept in the physicist's description of the world in fundamental terms that captures the everyday concept of something being a table or being a chair. But it still seems right to say that the entire behavior of those homely objects of furniture is completely determined by the fact that they are made up of the fundamental constituents of the world (molecules, atoms, subatomic particles) described by the physicist's theory, and that the behavior in question "follows from," in a causal sense, the lawlike behavior of the basic constituents of matter described in the physicist's laws.

The irreducibility of a theory of furniture to fundamental physics would be, of course, of little interest. But it is much more important to consider the claims that such disciplines as sociology, psychology, and even biology might be "irreducible" to physics, even though the objects of their study have their behavior "supervening" entirely on the nature of the world described by the physicist.

The issues of the nature (or natures) of reduction between theories and between disciplines, and the accounts offered of inter-theoretical relations that may be thought of as non-reductive, even though the one theory can be said to describe the facts that ground the truth of the other, are the subject matter of Section A of this volume.

In the history of science there are those infrequent episodes known as scientific revolutions. The change from the earth-entered astronomy of Ptolemy to the sun-centered theory of Copernicus, the transition in dynamics first from Aristotelian theory to Cartesian, and shortly afterwards from Cartesian to the great synthesis of the theory of motion and gravity of Newton, and the great theoretical revisions of the twentieth century from Newton's account of space, time, and motion to the radically novel theories of relativistic spacetime and of quantum dynamics, are frequently cited examples of scientific revolutions.

It has been claimed more than once that such radical scientific transitions require a special account on their own. They are, it is alleged, episodes quite unlike the usual modest and continuous development of scientific theory. It is sometimes claimed that these changes in scientific belief systems are so radical that dropping the older theory and accepting its replacement cannot be analyzed in the usual terms of a rational scientific response to new experimental data and modified theoretical conceptions. Instead, it is sometimes claimed, such revolutions in thought are something that are "arational" in nature, more like a religious conversion than the usual change of belief under slow empirical pressure.

Furthermore, it is sometimes argued that the older and newer accounts of the world are so radically distinct in the way that they conceptualize the world, that it is misleading to even speak of the same empirical evidence being brought to bear on the older and newer theories. Even our conception of the evidence changes, it is sometimes claimed, in a scientific revolution. Many arguments to this effect rest upon the idea that the meaning of the theoretical concepts is fixed by the total theory in which they occur. If two theories, then, are quite radically different from one another, it is argued that even those words that appear in common in the context of the two theories, even those words in the observational vocabulary, must differ in their meaning in the two theoretical contexts.

In addition to these claims, the way in which our views of what the universe is like in its most fundamental nature changes so radically and discontinuously during scientific revolutions, some have said, that the existence of these revolutions suggests that a model of science that views theories as smoothly converging to some ultimate theory that we can call "the truth" about the world, with each newer theory assimilating the insights of its predecessor but adding to these in a cumulative manner, must be discarded. Instead we must see our world-view as continuously subject to radical shock, with no evidence that science is smoothly "progressive" or cumulative in the past, and no hope for such historical continuity in the future.

Some responses to these views about scientific revolutions and their consequences for the methodology of science have argued that there is more continuity than the proponents of the extreme radical nature of scientific change during revolutionary periods have allowed. We can, it is argued, deal systematically with how our views about elements of the world change during such scientific transitions, rather than having to say that the proponents of the older and the newer theory "live in different worlds altogether."

Those focussing on the radical nature of scientific revolutions have sometimes espoused an "irrealism" about theories quite different from the kinds of irrealism discussed in the works in Volume 2 of this series. If we can expect such radical changes in what we take to be the fundamental nature of the world in the future as we have seen occur in science in the past, how can we take seriously our current theory's postulations about what there is in the world as seriously describing "what there really is" at all?

In response to such claims some methodologists have argued that even with the past history of radical change of view about how the world is made up, science does allow the methodologist to make reasonable claims to the effects that our current best available scientific pictures of how the world is constructed do, indeed, give us insights that legitimize our making claims to the effect that these theories tell us "how things really are." One may try to argue on scientific, historical, and philosophical grounds that none of the arguments given by those who emphasize the revolutionary aspects of some scientific changes prevent us from having an overall view of fundamental science as both cumulative and progressive.

Other philosophers of science have argued that the insights of those who have emphasized the radical changes in science experienced in scientific revolutions must be taken deeper to heart. They argue that we must, to be sure, give up once and for all the idea of science as progressive and cumulative, and must forgo the old idea of science as even searching for some final "truth" about the world. But, these methodologists argue, that does not mean, as some have suggested, that the decisions scientists make when they accept and reject theories, even fundamental theories and even when the theories are radically at variance with one another, must be matters of "faith" or "commitment." Instead, it is suggested, we need new models of rationality to make it clear how scientists can be construed as acting rationally in their decisions to opt for alternative fundamental scientific theories, despite the need to give up so many of the traditional ideas about rationality and truth we previously had built into our accounts of scientific decision making.

These are the matters discussed in the works in Section B.

Unity of Science as a Working Hypothesis

1. Introduction

1.1. The expression "Unity of Science" is often encountered, but its precise content is difficult to specify in a satisfactory manner. It is the aim of this paper to formulate a precise concept of Unity of Science; and to examine to what extent that unity can be attained.

A concern with Unity of Science hardly needs justification. We are guided especially by the conviction that Science of Science, i.e., the meta-scientific study of major aspects of science, is the natural means for counterbalancing specialization by promoting the integration of scientific knowledge. The desirability of this goal is widely recognized; for example, many universities have programs with this end in view; but it is often pursued by means different from the one just mentioned, and the conception of the Unity of Science might be especially suited as an organizing principle for an enterprise of this kind.

1.2. As a preliminary, we will distinguish, in order of increasing strength, three broad concepts of Unity of Science:

First, Unity of Science in the weakest sense is attained to the extent to which all the terms of science [1] are reduced to the terms of some one discipline (e.g., physics, or psychology). This concept of *Unity of Language* (12) may be replaced by a number of sub-concepts depending on the manner in which one specifies the notion of "reduction" involved. Certain authors, for example, construe reduction as the definition of the terms of science by means of those in the selected basic discipline (reduction by means of biconditionals (47)); and some of these require the definitions in question to be analytic, or "true in virtue of the meanings of the terms involved" (epistemological reduction);

AUTHORS' NOTE: We wish to express our thanks to C. G. Hempel for constructive criticism. The responsibility for any shortcomings is, however, exclusively ours.

1

others impose no such restriction upon the biconditionals effecting reduction. The notion of reduction we shall employ is a wider one, and is designed to include reduction by means of biconditionals as a special case.

Second, Unity of Science in a stronger sense (because it implies Unity of Language, whereas the reverse is not the case) is represented by *Unity of Laws* (12). It is attained to the extent to which the laws of science become reduced to the laws of some one discipline. If the ideal of such an all-comprehensive explanatory system were realized, one could call it *Unitary Science* (18, 19, 20, 80). The exact meaning of 'Unity of Laws' depends, again, on the concept of "reduction" employed.

Third, Unity of Science in the strongest sense is realized if the laws of science are not only reduced to the laws of some one discipline, but the laws of that discipline are in some intuitive sense "unified" or "connected." It is difficult to see how this last requirement can be made precise; and it will not be imposed here. Nevertheless, trivial realizations of "Unity of Science" will be excluded, for example, the simple conjunction of several branches of science does not *reduce* the particular branches in the sense we shall specify.

1.3. In the present paper, the term 'Unity of Science' will be used in two senses, to refer, first, to an ideal state of science, and, second, to a pervasive trend within science, seeking the attainment of that ideal.

In the first sense, 'Unity of Science' means the state of unitary science. It involves the two constituents mentioned above: unity of vocabulary, or "Unity of Language"; and unity of explanatory principles, or "Unity of Laws." That Unity of Science, in this sense, can be fully realized constitutes an over-arching meta-scientific hypothesis which enables one to see a unity in scientific activities that might otherwise appear disconnected or unrelated, and which encourages the construction of a unified body of knowledge.

In the second sense, Unity of Science exists as a trend within scientific inquiry, whether or not unitary science is ever attained, and notwithstanding the simultaneous existence, (and, of course, legitimacy) of other, even *incompatible*, trends.

1.4. The expression 'Unity of Science' is employed in various other senses, of which two will be briefly mentioned in order to distinguish them from the sense with which we are concerned. In the first place, what is sometimes referred to is something that we may call the *Unity of Method* in science. This might be represented by the thesis that all the empirical sciences employ the same standards of explanation, of significance, of evidence, etc.

In the second place, a radical reductionist thesis (of an alleged "logical," not an empirical kind) is sometimes referred to as the thesis of the Unity of Science. Sometimes the "reduction" asserted is the definability of all the terms of science in terms of *sensationalistic predicates* (10); sometimes the notion of "reduction" is wider (11) and predicates referring to *observable qualities of physical things* are taken as basic (12). These theses are epistemological ones, and ones which today appear doubtful. The epistemological uses of the terms 'reduction', 'physicalism', 'Unity of Science', etc., should be carefully distinguished from the use of these terms in the present paper.

2. Unity of Science and Micro-Reduction

2.1. In this paper we shall employ a concept of reduction introduced by Kemeny and Oppenheim in their paper on the subject (47), to which the reader is referred for a more detailed exposition. The principal requirements may be summarized as follows: given two theories T_1 and T_2, T_2 is said to be *reduced* to T_1 if and only if:

(1) The vocabulary of T_2 contains terms not in the vocabulary of T_1.

(2) Any observational data explainable by T_2 are explainable by T_1.

(3) T_1 is at least as well systematized as T_2. (T_1 is normally more complicated than T_2; but this is allowable, because the reducing theory normally explains more than the reduced theory. However, the "ratio," so to speak, of simplicity to explanatory power should be at least as great in the case of the reducing theory as in the case of the reduced theory.)[2]

Kemeny and Oppenheim also define the reduction of a branch of science B_2 by another branch B_1 (e.g., the reduction of chemistry to physics). Their procedure is as follows: take the accepted theories of B_2 at a given time t as T_2. Then B_2 is reduced to B_1 at time t if and only if there is some theory T_1 in B_1 at t such that T_1 reduces T_2 (47). Analogously, if some of the theories of B_2 are reduced by some T_1 belonging to branch B_1 at t, we shall speak of a *partial reduction* of B_2 to B_1 at t. This approach presupposes (1) the familiar assumption that some division of the total vocabulary of both branches into theoretical and observational terms is given, and (2) that the two branches have the same observational vocabulary.

2.2. The essential feature of a micro-reduction is that the branch B_1 deals with the parts of the objects dealt with by B_2. We must suppose that corresponding to each branch we have a specific universe of discourse U_{Bi};[3] and that we have a part-whole relation, Pt (75; 76, especially p. 91). Under the following conditions we shall say that the reduction of B_2 to B_1[4] is a micro-reduction: B_2 is reduced to B_1; and the objects in the universe of discourse of B_2 are wholes which possess a decomposition (75; 76, especially p. 91) into proper parts all of which belong to the universe of discourse of B_1. For example, let us suppose B_2 is a branch of science which has multicellular living things as its universe of discourse. Let B_1 be a branch with cells as its universe of discourse. Then the things in the universe of discourse of B_2 can be decomposed into proper parts belonging to the universe of discourse of B_1. If, in addition, it is the case that B_1 reduces B_2 at the time t, we shall say that *B_1 micro-reduces B_2* at time t.

We shall also say that a branch B_1 is a *potential* micro-reducer of a branch B_2 if the objects in the universe of discourse of B_2 are wholes which possess a decomposition into proper parts all of which belong to the universe of discourse of B_1. The definition is the same as the definition of 'micro-reduces' except for the omission of the clause 'B_2 is reduced to B_1.'

Any micro-reduction constitutes a step in the direction of *Unity of Language* in science. For, if B_1 reduces B_2, it explains everything that B_2 does (and normally, more besides). Then, even if we cannot define in B_1 analogues for some of the theoretical terms of B_2, we can use *B_1 in place of B_2*. Thus any reduction, in the sense explained, permits a "reduction" of the total vocabulary of science by making it possible to dispense with some terms.[5] Not every reduction moves in the direction of Unity of Science; for instance reductions *within* a branch lead to a simplification of the vocabulary of science, but they do not necessarily lead in the direction of Unity of Science as we have characterized it (although they may at times fit into that trend). However, micro-reductions, and even partial micro-reductions, insofar as they permit us to replace some of the terms of one branch of science by terms of another, do move in this direction.

Likewise, the micro-reduction of B_2 to B_1 moves in the direction of

Unity of Laws; for it "reduces" the total number of scientific laws by making it possible, in principle, to dispense with the laws of B_2 and explain the relevant observations by using B_1.

The relations 'micro-reduces' and 'potential micro-reducer' have very simple properties: (1) they are transitive (this follows from the transitivity of the relations 'reduces' and 'Pt'); (2) they are irreflexive (no branch can micro-reduce itself); (3) they are asymmetric (if B_1 micro-reduces B_2, B_2 never micro-reduces B_1). The two latter properties are not purely formal; however, they require for their derivation only the (certainly true) empirical assumption that there does not exist an infinite descending chain of proper parts, i.e., a series of things $x_1, x_2, x_3 \ldots$ such that x_2 is a proper part of x_1, x_3 is a proper part of x_2, etc.

The just-mentioned formal property of the relation 'micro-reduces'— its transitivity—is of great importance for the program of Unity of Science. It means that micro-reductions have a cumulative character. That is, if a branch B_3 is micro-reduced to B_2, and B_2 is in turn micro-reduced to B_1, then B_3 is automatically micro-reduced to B_1. This simple fact is sometimes overlooked in objections [6] to the theoretical possibility of attaining unitary science by means of micro-reduction. Thus it has been contended that one manifestly cannot explain human behavior by reference to the laws of atomic physics. It would indeed be fantastic to suppose that the simplest regularity in the field of psychology could be explained directly—i.e., "skipping" intervening branches of science— by employing subatomic theories. But one may believe in the attainability of unitary science without thereby committing oneself to this absurdity. It is not absurd to suppose that psychological laws may eventually be explained in terms of the behavior of individual neurons in the brain; that the behavior of individual cells—including neurons— may eventually be explained in terms of their biochemical constitution; and that the behavior of molecules—including the macro-molecules that make up living cells—may eventually be explained in terms of atomic physics. If this is achieved, then psychological laws will have, in principle, been reduced to laws of atomic physics, although it would nevertheless be hopelessly impractical to try to derive the behavior of a single human being directly from his constitution in terms of elementary particles.

2.3. Unitary science certainly does not exist today. But will it ever be attained? It is useful to divide this question into two subquestions:

Paul Oppenheim and Hilary Putnam

(1) If unitary science can be attained at all, how can it be attained? (2) Can it be attained at all?

First of all, there are various abstractly possible ways in which unitary science might be attained. However, it seems very doubtful, to say the least, that a branch B_2 could be reduced to a branch B_1, if the things in the universe of discourse of B_2 are not themselves in the universe of discourse of B_1 and also do not possess a decomposition into parts in the universe of discourse of B_1. ("They don't speak about the same things.")

It does not follow that B_1 must be a potential micro-reducer of B_2, i.e., that all reductions are micro-reductions.

There are many cases in which the reducing theory and the reduced theory belong to the same branch, or to branches with the same universe of discourse. When we come, however, to branches with different universes—say, physics and psychology—it seems clear that the possibility of reduction depends on the existence of a structural connection between the universes via the 'Pt' relation. Thus one cannot plausibly suppose—for the present at least—that the behavior of inorganic matter is explainable by reference to psychological laws; for inorganic materials do not consist of living parts. One supposes that psychology may be reducible to physics, but not that physics may be reducible to psychology!

Thus, the only method of attaining unitary science that appears to be seriously available at present is micro-reduction.

To turn now to our second question, can unitary science be attained? We certainly do not wish to maintain that it has been *established* that this is the case. But it does not follow, as some philosophers seem to think, that a tentative acceptance of the hypothesis that unitary science can be attained is therefore a mere "act of faith." We believe that this hypothesis is *credible*; [7] and we shall attempt to support this in the latter part of this paper, by providing empirical, methodological, and pragmatic reasons in its support. We therefore think the assumption that unitary science can be attained through cumulative micro-reduction recommends itself as a *working hypothesis*.[8] That is, we believe that it is in accord with the standards of reasonable scientific judgment to tentatively accept this hypothesis and to work on the assumption that further progress can be made in this direction, without claiming that its truth has been established, or denying that success may finally elude us.

3. Reductive Levels

3.1. As a basis for our further discussion, we wish to consider now the possibility of ordering branches in such a way as to indicate the major potential micro-reductions standing between the present situation and the state of unitary science. The most natural way to do this is by their universes of discourse. We offer, therefore, a system of reductive levels so chosen that a branch with the things of a given level as its universe of discourse will always be a potential micro-reducer of any branch with things of the next higher level (if there is one) as its universe of discourse.

Certain conditions of adequacy follow immediately from our aim. Thus:

(1) There must be several levels.

(2) The number of levels must be finite.

(3) There must be a unique lowest level (i.e., a unique "beginner" under the relation 'potential micro-reducer'); this means that success at transforming all the *potential* micro-reductions connecting these branches into *actual* micro-reductions must, *ipso facto*, mean reduction to a single branch.

(4) Any thing of any level except the lowest must possess a decomposition into things belonging to the next lower level. In this sense each level, will be as it were a "common denominator" for the level immediately above it.

(5) Nothing on any level should have a part on any higher level.

(6) The levels must be selected in a way which is "natural" [9] and justifiable from the standpoint of present-day empirical science. In particular, the step from any one of our reductive levels to the next lower level must correspond to what is, scientifically speaking, a crucial step in the trend toward over-all physicalistic reduction.

The accompanying list gives the levels we shall employ; [10] the reader may verify that the six conditions we have listed are all satisfied.

6	Social groups
5	(Multicellular) living things
4	Cells
3	Molecules
2	Atoms
1	Elementary particles

Any whole which possesses a decomposition into parts all of which

7

are on a given level, will be counted as also belonging to that level. Thus each level includes all higher levels. However, the highest level to which a thing belongs will be considered the "proper" level of that thing.

This inclusion relation among our levels reflects the fact that scientific laws which apply to the things of a given level and to all combinations of those things also apply to all things of higher level. Thus a physicist, when he speaks about "all physical objects," is also speaking about living things—but not qua living things.

We maintain that each of our levels is *necessary* in the sense that it would be utopian to suppose that one might reduce all of the major theories or a whole branch concerned with any one of our six levels to a theory concerned with a lower level, *skipping* entirely the *immediately* lower level; and we maintain that our levels are *sufficient* in the sense that it would *not* be utopian to suppose that a major theory on any one of our levels *might* be directly reduced to the next lower level. (Although this is not to deny that it may be convenient, in special cases, to introduce intervening steps.)

However, this contention is significant only if we suppose some set of predicates to be associated with each of these levels. Otherwise, as has been pointed out,[11] trivial micro-reductions would be possible; e.g., we might introduce the property "Tran" (namely, the property of being an atom of a transparent substance) and then "explain the transparency of water in terms of properties on the atomic level," namely, by the hypothesis that all atoms of water have the property Tran. More explicitly, the explanation would consist of the statements

(a) $(x)(x$ is transparent $\equiv (y)(y$ is an atom of $x \supset \text{Tran}(y)))$
(b) $(x)(x$ is water $\supset (y)(y$ is an atom of $x \supset \text{Tran}(y)))$

To exclude such trivial "micro-reductions," we shall suppose that with each level there is associated a list of the theoretical predicates normally employed to characterize things on that level at present (e.g., with level 1, there would be associated the predicates used to specify spatio-temporal coordinates, mass-energy, and electric charge). And when we speak of a theory concerning a given level, we will mean not only a theory whose universe of discourse is that level, but one whose predicates belong to the appropriate list. Unless the hypothesis that theories concerning level $n + 1$ can be reduced by a theory concerning level n is restricted in this way, it lacks any clear empirical significance.

3.2. If the "part-whole" ('Pt') relation is understood in the wide sense, that x Pt y holds if x is spatially or temporally contained in y, then everything, continuous or discontinuous, belongs to one or another reductive level; in particular to level 1 (at least), since it is a whole consisting of elementary particles. However, one may wish to understand 'whole' in a narrower sense (as "structured organization of elements" [12]). Such a specialization involves two essential steps: (1) the construction of a calculus with such a narrower notion as its primitive concept, and (2) the definition of a particular 'Pt' relation satisfying the axioms of the calculus.

Then the problem will arise that some things do not belong to any level. Hence a theory dealing with such things might not be micro-reduced even if all the micro-reductions indicated by our system of levels were accomplished; and for this reason, unitary science might not be attained.

For a trivial example, "a man in a phone booth" is an aggregate of things on different levels which we would not regard as a whole in such a narrower sense. Thus, such an "object" does not belong to any reductive level; although the "phone booth" belongs to level 3 and the man belongs to level 5.

The problem posed by such aggregates is not serious, however. We may safely make the assumption that the behavior of "man in phone booths" (to be carefully distinguished from "men in phone booths") could be completely explained given (a) a complete physicochemical theory (i.e., a theory of levels up to 3, including "phone booths"), and (b) a complete individual psychology (or more generally, a theory of levels up to 5). With this assumption in force, we are able to say: If we can construct a theory that explains the behavior of all the objects in our system of levels, then it will also handle the aggregates of such objects.

4. The Credibility of Our Working Hypothesis

4.1. John Stuart Mill asserts (55, Book VI, Chapter 7) that since (in our wording) human social groups are wholes whose parts are individual persons, the "laws of the phenomena of society" are "derived from and may be resolved into the laws of the nature of individual man." In our terminology, this is to suggest that it is a logical truth that theories concerning social groups (level 6) can be micro-reduced by theories

9

concerning individual living things (level 5); and, *mutatis mutandis*, it would have to be a logical truth that theories concerning any other level can be micro-reduced by theories concerning the next lower level. As a consequence, what we have called the "working hypothesis" that unitary science can be attained would likewise be a logical truth.

Mill's contention is, however, not so much *wrong* as it is vague. What is one to count as "the nature of individual man"? As pointed out above (section 3.1) the question whether theories concerning a given reductive level can be reduced by a theory concerning the next lower level has empirical content only if the theoretical vocabularies are specified; that is, only if one associates with each level, as we have supposed to be done, a particular set of theoretical concepts. Given, e.g., a sociological theory T_2, the question whether there exists a true psychological theory T_1 *in a particular vocabulary* which reduces T_2 is an empirical question. Thus our "working hypothesis" is one that can only be justified on empirical grounds.

Among the factors on which the degree of credibility of any empirical hypothesis depends are (45, p. 307) the *simplicity* of the hypothesis, the *variety* of the evidence, its *reliability*, and, last but not least, the *factual* support afforded by the evidence. We proceed to discuss each of these factors.

4.2. As for the *simplicity* [13] of the hypothesis that unitary science can be attained, it suffices to consider the traditional alternatives mentioned by those who oppose it. "Hypotheses" such as Psychism and Neo-Vitalism assert that the various objects studied by contemporary science have special parts or attributes, unknown to present-day science, in addition to those indicated in our system of reductive levels. For example, men are said to have not only cells as parts; there is also an immaterial "psyche"; living things are animated by "entelechies" or "vital forces"; social groups are moved by "group minds." But, in none of these cases are we provided *at present* with postulates or coordinating definitions which would permit the derivation of testable predictions. Hence, the claims made for the hypothetical entities just mentioned lack any clear scientific meaning; and as a consequence, the question of supporting evidence cannot even be raised.

On the other hand, if the effort at micro-reduction should seem to fail, we cannot preclude the introduction of theories postulating presently unknown relevant parts or presently unknown relevant attributes for some or all of the objects studied by science. Such theories are perfectly admissible, provided they have genuine explanatory value. For example, Dalton's chemical theory of molecules might not be reducible to the best available theory of atoms at a given time if the latter theory ignores the existence of the electrical properties of atoms. Thus the hypothesis of micro-reducibility,[14] as the meaning is specified at a particular time, may be false because of the insufficiency of the theoretical apparatus of the reducing branch.

Of course, a new working hypothesis of micro-reducibility, obtained by enlarging the list of attributes associated with the lowest level, might then be correct. However, if there are presently unknown attributes of a more radical kind (e.g., attributes which are relevant for explaining the behavior of living, but not of non-living things), then no such simple "repair" would seem possible. In this sense, Unity of Science is an alternative to the view that it will eventually be necessary to *bifurcate* the conceptual system of science, by the postulation of new entities or new attributes unrelated to those needed for the study of inanimate phenomena.

4.3. The requirement that there be *variety* of evidence assumes a simple form in our present case. If all the past successes referred to a single pair of levels, then this would be poor evidence indeed that theories concerning each level can be reduced by theories concerning a lower level. For example, if all the past successes were on the atomic level, we should hardly regard as justified the inference that laws concerning social groups can be explained by reference to the "individual psychology" of the members of those groups. Thus, the first requirement is that one should be able to provide examples of successful micro-reductions between several pairs of levels, preferably between all pairs.

Second, within a given level what is required is, preferably, examples of different kinds, rather than a repetition of essentially the same example many times. In short, one wants good evidence that *all* the phenomena of the given level can be micro-reduced.

We shall present below a survey of the past successes in each level. This survey is, of course, only a sketch; the successful micro-reductions and projected micro-reductions in biochemistry alone would fill a large book. But even from this sketch it will be apparent, we believe, how great the variety of these successful micro-reductions is in both the respects discussed.

Paul Oppenheim and Hilary Putnam

4.4. Moreover, we shall, of course, present only evidence from authorities regarded as *reliable* in the particular area from which the theory or experiment involved is drawn.

4.5. The important factor *factual support* is discussed only briefly now, because we shall devote to it many of the following pages and would otherwise interrupt our presentation.

The first question raised in connection with any hypothesis is, of course, what *factual support* it possesses; that is, what confirmatory or disconfirmatory evidence is available. The evidence supporting a hypothesis is conveniently subdivided into that providing *direct* and that providing *indirect* factual support. By the direct factual support for a hypothesis we mean, roughly,[15] the proportion of confirmatory as opposed to disconfirmatory instances. By the indirect factual support, we mean the inductive support obtained from other well-confirmed hypotheses that lend credibility to the given hypothesis. While intuitively adequate quantitative measures of direct factual support have been worked out by Kemeny and Oppenheim,[16] no such measures exist for indirect factual support. The present paper will rely only on intuitive judgments of these magnitudes, and will not assume that quantitative explicata will be worked out.

As our hypothesis is that theories of each reductive level can be micro-reduced by theories of the next lower level, a "confirming instance" is simply any successful micro-reduction between any two of our levels. The *direct* factual support for our hypothesis is thus provided by the *past* successes at reducing laws about the things on each level by means of laws referring to the parts on lower (usually, the next lower) levels. In the sequel, we shall survey the past successes with respect to each pair of levels.

As *indirect* factual support, we shall cite evidence supporting the hypothesis that each reductive level is, in evolution and ontogenesis (in a wide sense presently to be specified) prior to the one above it. The hypothesis of *evolution* means here that (for n = 1 . . . 5) there was a time when there were things of level n, but no things of any higher level. This hypothesis is highly speculative on levels 1 and 2; fortunately the micro-reducibility of the molecular to the atomic level and of the atomic level to the elementary particle level is relatively well established on other grounds.

Similarly, the hypothesis of ontogenesis is that, in certain cases, for any particular object on level n, there was a time when it did not exist, but when some of its parts on the next lower level existed; and that it developed or was causally produced out of these parts.[17]

The reason for our regarding evolution and ontogenesis as providing indirect factual support for the Unity of Science hypothesis may be formulated as follows:

Let us, as is customary in science, assume causal determination as a guiding principle; i.e., let us assume that things that appear later in time can be accounted for in terms of things and processes at earlier times. Then, if we find that there was a time when a certain whole did not exist, and that things on a lower level came together to form that whole, it is very natural to suppose that the characteristics of the whole can be causally explained by reference to these earlier events and parts; and that the theory of these characteristics can be micro-reduced by a theory involving only characteristics of the parts.

For the same reason, we may cite as further indirect factual support for the hypothesis of empirical Unity of Science the various successes at synthesizing things of each level out of things on the next lower level. Synthesis strongly increases the evidence that the characteristics of the whole in question are causally determined by the characteristics, including spatio-temporal arrangement, of its parts by showing that the object is produced, under controlled laboratory conditions, whenever parts with those characteristics are arranged in that way.

The consideration just outlined seems to us to constitute an argument against the view that, as objects of a given level combine to form wholes belonging to a higher level, there appear certain new phenomena which are "emergent" (35, p. 151; 76, p. 93) in the sense of being forever irreducible to laws governing the phenomena on the level of the parts. What our argument opposes is not, of course, the obviously true statement that there are many phenomena which are not reducible by currently available theories pertaining to lower levels; our working hypothesis rejects merely the claim of absolute irreducibility, unless such a claim is supported by a theory which has a sufficiently high degree of credibility; thus far we are not aware of any such theory. It is not sufficient, for example, simply to advance the claim that certain phenomena considered to be specifically human, such as the use of verbal language, in an abstract and generalized way, can never be explained on the basis of neurophysiological theories, or to make the claim that this conceptual

13

capacity distinguishes man in principle and not only in degree from non-human animals.

4.6. Let us mention in passing certain *pragmatic* and *methodological* points of view which speak in favor of our working hypothesis:

(1) It is of *practical* value, because it provides a good synopsis of scientific activity and of the relations among the several scientific disciplines.

(2) It is, as has often been remarked, *fruitful* in the sense of stimulating many different kinds of scientific research. By way of contrast, belief in the *irreducibility* of various phenomena has yet to yield a single accepted scientific theory.

(3) It corresponds *methodologically* to what might be called the "Democritean tendency" in science; that is, the pervasive methodological tendency [18] to try, insofar as is possible, to explain apparently dissimilar phenomena in terms of qualitatively identical parts and their spatio-temporal relations.

5. Past Successes at Each Level

5.1. By comparison with what we shall find on lower levels, the micro-reduction of level 6 to lower ones has not yet advanced very far, especially in regard to human societies. This may have at least two reasons: First of all, the body of well established theoretical knowledge on level 6 is still rather rudimentary, so that there is not much to be micro-reduced. Second, while various precise theories concerning certain special types of phenomena on level 5 have been developed, it seems as if a good deal of further theoretical knowledge concerning other areas on the same level will be needed before reductive success on a larger scale can be expected.[19] However, in the case of certain very primitive groups of organisms, astonishing successes have been achieved. For instance, the differentiation into social castes among certain kinds of insects has been tentatively explained in terms of the secretion of so-called social hormones (3).

Many writers [20] believe that there are some laws common to all forms of animal association, including that of humans. Of greater potential relevance to such laws are experiments dealing with "pecking order" among domestic fowl (29). In particular, experiments showing that the social structure can be influenced by the amount of male hormone in individual birds suggest possible parallels farther up the evolutionary scale.

With respect to the problems of human social organization, as will be seen presently, two things are striking: (1) the most developed body of theory is undoubtedly in the field of economics, and this is at present entirely micro-reductionistic in character; (2) the main approaches to social theory are all likewise of this character. (The technical term 'micro-reduction' is not, of course, employed by writers in these fields. However, many writers have discussed "the Principle of Methodological Individualism"; [21] and this is nothing more than the special form our working hypothesis takes in application to human social groups.)

In economics, if very weak assumptions are satisfied, it is possible to represent the way in which an individual orders his choices by means of an individual preference function. In terms of these functions, the economist attempts to explain group phenomena, such as the market, to account for collective consumer behavior, to solve the problems of welfare economics, etc. As theories for which a micro-reductionistic derivation is accepted in economics we could cite all the standard macro-theories; e.g., the theories of the business cycle, theories of currency fluctuation (Gresham's law to the effect that bad money drives out good is a familiar example), the principle of marginal utility, the law of demand, laws connecting change in interest rate with changes in inventory, plans, equipment, etc. The relevant point is while the economist is no longer dependent on the oversimplified assumption of "economic man," the explanation of economic phenomena is still in terms of the preferences, choices, and actions available to *individuals*.

In the realm of *sociology*, one can hardly speak of any major theory as "accepted." But it is of interest to survey some of the major theoretical approaches from the standpoint of micro-reduction.

On the one hand, there is the *economic determinism* represented by Marx and Veblen. In the case of Marx the assumptions of classical economics are openly made: Individuals are supposed—at least on the average, and in the long run—to act in accordance with their material interests. From this assumption, together with a theory of the business cycle which, for all its undoubted originality, Marx based on the classical laws of the market, Marx derives his major laws and predictions. Thus Marxist sociology is micro-reductionistic in the same sense as classical economics, and shares the same basic weakness (the assumption of "economic man").

Veblen, although stressing class interests and class divisions as did

15

Marx, introduces some non-economic factors in his sociology. His account is ultimately in terms of individual psychology; his hypothesis of "conspicuous consumption" is a brilliant—and characteristic—example.

Max Weber produced a sociology strongly antithetical to Marx's. Yet each of his explanations of group phenomena is ultimately in terms of individual psychology; e.g., in his discussion of political parties, he argues that people *enjoy* working under a "charismatic" leader, etc.

Indeed the psychological (and hence micro-reductionistic) character of the major sociologies (including those of Mannheim, Simmel, etc., as well as the ones mentioned above (54, 86, 94, 103)) is often recognized. Thus one may safely say, that while there is no one accepted sociological theory, all of these theoretical approaches represent attempted micro-reductions.

5.2. Since Schleiden and Schwann (1838/9), it is known that all living things consist of cells. Consequently, explaining the laws valid on level 5 by those on the cell level means micro-reducing all phenomena of plants and animals to level 4.

As instances of past successes in connection with level 5 we have chosen to cite, in preference to other types of example, micro-reductions and projected micro-reductions dealing with central nervous systems as wholes and nerve cells as parts. Our selection of these examples has not been determined by anthropocentrism. First of all, substantially similar problems arise in the case of multicellular animals, as nearly all of them possess a nervous system; and, second, the question of micro-reducing those aspects of behavior that are controlled by the central nervous system in man and the higher animals is easily the most significative (85, p. 1) one at this level, and therefore most worth discussing.

Very great activity is, in fact, apparent in the direction of micro-reducing the phenomena of the central nervous system. Much of this activity is very recent; and most of it falls under two main headings: neurology, and the *logical design of nerve nets*. (Once again, the technical term 'micro-reduction' is not actually employed by workers in these fields. Instead, one finds widespread and lasting discussion concerning the advantages of "molecular" versus "molar" [22] explanations, and concerning "reductionism." [23])

Theories constructed by neurologists are the product of highly detailed experimental work in neuroanatomy, neurochemistry, and neurophysiology, including the study of electric activity of the nervous system; e.g., electroencephalography.[24]

As a result of these efforts, it has proved possible to advance more or less hypothetical explanations on the cellular level for such phenomena as association, memory, motivation, emotional disturbance, and some of the phenomena connected with learning, intelligence, and perception. For example, a theory of the brain has been advanced by Hebb (32) which accounts for all of the above-mentioned phenomena. A classical psychological law, the Weber-Fechner law (insofar as it seems to apply), has likewise been micro-reduced, as a result of the work of Hoagland (36).

We turn now to *the logical design of nerve nets*: The logician Turing[25] proposed (and solved) the problem of giving a characterization of computing machines in the widest sense—mechanisms for solving problems by effective series of logical operations. This naturally suggests the idea of seeing whether a "Turing machine" could consist of the elements used in neurological theories of the brain; that is, whether it could consist of a network of neurons. Such a nerve network could then serve as a hypothetical model for the brain.

Such a network was first constructed by McCulloch and Pitts.[26] The basic element is the neuron, which, at any instant, is either *firing* or *not firing* (quiescent). On account of the "all or none" character of the activity of this basic element, the *nerve net* designed by McCulloch and Pitts constitutes, as it were, a digital computer. The various relations of propositional logic can be represented by instituting suitable connections between neurons; and in this way the hypothetical net can be "programmed" to solve any problem that will yield to a predetermined sequence of logical or mathematical operations. McCulloch and Pitts employ approximately 10^4 elements in their net; in this respect they are well below the upper limit set by neurological investigation, since the number of neurons in the brain is estimated to be of the order of magnitude of 10^{10}. In other respects, their model was, however, unrealistic: no allowance is made for time delay, or for random error, both of which are important features of all biological processes.

Nerve nets incorporating both of these features have been designed by von Neumann. Von Neumann's model employs bundles of nerves rather than single nerves to form a network; this permits the simultaneous performance of each operation as many as 20,000 times as a

17

check against error. This technique of constructing a computer is impractical at the level of present-day technology, von Neumann admits, "but quite practical for a perfectly conceivable, more advanced technology, and for the natural relay-organs (neurons). I.e., it merely calls for micro-componentry which is not at all unnatural as a concept on this level" (97, p. 87). Still further advances in the direction of adapting these models to neurological data are anticipated. In terms of such nerve nets it is possible to give hypothetical micro-reductions for *memory, exact thinking, distinguishing similarity or dissimilarity of stimulus patterns, abstracting of "essential"* components of a stimulus pattern, recognition of shape regardless of form and of chord regardless of pitch (phenomena of great importance in Gestalt psychology (5, pp. 128, 129, 152)), *purposeful behavior* as controlled by negative feedback, *adaptive behavior,* and *mental disorders.*

It is the task of the neurophysiologist to test these models by investigating the existence of such nets, scanning units, reverberating networks, and pathways of feedback, and to provide physiological evidence of their functioning. Promising studies have been made in this respect.

5.3. As past successes in connection with level 4 (i.e. as cases in which phenomena involving whole cells [27] have been explained by theories concerning the molecular level) we shall cite micro-reductions dealing with three phenomena that have a ˙fundamental character for all of biological science: the *decoding, duplication,* and *mutation* of the genetic information that is ultimately responsible for the development and maintenance of order in the cell. Our objective will be to show that at least one well-worked-out micro-reducing theory has been advanced for each phenomenon.[28] (The special form taken by our working hypothesis on this level is "methodological mechanism.")

Biologists have long had good evidence indicating that the genetic information in the cell's nucleus—acting as an "inherited message"—exerts its control over cell biochemistry through the production of specific protein catalysts (enzymes) that mediate particular steps (reactions) in the chemical order that is the cell's life. The problem of *"decoding"* the control information in the nucleus thus reduces to how the specific molecules that comprise it serve to specify the construction of specific protein catalysts. The problem of *duplication* (one aspect of the over-all problem of inheritance) reduces to how the molecules of genetic material can be copied—like so many "blueprints."

18

And the problem of mutation (elementary step in the evolution of new inheritable messages) reduces to how "new" forms of the genetic molecules can arise.

In the last twenty years evidence has accumulated implicating desoxyribose nucleic acid (DNA) as the principal "message-carrying" molecule and constituting the genetic material of the chromosomes. Crick and Watson's [29] brilliant analysis of DNA structure leads to powerful micro-reducing theories that explain the decoding and duplication of DNA. It is known that the giant molecules that make up the nucleic acids have, like proteins (49, 66, 67), the structure of a backbone with side groups attached. But, whereas the proteins are polypeptides, or chains of amino-acid residues (slightly over 20 kinds of amino acids are known); the nucleic acids have a phosphate-sugar backbone, and there are only 4 kinds of side groups all of which are nitrogen bases (purines and pyrimidines). Crick and Watson's model contains a pair of DNA chains wound around a common axis in the form of two interlocking helices. The two helices are held together (forming a helical "ladder") by hydrogen bonds between pairs of the nitrogen bases, one belonging to each helix. Although 4 bases occur as side groups only 2 of 16 conceivable pairings are possible, for steric reasons. These 2 pairs of bases recur along the length of the DNA molecule and thus invite a picturesque analogy with the dots and dashes of the Morse code. They can be arranged in any sequence: there is enough DNA in a single cell of the human body to encode in this way 1000 large textbooks. The model can be said to imply that the genetic "language" of the inherited control message is a "language of surfaces": the information in DNA structure is decoded as a sequence of amino acids in the proteins which are synthesized under ultimate DNA control. The surface structure of the DNA helix, dictated by the sequence of base pairs, specifies like a template [30] the sequence of amino acids laid down end to end in the fabrication of polypeptides.

Watson and Crick's model immediately suggests how the DNA might produce an exact copy of itself—for transmission as an inherited message to the succeeding generation of cells. The DNA molecule, as noted above, consists of two interwoven helices, each of which is the complement of the other. Thus each chain may act as a mold on which a complementary chain can be synthesized. The two chains of a DNA molecule need only unwind and separate. Each begins to build a new complement onto itself, as loose units, floating in the cell, attach themselves to the bases in the single DNA chain. When the process is completed, there are two pairs of chains where before there was only one! [31]

Mutation of the genetic information has been explained in a molecular (micro-reduction) theory advanced some years ago by Delbrück.[32] Delbrück's theory was conceived long before the newer knowledge of DNA was available; but it is a very general model in no way vitiated by Crick and Watson's model of the particular molecule constituting the genetic material. Delbrück, like many others, assumed that the gene is a single large "nucleo-protein" molecule. (This term is used for macromolecules, such as viruses and the hypothetical "genes," which consist of protein and nucleic acid. Some recent theories even assume that an entire chromosome is a single such molecule.) According to Delbrück's theory, different quantum levels within the atoms of the molecule correspond to different hereditary characteristics. A mutation is simply a quantum jump of a rare type (i.e., one with a high activation energy). The observed variation of the spontaneous mutation rate with temperature is in good quantitative agreement with the theory.

Such hypotheses and models as those of Crick and Watson, and of Delbrück, are at present far from sufficient for a complete micro-reduction of the major biological generalization, e.g., evolution and general genetic theory (including the problem of the control of development). But they constitute an encouraging start towards this ultimate goal and, to this extent, an indirect support for our working hypothesis.

5.4. Only in the twentieth century has it been possible to micro-reduce to the atomic and in some cases directly to the subatomic level most of the *macro-physical* aspects of matter (e.g., the high fluidity of water, the elasticity of rubber, and the hardness of diamond) as well as the *chemical* phenomena of the elements, i.e. those changes of the peripheral electrons which leave the nucleus unaffected. In particular, electronic theories explain, e.g., the laws governing valence, the various types of bonds, and the "resonance" of molecules between several equivalent electronic structures. A complete explanation of these phenomena and those of the Periodic Table is possible only with the help of Pauli's exclusion principle which states in one form that no two electrons of the same atom can be alike in all of 4 "quantum numbers." While some molecular laws are not yet micro-reduced, there is every hope that further successes will be obtained in these respects. Thus Pauling (63, 64) writes:

20

There are still problems to be solved, and some of them are great problems—an example is the problem of the detailed nature of catalytic activity. We can feel sure, however, that this problem will in the course of time be solved in terms of quantum theory as it now exists: there seems little reason to believe that some fundamental new principle remains to be discovered in order that catalysis be explained (64).

5.5. Micro-reduction of level 2 to level 1 has been mentioned in the preceding section because many molecular phenomena are at present (skipping the atomic level) explained with reference to laws of elementary particles.[33] Bohr's basic (and now somewhat outdated) model of the atom as a kind of "solar system" of elementary particles is today part of everyone's conceptual apparatus; while the mathematical development of theory in its present form is formidable indeed! Thus we shall not attempt to give any details of this success. But the high rate of progress in this field certainly gives reason to hope that the unsolved problems, especially as to the forces that hold the nucleus together, will likewise be explained in terms of an elementary particle theory.

6. Evolution, Ontogenesis, and Synthesis

6.1. As pointed out in section 4.5, *evolution* provides indirect factual support for the working hypothesis that unitary science is attainable. Evolution (in the present sense) is an over-all phenomenon involving all levels, from 1 through 6; the mechanisms of chance variation and "selection" operate throughout in ways characteristic for the evolutionary level involved.[34] Time scales have, indeed, been worked out by various scientists showing the times when the first things of each level first appeared.[35] (These times are, of course, the less hypothetical the higher the level involved.) But even if the hypothesis of evolution should fail to hold in the case of certain levels, it is important to note that whenever it does hold—whenever it can be shown that things of a given level existed before things of the next higher level came into existence—some degree of indirect support is provided to the particular special case of our working hypothesis that concerns those two levels.

The hypothesis of "evolution" is most speculative insofar as it concerns levels 1 to 3. Various cosmological hypotheses are at present undergoing lively discussion.[36] According to one of these, strongly urged by Gamow (24, 25, 26), the first nuclei did not form out of elementary particles until five to thirty minutes after the start of the universe's expansion; molecules may not have been able to exist until considerably later. Most present-day cosmologists still subscribe to such evolutionary views of the universe; i.e., there was a "zero point" from which the evolution of matter began, with diminishing density through expansion. However, H. Bondi, T. Gold, and F. Hoyle have advanced a conflicting idea, the "steady state" theory, according to which there is no "zero point" from which the evolution of matter began; but matter is continuously created, so that its density remains constant in spite of expansion. There seems to be hope that these rival hypotheses will be submitted to specific empirical tests in the near future. But, fortunately, we do not have to depend on hypotheses that are still so highly controversial: as we have seen, the micro-reducibility of molecular and atomic phenomena is today not open to serious doubt.

Less speculative are theories concerning the origin of life (transition from level 3 to level 4). Calvin (9; Fox, 22) points out that four mechanisms have been discovered which lead to the formation of amino acids and other organic materials in a mixture of gases duplicating the composition of the primitive terrestrial atmosphere.[37] These have, in fact, been tested experimentally with positive results. Many biologists today accept with Oparin (61) the view that the evolution of life as such was not a single chance event but a long process possibly requiring as many as two billion years, until precellular living organisms first appeared.

According to such views, "chemical evolution" gradually leads in an appropriate environment to evolution in the familiar Darwinian sense. In such a process, it hardly has meaning to speak of a point at which "life appeared." To this day controversies exist concerning the "dividing line" between living and non-living things. In particular, viruses are classified by some biologists as *living*, because they exhibit *self-duplication* and *mutability*; but most biologists refuse to apply the term to them, because viruses exhibit these characteristic phenomena of life only due to activities of a living cell with which they are in contact. But, wherever one draws the line,[38] non-living molecules preceded primordial living substance, and the latter evolved gradually into highly organized living units, the unicellular ancestors of all living things. The "first complex molecules endowed with the faculty of reproducing their own kind" must have been synthesized—and with them the beginning of evolution in the Darwinian sense—a few billion years ago, Goldschmidt

(27, p. 84) asserts: "all the facts of biology, geology, paleontology, bio-chemistry, and radiology not only agree with this statement but actually prove it."

Evolution at the next two levels (from level 4 to level 5, and from 5 to 6) is not speculative at all, but forms part of the broad line of Darwinian evolution, so well marked out by the various kinds of evidence referred to in the statement just quoted. The line of development is again a continuous one; [39] and it is to some extent arbitrary (as in the case of "living" versus "non-living") to give a "point" at which true multicellulars first appeared, or at which an animal is "social" rather than "solitary." But in spite of this arbitrariness, it is safe to say that:

(a) Multicellulars evolved from what were originally competing single cells; the "selection" by the environment was in this case determined by the superior survival value of the cooperative structure.[40]

(b) Social animals evolved from solitary ones for similar reasons; and, indeed, there were millions of years during which there were only solitary animals on earth, and not yet their organizations into social structures.[41]

6.2. To illustrate ontogenesis, we must show that particular things of a particular level have arisen out of particular things of the next lower level. For example, it is a consequence of most contemporary cosmological theories—whether of the evolutionary or of the "steady state" type—that each existent atom must have originally been formed by a union of elementary particles. (Of course an atom of an element may subsequently undergo "transmutation.") However, such theories are extremely speculative. On the other hand, the chemical union of atoms to form molecules is commonplace in nature.

Coming to the higher levels of the reductive hierarchy, we have unfortunately a hiatus at the level of cells. Individual cells do not, as far as our observations go, ever develop out of individual molecules; on the contrary, "cells come only from cells," as Virchow stated about one hundred years ago. However, a characteristic example of ontogenesis of things of one level out of things of the next lower level is afforded by the development of multicellular organisms through the process of mitosis and cell division. All the hereditary characteristics of the organism are specified in the "genetic information" carried in the chromosomes of each individual cell, and are transmitted to the resultant organism through cell division and mitosis.

A more startling example of ontogenesis at this level is provided by the *slime molds* studied by Bonner (3). These are isolated amoebae; but, at a certain stage, they "clump" together chemotactically and form a simple multicellular organism, a sausage-like "slug"! This "slug" crawls with comparative rapidity and good coordination. It even has senses of a sort, for it is attracted by light.

As to the level of social groups, we have some ontogenetic data, however slight; for children, according to the well-known studies of Piaget (70, 71) (and other authorities on child behavior), acquire the capacity to cooperate with one another, to be concerned with each other's welfare, and to form groups in which they treat one another as peers, only after a number of years (not before seven years of age, in Piaget's studies). Here one has in a rudimentary form what we are looking for: the ontogenetic development of progressively more social behavior (level 6) by what begin as relatively "egocentric" and unsocialized individuals (level 5).

6.3. *Synthesis* affords factual support for micro-reduction much as ontogenesis does; however, the evidence is better because synthesis usually takes place under *controlled* conditions. Thus it enables one to show that one can obtain an object of the kind under investigation *invariably* by instituting the appropriate causal relations among the parts that go to make it up. For this reason, we may say that success in synthesizing is as strong evidence as one can have for the possibility of micro-reduction, short of actually finding the micro-reducing theory.

To begin on the lowest level of the reductive hierarchy, that one can obtain an atom by bringing together the appropriate elementary particles is a basic consequence of elementary nuclear physics. A common example from the operation of atomic piles is the synthesis of deuterium. This proceeds as one bombards protons (in, e.g., hydrogen gas) with neutrons.

The synthesis of a molecule by chemically uniting atoms is an elementary laboratory demonstration. One familiar example is the union of oxygen and hydrogen gas. Under the influence of an electric spark one obtains the appearance of H_2O molecules.

The next level is that of life. "On the borderline" are the viruses. Thus success at synthesizing a virus out of non-living macro-molecules would count as a first step to the synthesis of cells (which at present seems to be an achievement for the far distant future).

While success at synthesizing a virus out of atoms is not yet in sight, synthesis out of non-living highly complex macro-molecules has been accomplished. At the University of California Virus Laboratory (23), protein obtained from viruses has been mixed with nucleic acid to obtain active virus. The protein does not behave like a virus—it is completely non-infectious. However, the reconstituted virus has the same structure as "natural" virus, and will produce the tobacco mosaic disease when applied to plants. Also new "artificial" viruses have been produced by combining the nucleic acid from one kind of virus with the protein from a different kind. Impressive results in synthesizing proteins have been accomplished: e.g., R. B. Woodward and C. H. Schramm (107; see also Nogushi and Hayakawa, 60; and Oparin, 61) have synthesized "protein analogues"—giant polymers containing at least 10,000 amino-acid residues.

At the next level, no one has of course synthesized a whole multi-cellular organism out of individual cells; but here too there is an impressive partial success to report. Recent experiments have provided detailed descriptions of the manner in which cells organize themselves into whole multicellular tissues. These studies show that even isolated whole cells, when brought together in random groups, could effectuate the characteristic construction of such tissues.[42] Similar phenomena are well known in the case of sponges and fresh-water polyps.

Lastly, the "synthesis" of a new social group by bringing together previously separated individuals is extremely familiar; e.g., the organization of new clubs, trade unions, professional associations, etc. One has even the deliberate formation of whole new societies, e.g., the formation of the Oneida community of utopians, in the nineteenth century, or of the state of Israel by Zionists in the twentieth.

There have been experimental studies in this field; among them, the pioneer work of Kurt Lewin and his school is especially well known.[43]

7. Concluding Remarks

The possibility that all science may one day be reduced to micro-physics (in the sense in which chemistry seems today to be reduced to it), and the presence of a unifying trend toward micro-reduction running through much of scientific activity, have often been noticed both by specialists in the various sciences and by meta-scientists. But these opinions have, in general, been expressed in a more or less vague manner and without very deep-going justification. It has been our aim, first, to provide precise definitions for the crucial concepts involved, and, second, to reply to the frequently made accusations that belief in the attainability of unitary science is "a mere act of faith." We hope to have shown that, on the contrary, a tentative acceptance of this belief, an acceptance of it as a working hypothesis, is *justified*, and that the hypothesis is *credible*, partly on methodological grounds (e.g., the simplicity of the hypothesis, as opposed to the bifurcation that rival suppositions create in the conceptual system of science), and partly because there is really a large mass of direct and indirect evidence in its favor.

The idea of reductive levels employed in our discussion suggests what may plausibly be regarded as a *natural order of sciences*. For this purpose, it suffices to take as "fundamental disciplines" the branches corresponding to our levels. It is understandable that many of the well-known orderings of things [44] have a rough similarity to our reductive levels, and that corresponding orderings of sciences are more or less similar to our order of 6 "fundamental disciplines." Again, several successive levels may be grouped together (e.g. physics today conventionally deals at least with levels 1, 2, and 3; just as biology deals with at least levels 4 and 5). Thus we often encounter a division into simply physics, biology, and social sciences. But these other efforts to solve a problem which goes back to ancient times [45] have apparently been made on more or less intuitive grounds; it does not seem to have been realized that these orderings are "natural" in a deeper sense, of being based on the relation of *potential micro-reducer* obtaining between the branches of science.

It should be emphasized that these six "fundamental disciplines" are, largely, fictitious ones (e.g., there is no actual branch whose universe of discourse is *strictly* molecules and combinations thereof). If one wishes a less idealized approach, one may utilize a concept in semantical information theory which has been defined by one of us (3). This is the semantical functor: 'the amount of information the statement S contains about the class C' (or, in symbols: $inf(S, C)$). Then one can characterize any theory S (or any branch, if we are willing to identify a branch with a conjunction of theories) by a sextuple: namely, $inf(S, level 1)$, $inf(S, level 2) \ldots inf(S, level 6)$. This sextuple can be regarded as the "locus" of the branch S in a six-dimensional space. The axes are the loci of the imaginary "fundamental disciplines" just referred to; any real branch (e.g., present-day biology) will probably have a position not quite on any axis, but nearer to one than to the others. Whereas the orderings to which we referred above generally begin with the historically given branches, the procedure just described reverses this tendency. First a continuous order is defined in which any imaginable branch can be located; then one investigates the relations among the actually existing branches. These positions may be expected to change with time; e.g., as micro-reduction proceeds, "biology" will occupy a position closer to the "level 1" axis, and so will all the other branches. The continuous order may be described as "Darwinian" rather than "Linnean"; it derives its naturalness, not from agreement with intuitive or customary classifications, but from its high systematic import in the light of the hypothesis that Unity of Science is attainable.

NOTES

[1] Science, in the wider sense, may be understood as including the formal disciplines, mathematics, and logic, as well as the empirical ones. In this paper, we shall be concerned with science only in the sense of empirical disciplines, including the socio-humanistic ones.

[2] By a "theory" (in the widest sense) we mean any hypothesis, generalization, or law (whether deterministic or statistical), or any conjunction of these; likewise by "phenomena" (in the widest sense) we shall mean either particular occurrences or theoretically formulated general patterns. Throughout this paper, "explanation" ("explainable" etc.) is used as defined in Hempel and Oppenheim (35). As to "explanatory power," there is a definite connection with "systematic power." See Kemeny and Oppenheim (46, 47).

[3] If we are willing to adopt a "Taxonomic System" for classifying all the things dealt with by science, then the various classes and subclasses in such a system could represent the possible "universes of discourse." In this case, the U_{Bi} of any branch would be associated with the extension of a taxonomic term in the sense of Oppenheim (62).

[4] Henceforth, we shall as a rule omit the clause 'at time t'.

[5] Oppenheim (62, section 3) has a method for measuring such a reduction.

[6] Of course, in some cases, such "skipping" does occur in the process of micro-reduction, as shall be illustrated later on.

[7] As to degree of credibility, see Kemeny and Oppenheim (45, especially p. 307).

[8] The "acceptance, as an overall fundamental working hypothesis, of the reduction theory, with physical science as most general, to which all others are reducible; with biological science less general; and with social science least general of all," has been emphasized by Hockett (37, especially p. 571).

[9] As to natural, see Hempel (33, p. 52), and Hempel and Oppenheim (34, pp. 107, 110).

[10] Many well known hierarchical orders of the same kind (including some compatible with ours) can be found in modern writings. It suffices to give the following quotation from an article by L. von Bertalanffy (95, p. 164): "Reality, in the modern conception, appears as a tremendous hierarchical order of organized entities, leading, in a superposition of many levels, from physical and chemical to biological and sociological systems. Unity of Science is granted, not by an utopian reduction of all

sciences to physics and chemistry, but by the structural uniformities of the different levels of reality." As to the last sentence, we refer in the last paragraph of section 2.2 to the problem noted. Von Bertalanffy has done pioneer work in developing a General System Theory which, in spite of some differences of emphasis, is an interesting contribution to our problem.

[11] The following example is a slight modification of the one given in Hempel and Oppenheim (35, p. 148). See also Rescher and Oppenheim (76, pp. 93, 94).

[12] See Rescher and Oppenheim (76, p. 100), and Rescher (75). Of course, nothing is intrinsically a "true" whole; the characterization of certain things as "wholes" is always a function of the point of view, i.e. of the particular 'Pt' relation selected. For instance, if a taxonomic system is given, it is very natural to define 'Pt' so that the "wholes" will correspond to the things of the system. Similarly for aggregate see Rescher and Oppenheim (76, p. 90, n. 1).

[13] See Kemeny and Oppenheim (47, n. 6). A suggestive characterization of simplicity in terms of the "entropy" of a theory has been put forward by Rothstein (78). Using Rothstein's terms, we may say that any micro-reduction moves in the direction of lower entropy (greater organization).

[14] The statement that B_2 is *micro-reducible* to B_1 means (according to the analysis we adopt here) that some *true* theory belonging to B_1—i.e., some true theory with the appropriate vocabulary and universe of discourse, whether accepted or not, and whether it is ever even written down or not—micro-reduces every true theory of B_2. This seems to be what people have in mind when they assert that a given B_2 may not be reduced to a given B_1 at a certain time, but may nonetheless be reducible (micro-reducible) to it.

[15] See Kemeny and Oppenheim (45, p. 307); also for "related concepts," like Carnap's "degree of confirmation" see Carnap (13).

[16] As to degree of credibility see Kemeny and Oppenheim (45, especially p. 307).

[17] Using a term introduced by Kurt Lewin (48), we can also say in such a case: any particular object on level n is *genidentical* with these parts.

[18] Though we cannot accept Sir Arthur Eddington's idealistic implications, we quote from his *Philosophy of Physical Science* (17, p. 125): "I conclude therefore that our engrained form of thought is such that we shall not rest satisfied until we are able to represent all physical phenomena as an interplay of a vast number of structural units intrinsically alike. All the diversity of phenomena will be then seen to correspond to different forms of relatedness of these units or, as we should usually say, different configurations."

[19] M. Scriven has set forth some suggestive considerations on this subject in his essay, "A Possible Distinction between Traditional Scientific Disciplines and the Study of Human Behavior" (79).

[20] See e.g. Kartman (43), with many quotations, references, and notes, some of them micro-reductionistic.

[21] This term has been introduced by F. A. Hayek (31). See also Watkins (98, especially pp. 729–732) and Watkins (99). We owe valuable information in economics to W. J. Baumol, Princeton University.

[22] This distinction, first made by C. D. Broad (6, p. 616), adopted by E. C. Tolman (90), C. L. Hull (39), and others, is still in use, in spite of objections against this terminology.

[23] This is the form our working hypothesis takes on this level in this field. See in this connection the often quoted paper by K. MacCorquodale and P. E. Meehl, "On a Distinction between Hypothetical Constructs and Intervening Variables" (52), and some of the discussions in the "Symposium on the Probability Approach in Psychology" (73), as well as references therein, to H. Feigl, W. Koehler, D. Krech, and C. C. Pratt.

[24] As to neuroanatomy, see e.g. W. Penfield (69); as to neurochemistry, see e.g.

Rosenblueth (77, especially Chapter 26 for acetylcholine and the summaries on pp. 134–135, 274–275); as to *The Electric Activity of the Nervous System*, see the book of this title by Brazier (5). See this last book also for neuroanatomy, neurophysiology, neurochemistry. See Brazier (5, pp. 128, 129, 152) for micro-reduction of Gestalt phenomena mentioned below.

²⁵ Turing (91, 92). For an excellent popular presentation, see Kemeny (44).

²⁶ See the often quoted paper by McCulloch and Pitts (53), and later publications by these authors, as well as other papers in this field in the same *Bulletin of Mathematical Biophysics*, e.g. by N. Rashevsky. See also Platt (72) for a "complementary approach which might be called amplifier theory." For more up to date details, see Shannon and McCarthy's (82) *Automata Studies*, including von Neumann's model, discussed by him (82, pp. 43–98).

²⁷ Throughout this paper, "cell" is used in a wide sense, i.e., "Unicellular" organism or single cell in a multicellular organism.

²⁸ For more details and much of the following, see Simpson, Pittendrigh and Tiffany (87), Goldschmidt (28), and Horowitz (38). For valuable suggestions we are indebted to C. S. Pittendrigh who also coined the terms "message carrying molecule" and "languages of surface" used in our text.

²⁹ See in reference to the following discussion Watson and Crick (100), also (101), and (102), and Crick (15).

³⁰ Pauling and Delbrück (68). A micro-reducing theory has been proposed for these activities using the "lock-key" model. See Pauling, Campbell and Pressman (65), and Burnet (8).

³¹ For a mechanical model, see von Neumann (96) and Jacobson (40).

³² See Timoféeff-Ressovsky (89, especially pp. 108–138). It should, however, be noted that since Delbrück's theory was put forward, his model has proved inadequate for explaining genetic facts concerning mutation. And it is reproduced here only as a historical case of a micro-reducing theory that, in its day, served valuable functions.

³³ We think that, throughout this paper, our usage of thing language also on this level is admissible in spite of well-known difficulties and refer e.g. to Born (4), and Johnson (42).

³⁴ See e.g. Broad (6, especially p. 93), as to "a general tendency of one order to combine with each other under suitable conditions to form complexes of the next order." See also Blum (1, and 2, especially p. 608); Needham (59, especially pp. 184–185); and Dodd (16).

³⁵ This wording takes care of "regression," a reversal of trend, illustrated e.g. by parasitism.

³⁶ For a clear survey of cosmological hypotheses see the 12 articles published in the issue of *Scientific American* cited under Gamow (26).

³⁷ Perhaps the most sensational method is an experiment suggested by H. C. Urey and made by S. L. Miller (56, 57), according to which amino acids are formed when an electric discharge passes through a mixture of methane, hydrogen, ammonia, and water.

³⁸ "Actually life has many attributes, almost any one of which we can reproduce in a nonliving system. It is only when they all appear to a greater or lesser degree in the same system simultaneously that we call it living" (Calvin, 9, p. 252). Thus the dividing line between "living" and "non-living" is obtained by transforming an underlying "multidimensional concept of order" (see Hempel and Oppenheim, 34, pp. 65–77), in a more or less arbitrary way, into a dichotomy. See also Stanley (88, especially pp. 15 and 16 of the reprint of this article).

³⁹ See note 38 above.

⁴⁰ For details, see Lindsey (50, especially pp. 136–139, 152–153, 342–344). See also Burkholder (7).

⁴¹ See e.g. the publications (104, 105, 106) by Wheeler. See also Haskins (30,

especially pp. 30–36). Since we are considering evolution on level 6 as a whole, we can refrain from discussing the great difference between, on the one hand, chance mutations, natural selection, and "instinctive" choices and, on the other hand, the specific faculty of man of consciously and willfully directing social evolution in time stretches of specifically small orders of magnitude (see Zilsel, 108).

⁴² See Moscana (58) and his references, especially to work by the same author and by Paul Weiss.

⁴³ See Lippitt (51). For recent experiments, see Sherif and Sherif (84, Chapters 6 and 9), and Sherif (83).

⁴⁴ See note 10 above.

⁴⁵ For details, see Flint (21), and Vannerus (93). Auguste Comte in his *Cours de Philosophie Positive*, Première et Deuxième Leçons (14), has given a hierarchical order of 6 "fundamental disciplines" which, independently from its philosophical background, is amazingly modern in many respects, as several contemporary authors recognize.

REFERENCES

1. Blum, H. F. *Time's Arrow and Evolution*. Princeton: Princeton Univ. Press, 1951.
2. Blum, H. F. "Perspectives in Evolution," *American Scientist*, 43:595–610 (1955).
3. Bonner, J. T. *Morphogenesis*. Princeton: Princeton Univ. Press, 1952.
4. Born, M. "The Interpretation of Quantum Mechanics," *British Journal for the Philosophy of Science*, 3:95–106 (1953).
5. Brazier, M. A. B. *The Electric Activity of the Nervous System*. London: Sir Isaac Pitman & Sons, Ltd., 1951.
6. Broad, C. D. *The Mind and its Place in Nature*. New York: Harcourt, Brace, 1925.
7. Burkholder, P. R. "Cooperation and Conflict among Primitive Organisms," *American Scientist*, 40:601–631 (1952).
8. Burnet, M. "How Antibodies are Made," *Scientific American*, 191:74–78 (November 1954).
9. Calvin, M. "Chemical Evolution and the Origin of Life," *American Scientist*, 44:248–263 (1956).
10. Carnap, R. *Der logische Aufbau der Welt*. Berlin-Schlachtensee: Im Weltkreis-Verlag, 1928. Summary in N. Goodman, *The Structure of Appearances*, pp. 114–146. Cambridge: Harvard Univ. Press, 1951.
11. Carnap, R. "Testability and Meaning," *Philosophy of Science*, 3:419–471 (1936), and 4:2–40 (1937). Reprinted by Graduate Philosophy Club, Yale University, New Haven, 1950.
12. Carnap, R. *Logical Foundations of the Unity of Science, International Encyclopedia of Unified Science*. Vol. I, pp. 42–62. Chicago: Univ. of Chicago Press, 1938.
13. Carnap, R. *Logical Foundations of Probability*. Chicago: Univ. of Chicago Press, 1950.
14. Comte, Auguste. *Cours de Philosophie Positive*. 6 Vols. Paris: Bachelier, 1830–42.
15. Crick, F. H. C. "The Structure of Hereditary Material," *Scientific American*, 191: 54–61 (October 1954).
16. Dodd, S. C. "A Mass-Time Triangle," *Philosophy of Science*, 11:233–244 (1944).
17. Eddington, Sir Arthur. *The Philosophy of Physical Science*. Cambridge: Cambridge University Press, 1949.
18. Feigl, H. "Logical Empiricism," in D. D. Runes (ed.), *Twentieth Century*

Philosophy, pp. 371–416. New York: Philosophical Library, 1943. Reprinted in H. Feigl and W. Sellars (eds.), Readings in Philosophical Analysis. New York: Appleton-Century-Crofts, 1949.

19. Feigl, H. "Unity of Science and Unitary Science," in H. Feigl and M. Brodbeck (eds.), Readings in the Philosophy of Science, pp. 382–384. New York: Appleton-Century-Crofts, 1953.

20. Feigl, H. "Functionalism, Psychological Theory and the Uniting Sciences: Some Discussion Remarks," Psychological Review, 62:232–235 (1955).

21. Flint, R. Philosophy as Scientia Scientiarum and the History of the Sciences. New York: Scribner, 1904.

22. Fox, S. W. "The Evolution of Protein Molecules and Thermal Synthesis of Biochemical Substances," American Scientist, 44:347–359 (1956).

23. Fraenkel-Conrat, H. "Rebuilding a Virus," Scientific American, 194:42–47 (June 1956).

24. Gamow, G. "The Origin and Evolution of the Universe," American Scientist, 39:393–406 (1951).

25. Gamow, G. The Creation of the Universe. New York: Viking Press, 1952.

26. Gamow, G. "The Evolutionary Universe," Scientific American, 195:136–154 (September 1956).

27. Goldschmidt, R. B. "Evolution, as Viewed by One Geneticist," American Scientist, 40:84–98 (1952).

28. Goldschmidt, R. B. Theoretical Genetics. Berkeley and Los Angeles: Univ. of California Press, 1955.

29. Guhl, A. M. "The Social Order of Chickens," Scientific American, 194:42–46 (February 1956).

30. Haskins, C. P. Of Societies and Man. New York: Norton & Co., 1951.

31. Hayek, F. A. Individualism and the Economic Order. Chicago: Univ. of Chicago Press, 1948.

32. Hebb, D. O. The Organization of Behavior. New York: Wiley, 1949.

33. Hempel, C. G. Fundamentals of Concept Formation in the Empirical Sciences, Vol. II, No. 7 of International Encyclopedia of Unified Science. Chicago: Univ. of Chicago Press, 1952.

34. Hempel, C. G., and P. Oppenheim. Der Typusbegriff im Lichte der neuen Logik; wissenschaftstheoretische Untersuchungen zur Konstitutionsforschung und Psychologie. Leiden: A. W. Sythoff, 1936.

35. Hempel, C. G., and P. Oppenheim. "Studies in the Logic of Explanation," Philosophy of Science, 15:135–175 (1948).

36. Hoagland, H. "The Weber-Fechner Law and the All-or-None Theory," Journal of General Psychology, 3:351–373 (1930).

37. Hockett, C. H. "Biophysics, Linguistics, and the Unity of Science," American Scientist, 36:558–572 (1948).

38. Horowitz, N. H. "The Gene," Scientific American, 195:78–90 (October 1956).

39. Hull, C. L. Principles of Animal Behavior. New York: D. Appleton-Century, Inc., 1943.

40. Jacobson, H. "Information, Reproduction, and the Origin of Life," American Scientist, 43:119–127 (1955).

41. Jeffress, L. A. Cerebral Mechanisms in Behavior; the Hixon Symposium. New York: Wiley, 1951.

42. Johnson, M. "The Meaning of Time and Space in Philosophies of Science," American Scientist, 39:412–431 (1951).

43. Kartman, L. "Metaphorical Appeals in Biological Thought," American Scientist, 44:296–301 (1956).

44. Kemeny, J. G. "Man Viewed as a Machine," Scientific American, 192:58–66 (April 1955).

45. Kemeny, J. G., and P. Oppenheim. "Degree of Factual Support," *Philosophy of Science*, 19:307–324 (1952).
46. Kemeny, J. G., and P. Oppenheim. "Systematic Power," *Philosophy of Science*, 22:27–33 (1955).
47. Kemeny, J. G., and P. Oppenheim. "On Reduction," *Philosophical Studies*, 7:6–19 (1956).
48. Lewin, Kurt. *Der Begriff der Genese*. Berlin: Verlag von Julius Springer, 1922.
49. Linderstrom-Lang, K. U. "How is a Protein Made?" *American Scientist*, 41: 100–106 (1953).
50. Lindsey, A. W. *Organic Evolution*. St. Louis: C. V. Mosbey Company, 1952.
51. Lippitt, R. "Field Theory and Experiment in Social Psychology," *American Journal of Sociology*, 45:26–79 (1939).
52. MacCorquodale, K., and P. E. Meehl. "On a Distinction Between Hypothetical Constructs and Intervening Variables," *Psychological Review*, 55:95–105 (1948).
53. McCulloch, W. S., and W. Pitts. "A Logical Calculus of the Ideas Immanent in Nervous Activity," *Bulletin of Mathematical Biophysics*, 5:115–133 (1943).
54. Mannheim, K. *Ideology and Utopia*. New York: Harcourt, Brace, 1936.
55. Mill, John Stuart. *System of Logic*. New York: Harper, 1848 (1st ed. London, 1843).
56. Miller, S. L. "A Production of Amino Acids Under Possible Primitive Earth Conditions," *Science*, 117:528–529 (1953).
57. Miller, S. L. "Production of Some Organic Compounds Under Possible Primitive Earth Conditions," *Journal of the American Chemical Society*, 77:2351–2361 (1955).
58. Moscana, A. "Development of Heterotypic Combinations of Dissociated Embryonic Chick Cells," *Proceedings of the Society for Experimental Biology and Medicine*, 92:410–416 (1956).
59. Needham, J. *Time*. New York: Macmillan, 1943.
60. Nogushi, J., and T. Hayakawa. Letter to the Editor, *Journal of the American Chemical Society*, 76:2846–2848 (1954).
61. Oparin, A. I. *The Origin of Life*. New York: Macmillan, 1938 (Dover Publications, Inc. edition, 1953).
62. Oppenheim, P. "Dimensions of Knowledge," *Revue Internationale de Philosophie*, Fascicule 40, Section 7 (1957).
63. Pauling, L. "Chemical Achievement and Hope for the Future," *American Scientist*, 36:51–58 (1948).
64. Pauling, L. "Quantum Theory and Chemistry," *Science*, 113:92–94 (1951).
65. Pauling, L., D. H. Campbell, and D. Pressmann. "The Nature of Forces between Antigen and Antibody and of the Precipitation Reaction," *Physical Review*, 63:203–219 (1943).
66. Pauling, L., and R. B. Corey. "Two Hydrogen-Bonded Spiral Configurations of the Polypeptide Chain," *Journal of the American Chemical Society*, 72:5349 (1950).
67. Pauling, L., and R. B. Corey. "Atomic Coordination and Structure Factors for Two Helical Configurations," *Proceedings of the National Academy of Science* (U.S.), 37:235 (1951).
68. Pauling, L., and M. Delbrück. "The Nature of Intermolecular Forces Operative in Biological Processes," *Science*, 92:585–586 (1940).
69. Penfield, W. "The Cerebral Cortex and the Mind of Man," in P. Laslett (ed.), *The Physical Basis of Mind*, pp. 56–64. Oxford: Blackwell, 1950.
70. Piaget, J. *The Moral Judgment of the Child*. London: Kegan Paul, Trench, Trubner and Company, Ltd., 1932.
71. Piaget, J. *The Language and Thought of the Child*. London: Kegan Paul, Trench, Trubner and Company; New York: Harcourt, Brace, 1926.

72. Platt, J. R. "Amplification Aspects of Biological Response and Mental Activity," American Scientist, 44:180–197 (1956).
73. Probability Approach in Psychology (Symposium), Psychological Review, 62: 193–242 (1955).
74. Rashevsky, N. Papers in general of Rashevsky, published in the Bulletin of Mathematical Biophysics, 5 (1943).
75. Rescher, N. "Axioms of the Part Relation," Philosophical Studies, 6:8–11 (1955).
76. Rescher, N. and P. Oppenheim. "Logical Analysis of Gestalt Concepts," British Journal for the Philosophy of Science, 6:89–106 (1955).
77. Rosenblueth, A. The Transmission of Nerve Impulses at Neuroeffector Junctions and Peripheral Synapses. New York: Technological Press of MIT and Wiley, 1950.
78. Rothstein, J. Communication, Organization, and Science. Indian Hills, Colorado: Falcon's Wing Press, 1957.
79. Scriven, M. "A Possible Distinction between Traditional Scientific Disciplines and the Study of Human Behavior," in H. Feigl and M. Scriven (eds.), Vol. I, Minnesota Studies in the Philosophy of Science, pp. 330–339. Minneapolis: Univ. of Minnesota Press, 1956.
80. Sellars, W. "A Semantical Solution of the Mind-Body Problem," Methodos, 5:45–84 (1953).
81. Sellars, W. "Empiricism and the Philosophy of Mind," in H. Feigl and M. Scriven (eds.), Minnesota Studies in the Philosophy of Science, Vol. I, pp. 253–329. Minneapolis: Univ. of Minnesota Press, 1956.
82. Shannon, C. E., and J. McCarthy (eds.), Automata Studies. Princeton: Princeton Univ. Press, 1956.
83. Sherif, M. "Experiments in Group Conflict," Scientific American, 195:54–58 (November 1956).
84. Sherif, M., and C. W. Sherif. An Outline of Social Psychology. New York: Harper, 1956.
85. Sherrington, Charles. The Integrative Action of the Nervous System. New Haven: Yale Univ. Press, 1948.
86. Simmel, G. Sociologie. Leipzig: Juncker und Humblot, 1908.
87. Simpson, G. G., C. S. Pittendrigh, and C. H. Tiffany. Life. New York: Harcourt, Brace, 1957.
88. Stanley, W. M. "The Structure of Viruses," reprinted from publication No. 14 of the American Association for the Advancement of Science, The Cell and Protoplasm, pp. 120–135 (reprint consulted) (1940).
89. Timoféeff-Ressovsky, N. W. Experimentelle Mutationsforschung in der Vererbungslehre. Dresden und Leipzig: Verlag von Theodor Steinkopff, 1937.
90. Tolman, E. C. Purposive Behavior in Animals and Men. New York: The Century Company, 1932.
91. Turing, A. M. "On Computable Numbers, With an Application to the Entscheidungsproblem," Proceedings of the London Mathematical Society, Ser. 2, 42:230–265 (1936).
92. Turing, A. M. "A Correction," Proceedings of the London Mathematcial Society, Ser. 2, 43:544–546 (1937).
93. Vannerus, A. Vetenskapssystematik. Stockholm: Aktiebolaget Ljus, 1907.
94. Veblen, T. The Theory of the Leisure Class. London: Macmillan, 1899.
95. Von Bertalanffy, L. "An Outline of General System Theory," The British Journal for the Philosophy of Science, 1:134–165 (1950).
96. Von Neumann, John. "The General and Logical Theory of Automata," in L. A. Jeffress (ed.), Cerebral Mechanisms in Behavior; The Hixon Symposium, pp. 20–41. New York: John Wiley and Sons, Inc., 1951.

Paul Oppenheim and Hilary Putnam

97. Von Neumann, John. "Probabilistic Logics and the Synthesis of Reliable Organisms from Unreliable Components," in C. E. Shannon and J. McCarthy (eds.), *Automata Studies*. Princeton: Princeton Univ. Press, 1956.
98. Watkins, J. W. N. "Ideal Types and Historical Explanation," in H. Feigl and M. Brodbeck (eds.), *Readings in the Philosophy of Science*, pp. 723–743. New York: Appleton-Century-Crofts, 1953.
99. Watkins, J. W. N. "A Reply," *Philosophy of Science*, 22:58–62 (1955).
100. Watson, J. D., and F. H. C. Crick. "The Structure of DNA," *Cold Spring Harbor Symposium on Quantitative Biology*, 18:123–131 (1953).
101. Watson, J. D., and F. H. C. Crick. "Molecular Structure of Nucleic Acids—A Structure for Desoxyribosenucleic Acid," *Nature*, 171:737–738 (1953).
102. Watson, J. D., and F. H. C. Crick. "Genetical Implications of the Structure of Desoxyribosenucleic Acid," *Nature*, 171:964–967 (1953).
103. Weber, M. *The Theory of Social and Economic Organization*, translated by A. M. Henderson and T. Persons. New York: Oxford Univ. Press, 1947.
104. Wheeler, W. M. *Social Life Among the Insects*. New York: Harcourt, Brace, 1923.
105. Wheeler, W. M. *Emergent Evolution and the Development of Societies*. New York: Norton & Co., 1928.
106. Wheeler, W. M. "Animal Societies," *Scientific Monthly*, 39:289–301 (1934).
107. Woodward, R. B., and C. H. Schramm. Letter to the Editor, *Journal of the American Chemical Society*, 69:1551 (1947).
108. Zilsel, E. "History and Biological Evolution," *Philosophy of Science*, 7:121–128 (1940).

Types of Inter-Theoretic Reduction[*]

by LAWRENCE SKLAR

I

The history of scientific theorising presents us with a number of notable accomplishments conveniently grouped together by methodologists as cases of 'the reduction of one theory to another'. The members of this family are, however, by no means identical twins. That is, despite similarities among all the various cases of reduction, a number of distinctive sub-types, whose members all resemble each other far more than they do any members of other sub-types, can be discerned. My aim is to provide the beginnings of a 'taxonomy' of inter-theoretical reductions. This will, of necessity, involve the examination of a number of extant theories of reduction, and, in some cases, their rejection or modification.

The philosopher's interest in reduction in the natural sciences is often motivated to a large extent by the hope that a sufficient understanding of reductions of this sort will provide insight into traditionally philosophical problems, such as the mind-body problem and the complex of issues concerning phenomenalism. I shall avoid concerning myself with these philosophical issues as much as possible, only allowing myself a few remarks on the possible relevance of the results obtained in the body of the paper to current philosophical speculation at the end of my more-or-less 'descriptive' methodological survey.

I also wish to avoid the complex of issues in the foundations of mathematics often labelled problems of reduction. Whatever is meant by the claim that arithmetic is reducible to logic, or more conservatively to set-theory, it is not that arithmetic is related to the more fundamental theory in anything like the way in which a physical theory is related to a theory to which it reduces.[1]

---

* Received 11.ii.66
[1] For a discussion of the nature of reduction in mathematics, cf. N. Goodman, *The Structure of Appearance*, Cambridge, Mass., 1951, ch. I and P. Benacerraf, 'What Numbers Could Not Be', *Phil. Rev.* **74,** (1965) 47-73. It is Goodman's contention, in the reference cited, that the type of reduction appropriate for mathematics is adequate as an account of much philosophical reduction as well.

2

In what follows we shall be concerned with explicating the relations which hold between two theories, T1 and T2, when the latter is said to have been reduced to the former. As a matter of convenience I shall often deal with the expressions of the theories in some natural language rather than with the 'quasi-linguistic' theories themselves, thus identifying theories with their expressions, theoretical concepts with descriptive general terms, etc. While risky in the context of some discussions, this deliberate ambiguity will probably not lead to difficulties in the problems we will be considering.

We may begin our taxonomy with a division of reductions into those cases designated by Nagel 'homogeneous' and those designated 'inhomogeneous'. Two theories will be said to be *homogeneous* if they share the same conceptual apparatus, and *inhomogeneous* if one contains a concept not found in the other. Actually we shall only be concerned to distinguish those cases in which the concepts of the reduced theory are a subset of those of the reducing theory (*homogeneous reduction*) from those cases in which the reduced theory contains some concept not present in the reducing (*inhomogeneous reduction*).

A particularly simple case of homogeneous reduction is that in which the reduced theory is simply deducible from the reducing theory. In such a case the problem of reduction presents no methodological difficulties not already encountered in an attempt to understand the 'internal' structure of a theory. Indeed, why call the relationship a reduction at all, where the reduced theory is, properly speaking, only a fragment of the reducing, developable from it by mere deductive reasoning? Had the reducing theory been developed first, and the reduced subsequently derived from it by a chain of deductive inference, we would not be tempted to describe the relationship between the theories as one of reduction. Where, however, the 'smaller' theory is developed first, and only later shown to be a consequence of a 'larger', newer theory, we are inclined to speak of a reduction having taken place, since the reduced theory had its independent existence prior to either the development of the reducing theory or to the demonstration of the deductive relationship between the two theories.

Matters are not, however, quite this simple even in these very rudimentary cases of reduction. If one looks for examples of reduction from the history of science, strictly derivational reductions are few and far between. One can *construe* various relationships of a strictly derivational sort as reductions, but an examination of cases of reduction pre-analytically so-called, shows that even in the case of homogeneous theories reduction is very rarely derivation.

Consider, for example, the reduction of the Galilean laws of falling bodies or the Keplerian laws of planetary motion to Newtonian mechanics. As has been often pointed out, the reduced theory cannot be derived from the reducing, despite the fact that the concepts of the reduced are certainly a subset of the concepts of the larger theory. What can be derived from the reducing theory is an *approximation* to the reduced, where this notion of approximation is suitably relativised to a degree of accuracy, a range of values of the independent parameters, etc. One might think that the modification of the account of reduction from the strictly to the approximately derivational is a matter of no great philosophical interest, although it might be a matter of some complexity; we all know what 'approximately' means, but just try to *say* what it means for the general case. But certain recent remarks by philosophers about reduction show that somewhat more needs to be done to clarify the matter of approximate derivational reduction.

Consider, for example, the following argument: If T2 reduces to T1, then T1 explains T2. But suppose that the laws deducible from T1 are at best approximations to the laws of T2. To say that A is at best an approximation to B is to say that A and B differ, and that they are, in fact, incompatible. But now consider the so-called deductive model of scientific explanation. According to this account it is a necessary condition of one theory's explaining another that the latter can be deduced from the former. But since one theory can explain another incompatible with a theory deducible from the explanans theory, we can see that the deductive account of explanation cannot possibly be correct.[1]

What are we to say to such an argument? Perhaps the following is a useful approach. First, it is undoubtedly true that in some cases of reduction, pre-analytically so-called, the reduced theory is incompatible with the reducing. Approximate derivational reductions are of this sort and, as we shall see, so are reductions of a rather different kind. A fundamental distinction appears to be necessary between reductions in which the reduced theory is *retained* as correct subsequent to the reduction, or even better confirmed by its being reducible to a well confirmed more general theory, and those reductions in which the reduced theory is instead replaced by the theory to which it reduces. The difference between these two cases is so radical that it might be better to restrict the extension of the term 'reduction' so as to exclude replacements as cases of reduction. Two considerations militate against this move, however, useful as it would be in eliminating possible confusion. First, many important cases

---

[1] P. K. Feyerabend, 'Explanation, Reduction and Empiricism', in H. Feigl and G. Maxwell, (eds.), *Minnesota Studies in the Philosophy of Science*, 3, 43-52.

of reduction in science, pre-analytically so-called, are clearly replacement-reductions. Second, there are many important ways in which replacement-reductions are like those reductions in which the reduced theory is retained, despite the obvious and important differences between the two kinds of reduction.

Although I think some recent authors have gone too far in claiming that all, or at least all significant, reductions are replacement-reductions,[1] I do not want to minimise the importance in science of such 'discarding' replacements in which the reduced theory is shown to be incorrect by the very act of its being reduced. But the conclusions to be drawn from this observation must be examined with care. Consider, for example, the use of the existence of replacement-reductions to attack the deductive theory of explanation. Does this attack hold up under scrutiny? I think not. But this is not easy to demonstrate, at least to everyone's satisfaction. For the issue is as to the valid application of the term 'explanation', and such *decisions* regarding usage are hardly amenable to deductive proof. What can be provided is a clarification of the point at issue and some justifications for a *recommendation* to apply the term 'explanation' in a particular way.

At least one plausible distinction to be drawn is that between one theory's *explaining* another, and the former's *explaining away* the latter. Is it plausible, for example, to maintain that quantum mechanics, assumed for the moment to be correct, *explains* an admittedly incorrect Newtonian mechanics? Surely not, for an incorrect theory cannot be explained on any grounds. ('Explain why you were late.' 'I wasn't.' 'I am well aware of that fact but I insist upon an explanation for your lateness anyway.') If we extend the extension of the term 'explanation' in scientific contexts, so as to allow the possibility of explaining incorrect theories we are forced by 'parity of discourse' to allow discourse in our ordinary language which we would intuitively regard as senseless.

What, then, *does* quantum mechanics explain, if it fails to explain Newtonian mechanics? Among other things it explains why Newtonian mechanics *seemed* to be correct; why it met with such apparent success for such a long period of time and under such thorough experimental scrutiny. ('Explain why you were late.' 'But I wasn't.' 'Ah! Now I see you are right. You weren't late. But why did everybody think you were?') If the distinction between explaining a theory and explaining its apparent success is drawn there is no reason whatever why we cannot admit that T2 is reducible to T1, admit that T2 is incompatible with T1, and yet

[1] See, for example, T. S. Kuhn, *The Structure of Scientific Revolutions*, Chicago, 1962, *passim*.

deny that T1 explains T2. Granting that many reductions are radical replacements of the reduced theory by the reducing one need not in any way loose one's confidence in a strictly deductive account of explanation.

3

The problem of reduction becomes more difficult when the theory to be reduced contains concepts not present in the reducing theory. One solution is simply to dispose of the reduced theory entirely, for a non-existent theory presents no difficulties of un-accounted-for concepts. But this radical course seems inappropriate in many cases. Often we feel inclined to accept the retention in some sense of the reduced theory, despite the fact that it contains concepts not present in the reducing, nor 'definable away' in terms of the concepts of the latter theory. In one type of degenerate case it will be possible to strictly deduce the reduced theory from the reducing, despite the presence in the former of 'obtrusive' concepts. We can, for example, deduce from a theory of the form 'p' a theory of the form 'p or q', even if 'q' contains terms not present in 'p'. But such 'trivialized' deductions will hardly do justice to the many cases of significant reduction in science in which the reduced theory contains a concept not present in the reducing.

Such inhomogeneous reductions are common. The theory of heredity speaks of genes, while this concept appears nowhere in the chemical side of biochemistry. Physical optics refers to light-waves, whereas the electromagnetic theory to which it reduces refers only to electric and magnetic fields, etc. How can it be possible for one theory to be reducible to another when it is patently clear that the sentences of the former could not be deduced, or even approximately deduced, from those of the latter, even allowing for linguistic transformations such as substitutions of synonymous terms and grammatical transformations? We shall see that at least two kinds of inhomogeneous reduction are possible: a weaker variety in which the retention of the reduced theory is retention only in a much weakened sense; and a stronger form in which the reduced theory remains well established even after the reduction, and may even be given added confirmation by it.

A model of reduction proposed as a *general* account of reduction by Kemeny and Oppenheim will provide us with some insight into the weaker version of inhomogeneous reduction.[1] Their theory, of an obviously positivistic and instrumentalistic nature, begins with a familiar division of the vocabulary of a theory into an observational and theoretical part.

[1] J. Kemeny and P. Oppenheim, 'On Reduction', *Phil. Studies*, **7**, (1952), 6-19.

Derivative'from this is a division of the sentences of the theory into observational—those having as their only essentially appearing non-logical terms observation terms—and theoretical—those containing at least one essential occurrence of a term from the theoretical vocabulary. Reasoning that the sole purpose of a theory is to provide observational consequences, one can then define an equivalence relation between theories by the notion that two theories are equivalent if the same class of observational sentences is derivable from each. But equivalence relations are symmetrical and reductive relations asymmetrical! The asymmetry is introduced through the notion of the *systematic power* of a theory. This is a measure of the ability of the theory to predict as wide a range of phenomena as possible from as little data as possible; and, although there exists no clear explication which is intuitively entirely satisfactory, interesting attempts in a direction toward it have been made.[1] Clearly if two theories are predictively equivalent and one is of greater systematic power than the other, then the more powerful theory is preferred as reducing theory. Again, if the class of deducible observational consequences of one theory is a proper subset of the consequences of another of greater or equal systematic power, then the more general theory is preferred. By means of these notions we can introduce an asymmetric relation of reduction on theories as follows: T_2 is reducible to T_1 if (*a*) all the observational sentences deducible from T_2 are deducible from T_1 as well, and (*b*) T_1 is of greater systematic power than T_2. The harmony of this account of reduction with the more general thesis that theories are 'instruments for the correlation of observations' is clear.

Three reasons militate against the acceptance of the Kemeny-Oppenheim account as a general analysis of reductions in science. Two of these are objections to the general analysis of the nature of theory presupposed by their account. As a consequence I shall be able only to state them rather than to argue for them. First, the distinction between observational and theoretical terms (these latter would be more appropriately designated 'non-observational') is one which has been much doubted in recent years. Without resorting to the stipulation of observational terms as terms in a phenomenal 'sense-datum' language, an epistemological radicalism now shunned by most philosophers of science, it seems difficult to provide anything like a division of the world into observables and non-observables. Even if we could make such an epistemological distinction it is not at all certain that a 'map' of it in terms of a linguistic distinction of kinds of *vocabulary* could be drawn.

---

[1] C. Hempel and P. Oppenheim, 'The Logic of Explanation', *Phil. of Science*, 15, (1948), 164, and J. Kemeny and P. Oppenheim, 'Systematic Power', *Phil. of Science*, (22, (1955), 27-35.

Second, the 'positivistic-instrumentalistic' view of the role of theories in science clearly suggested and presupposed by the Kemeny-Oppenheim account is itself a view of physical theories hard to defend. But without such a positivistic epistemology and metaphysics it is hard to see the plausibility of the Kemeny-Oppenheim account as an explication of the meaning of reduction in natural science.

Lastly, some empirical examination of science casts even more doubt upon this account of reduction. It is hard to find in the history of science any examples of reduction, intuitively so-called, which satisfy the account; or, more correctly, it is hard to find any examples which satisfy the Kemeny-Oppenheim account without satisfying the much more demanding accounts of strict derivational reduction or of the more rigid kind of inhomogeneous reduction we shall examine later as strong inhomogeneous reduction. That is, the model seems too *weak* to account for some types of reduction, failing to illuminate the strength of the connection between reduced and reducing theory; and too *strong* for the remaining cases, making demands upon the relation between reduced and reducing theories which are simply not satisfied.

To see the *weakness* of the model for some cases we need to explicate the stronger connection holding between the related theories. We have done this in the case of strict derivational reduction and shall proceed to the strict inhomogeneous reductions later on. To see the other side of the coin we need only note that the very cases the Kemeny-Oppenheim account was introduced to handle fail to meet its requirements. The account was originally introduced to provide for those cases of reduction in which the reduced theory was, in fact, replaced by the reducing. The authors seemed to have hoped that the incompatibility of the reduced and reducing theories in these cases could be isolated at the theoretical level, the level at which on their account no relation of equivalence or inclusion between the theories is required. But this is not true. For if the reducing and reduced theories made *no* difference in our observational predictions, in so far as they 'covered' the same type of phenomena, then what would have motivated us to *discard* the reduced theory in favour of the reducing? No matter how stringent one's requirements for a state of affairs to be 'observable' are, for example, Newtonian mechanics and special relativity simply do not predict the same observational results, or even *approximately* the same observational results, at least for a significant and easily specifiable range of experimental conditions.

To summarise, the Kemeny-Oppenheim account fails to hold for any reductions except those satisfying much more stringent and illuminating conditions. In other words, the only reductions which hold between

41

theories making the same observational predictions within the range of phenomena 'covered' by the reduced theory, are reductions which display an important relationship between the *theoretical* parts of the reduced and reducing theory.

An account of reduction derived from the Kemeny-Oppenheim account does, however, seem to do justice to the weak inhomogeneous reductions they intended to explicate. Let us suppose we are presented with two theories, the one, T_2, to be reduced to the other, T_1. In addition suppose that T_2 contains some concepts not present in T_1. How then can T_2 be reducible to T_1? Suppose that of the concepts contained in T_2 a certain proper subset consists of concepts shared with T_1. Let us call these concepts the *external* concepts of T_2 and T_1, designating the remaining concepts the concepts *internal* to their theories. This distinction is one to be relativised to the particular reduction in question and not to be identified with any variety of observational/theoretical distinction. The external concepts may be either observational or theoretical concepts, no matter how this latter distinction is drawn. Further, suppose we consider the sentences deducible from T_2 and from T_1 containing as essentially occurring descriptive terms only terms of the external sort. We call these the *external sentences* of the respective theories. In addition, suppose that within the joint domain of external sentences of the two theories there is a subset of sentences derivable both from T_1 and from T_2, or derivable from T_1 and such that approximations to them are derivable from T_2. Where these conditions are satisfied we shall say that T_2 is *partially* (or *approximately partially*) *reducible* to T_1.

Examples of partial reduction are easy to come by. If we first restrict our attention to sentences framable in purely kinematic concepts, and then further restrict our attention to the subset of sentences in this class dealing with sufficiently low velocities, we find for every sentence in this subset derivable from the relativistic theory there is a sentence approximative to it derivable from the Newtonian theory. It is only in this extremely weak sense that Newtonian mechanics is reducible to special relativity. We should note that a theory may be partially reducible to any number of other theories mutually incompatible both with regard to their observational consequences and with regard to their theoretical conceptual apparatus.

I might remark in passing that at least one philosopher would find fault in my maintaining that the replaced theory has any conceptual apparatus at all in common with the theory which supplants it.[1] While

---

[1] Cf. P. K. Feyerabend, op. cit. section 7.

admitting the force of this argument, partially based upon the notion that the meaning of a concept is based upon its role within the total theory in which it occurs and partially upon the radical 'conceptual shifts' necessary to go from, say, Newtonian space and time to relativistic space-time, I doubt its validity. But since I am not prepared to offer anything like even the beginnings of an analysis of the notion of the meaning of a theoretical term, I can only state my prejudice and admit my unease.

It is only reasonable that an account of reduction derived from an instrumentalistic view of the nature of theories is applicable in some cases. For what is it that makes us want to speak of some theoretical replacements as reductions—say the reductions of Newtonian mechanics to relativity theory and to quantum mechanics, or the reduction of hydrodynamics, with its presupposed continuous fluid media, to the atomic theory of fluids; and to speak of other replacements—say of the phlogiston by the oxygen of combustion or the caloric by the energetic theory of heat—as the mere discarding of one theory in favour of a better? It is only the survival of the older theory, in the reductive cases, as a 'useful instrument for prediction', despite its known falsity as a scientific theory. If reference to caloric, its presence and absence, its rate of flow, the capacity of bodies to absorb it, etc., were as useful to engineers—despite the fact that there is no caloric—as is the hydrodynamic theory to boat designers—despite the fact that there are no microscopically continuous fluid media—there is little doubt that we would speak of the reduction of the caloric theory to the energetic, rather than welcome its overthrow as we are now inclined to do. The relation of partial reduction is a very weak one and must be, as noted, relativised to many factors idiosyncratic to the particular reduction in question. But why not? After all, the theory reduced is not only incorrect but conceptually untenable as well. The only thing good about it is that it is a useful device for making predictions over a limited range of phenomena within its scope. It is only this feature which it shares with the reducing theory, and it is only this shared feature that is required by the notion of partial reduction.

4

More interesting then these cases of partial reduction are the cases of inhomogeneous reduction where the reduction is complete; that is, where the reduced theory is assimilated *in toto*, theoretical apparatus and all, to the reducing, and where the very fact of its being reducible serves to increase our confidence in its correctness. The reduction of the physical optical theory of light to the theory of electromagnetic radiation is of this nature and provides almost a paradigm of such a reduction.[1]

Nagel offers an interesting account of such inhomogeneous but total reductions.[2] He assumes, as I shall, that the concepts appearing in the reduced theory but absent from the reducing theory are not eliminable in terms of the concepts of the reducing theory by linguistic investigation alone. If such were the case, that is, of the obtrusive terms of the reduced theory were synonymous with expressions constructable out of the descriptive terms of the reducing theory, and if substitution for such terms of their equivalents allowed the derivational reduction of the reduced theory to the reducing, then the reduction would constitute a matter of linguistic insight and perhaps clever logical inference, but it would hardly require experimental justification and observational confirmation. But the reduction of, say, physical optics to electromagnetic theory is not of this sort, but is instead an important *factual* discovery of empirical science.

To account for such a reduction some connection must be found between the terms peculiar to the reduced theory and those found in the reducing, despite the fact that this connection cannot be supplied by assiduous use of dictionary, grammar book and logic text. Where can such a connection be found? Nagel rightly asserts that it must be an empirically justified scientific discovery. But what should its nature be? According to Nagel the connection between the two theories is to be found in the existence of a set of 'bridging' hypotheses or 'correlatory laws'. It is the discovery of these laws, containing concepts of the reducing theory *and* concepts peculiar to the reduced theory, which allows for the reduction of the one theory to the other. Originally Nagel insisted upon these correlatory hypotheses being universally quantified bi-conditionals, with one side of the bi-conditional containing as its only descriptive term one of the terms peculiar to the reduced theory. Subsequently he appears to have weakened this condition, allowing the laws to take other forms as well.[3]

But not just any set of correlatory laws will suffice to establish a reduction. For the purpose of the reduction is to show that given our

---

[1] Like nearly all examples, this one fails to be quite adequate. There are predictions of traditional physical optics incompatible with some of those of electromagnetic theory, e.g. the latter's prediction of the exponentially decaying penetration of electro-magnetic waves into the surface of a reflecting opaque object. But the reduced theory can in this case, as in others, be suitably modified, without loss of any of its idiosyncratic concepts, so that the modified version of the theory is properly totally reducible to the electromagnetic theory.

[2] E. Nagel, 'The Meaning of Reduction in the Natural Sciences', reprinted in A. Danto and S. Morgenbesser, (eds.), *Philosophy of Science*, New York, (1960), pp. 288-312 and E. Nagel, *The Structure of Science*, New York, 1961, ch. 11.

[3] E. Nagel, *The Structure of Science*, pp. 355-6.

reducing theory, *and* the correlatory laws, the independent assumption of the hypotheses of the reduced theory is no longer necessary. To meet this requirement we must insist that the correlatory laws meet the *condition of derivability*: that from the reducing theory and the correlatory laws the reduced theory can be derived by logic alone.[1] Let us call a body of 'bridging' hypotheses or 'correlatory laws' sufficient to allow us to deduce one theory from another with its aid a *correlatory theory* for the two theories, and the relation established by the existence of these laws a *correlation*. Its correlation an adequate analysis of total inhomogeneous reduction?

Unfortunately it is not. It is easy to show that correlation is insufficient for reduction, although necessary in a sense. First note its insufficiency. The Wiedemann-Franz law is a well-known law allowing one to infer the thermal conductive properties of a material from its electrical conductive properties. With it we can deduce from the laws of electrical conductivity of a material the laws of its thermal conductivity. But does this law establish the reduction of the theory of heat conduction to the theory of the conduction of electricity? No one has ever maintained that it does. What does explain *both* the electrical and thermal properties of matter, and the Wiedemann-Franz law as well, is the reduction of the macroscopic theory of matter to the theory of its atomic microscopic constitution. Although the correlation *points* to a reduction it does not *constitute* a reduction by itself.

But why doesn't this correlation count as a reduction? The answer seems clear. The derivation of the laws of the reduced theory, T_2, from those of the reducing, T_1, requires the introduction of the independent laws of the correlating set. Call their conjunction TC. But the fact that T_2 can be deduced from T_1 *and* TC is evidence only of the reducibility of T_2 to the now expanded theory, 'T_1 and TC'. This reduction is not inhomogeneous, since the conjunction of T_1 and TC contains all the concepts present in T_2; in fact, it is a simple derivational reduction. But this reduction is not the reduction of T_2 to T_1 alone originally sought for.

If we can find correlations which are not reductions, can we find reductions which display no set of correlatory laws? Not surprisingly we

---

[1] We might note that in its original form, with (1) the bridging postulates simple biconditionals containing each only a single one of the reduced theory's idiosyncratic terms, and (2) this only once on one side of the biconditional, and (3) with one bridging law for each of the terms peculiar to the reduced theory, the condition of derivability is unnecessary if we are interested only in reducing the reduced theory to *some* theory not containing the peculiar concepts in question. If we wish the reduction to hold to some *specified* theory not containing the peculiar concepts, or if our correlatory laws are not of the simple form described, however, the condition of derivability must be imposed as a separate requirement. See Nagel, ibid. p. 356 note.

can. Although we find books on the theory of physical optics, and books on the theory of electromagnetic radiation, we fail to find texts on the theory (necessary to account for the reduction of physical optics to electromagnetic theory on Nagel's account) of the correlation of light with electromagnetic radiation. There are, then, reductions without correlations.

But I have said that 'bridging' hypotheses are in some sense necessary to reductions, and I must now make that sense clear. What I shall propose is nothing which is surprising any longer, nor do I claim for it any originality. On the other hand it seems at least part of a correct account. It is that the place of correlatory laws is taken by empirically established *identifications* of two classes of entities.[1] Light waves are not correlated with electromagnetic waves, for they *are* electromagnetic waves. There are not two classes of entities, but only one; a class whose members may be referred to with equal correctness as light waves or as electromagnetic waves. The identification takes the form of a universally quantified conditional, or biconditional if the two classes are coextensive (which they are not in our example since there are electro-magnetic waves which are *not* light waves, although all light waves are electromagnetic waves); but it is a universalised conditional of a very special form.

The advantages of identification over 'mere' correlation have been dealt with at great length in recent literature, and I shall mention them quickly and without elaboration. I might note in passing that at least some of these advantages take on a particularly obvious aspect in the case of the reduction of the theory of the macroscopic properties of matter to the atomic theory, a reduction accomplished by the identification of macroscopic material substances with arrangements of atoms. In this case the very notion of a correlatory alternative to the identification seems peculiar. What would it be to *correlate* the piece of salt with an array of atoms at the same place and time rather than to identify the salt as an array of atoms?

Three distinct features of these identifications should be noted. First, these identifications are quite different from those identifications in which we identify some one individual referred to by one proper name with an individual referred to by a different name. Primarily this is because (1) the identifications are general, identifying whole *classes* of individuals, and (2) because the predicates used are not only meaningful, having sense as

---

[1] Notice of the special role of identificatory statements was originally made by philosophers concerned with the mind-body problem. See, for example, U. Place, 'Is Consciousness a Brain Process', *Brit. Jour. Psych.* **47,** (1956), 44-50; H. Feigl, 'The *Mental* and the *Physical*', in Feigl, Scriven and Maxwell (eds.), *Minnesota Studies in the Philosophy of Science,* **2,** pp. 439 ff.; J. J. C. Smart, 'Sensations and Brain Processes', *Phil. Review,* **68,** (1959), 141-56; H. Putnam, 'Minds and Machines', in S. Hook (ed.), *Dimensions of Mind,* New York, 1960, pp. 148-79.

well as denotation, but have their sense due to their role in a scientific theory of wide scope, which theory provides connection by means of laws between these predicates and many others of the same, higher or lower (i.e. 'more directly observational') level.

Second, the identifications, despite the fact that they are universally generalised conditionals, differ radically from ordinary physical laws in that a request for an explanation of them is inappropriate. You cannot, as many have pointed out, ask *why* light waves are electromagnetic waves. What has not been as fully pointed out is that the identificatory process hardly 'eliminates' responsibility for explanations previously present. To a large extent, and this is a notion which needs far more looking into, the identification simply *transfers* the need for explanation to some other part of the theory. True, we cannot ask why light waves are electromagnetic waves. But we surely wish to know why entities like electromagnetic waves display the properties which we formerly used as our *criteria* for the presence of light waves. And the answers to these questions may be complex indeed. In our example, for instance, it will require explaining why electromagnetic waves affect our ability to *see* physical objects, and this in turn requires the identification of eyes and objects with complex arrays of electromagnetically interactive atoms and their components.

Third, despite the ability of identifications to 'simplify' our ontology, their ability to 'eliminate' unnecessary entities must be dealt with cautiously. It is true that after the identification there is only one *class* of entities, where before we thought there were two. But there are still as many light waves in the world subsequent to the identification as there were before, and as much commitment to them in our theory. This should be contrasted with, for example, the replacement of the caloric theory by the energetic theory of heat. According to the latter theory there simply is no such stuff as caloric. According to the electromagnetic theory of light there may indeed be light waves, and there are—lots of them. A new theory may bring about ontological simplification either by absorbing the ontology of an existing theory, or by eliminating it.

5

Although I have committed myself only to offering a taxonomy of reductions, and have deliberately eschewed questions of a general metaphysical and epistemological nature, I cannot resist the temptation to conclude by reviewing one or two considerations suggested by the nature of identificatory reductions. These may be of interest to philosophers more concerned with the metaphysics and epistemology of identification, than with its role in intertheoretic reduction in the natural sciences.

(1) How do we know when one class of entities is to be identified with a class of entities differently designated, and when the behaviour of the entities in one class is merely to be correlated with the behaviour of the entities in the other? The answer seems to be, although I imagine it is fraught with difficulties, that you identify whenever you can. That is, unless one has grounds for doubting the correctness of the identification, one should assume its correctness, even if a mere correlation would be equally compatible with all known facts.

(2) Must entities be spatial in order to be identified? Perhaps, but you need not have known they were spatial to begin with. Genes are DNA molecules, and since molecules are spatial objects, so are genes. Prior to the identification no one *knew* that genes were spatial objects, although many of course suspected they were. In any case the identification need not be established by first determining the spatio-temporal coincidence of the objects to be identified as some have maintained.

(3) Since identicals are indiscernable (we need not worry about in-tensional properties), then if *a* is identical with *b*, and if *a* has property *P*, must not *b* have property *P*? Of course. Even if it sounds odd to speak of *b* as being *P* it may be true that it is. Who would have thought that the statement 'The gene for blue eyes weighs less than a pound' made good sense? Would Mendel have found it to offend his sense of ordinary language? Perhaps so, but it's true anyway.[1]

One fact which should be noted with more care than has been given it, however, is that when we wish to identify *a* with *b*, and when we *think* that *a* has property *P*, and when we will not tolerate asserting that *b* has *P*, one way out is simply to deny that *a* really has property *P* after all. A standard procedure for doing this seems to be to 'push' property *P* into the realm of the mental where, if you are a physical scientist, it won't bother you any more. Thus, consider the following:

A: 'Apples are arrays of atoms.'
B: 'But how can apples be arrays of atoms? After all, apples are *sweet* aren't they? And arrays of atoms can't really be *sweet*?'
A: 'But apples aren't *really* sweet. The sweetness is *in the mind*.'

I do not say this is a *good* argument on A's part. Certainly as it stands it simply will not do. But is not something like this at least part of the impetus behind the theory of 'secondary qualities'? Anyone wishing to use the analysis of identificatory reduction in the natural sciences to illuminate a

[1] H. Putnam, ibid. pp. 166-9.

theory of the relation of mind to brain will have to deal with this problem with care. For in this case there is no mental 'receptacle' for inconvenient properties.

(4) The fact that it is *things* which one identifies in inhomogeneous reductions leads to a number of puzzles. In our identificatory 'bridging' hypothesis the variables must be taken to range over things, for example light waves which are electromagnetic waves, and our predicates must be substantival, for example, '. . . is a light wave' or '. . . is an electromagnetic wave'. If we reframed our physical theories in a 'co-ordinate language' in which the variables ranged only over space-time points, or over their co-ordinates, and in which the predicates were all property-attributing of the points, or their co-ordinates, then no distinction could be drawn between an identification of, say, light waves with electromagnetic waves, and the 'mere' correlation of the appropriate properties of points or co-ordinates. Whether this is an argument against the adequacy of co-ordinate languages for physical theories, or against the tenability of the distinction between correlations and identifications is an interesting question. I have no desire to perpetuate a 'useless metaphysical' distinction between substantival and property-attributing predicates, but if the differences between correlations and identifications noted above are real differences, then a distinction of this sort, metaphysical as it may be, seems to be far from useless. Unfortunately I am completely unprepared to offer even the beginnings of an analysis into just what this distinction, and the concomitant distinction of generalisations into general identifications and correlations, amounts to.

(5) A warning must be given against the two casual extension of the notion of identification to 'entities' of categories other than that of substance. It is all very well to identify light waves with electromagnetic waves, salt crystals with arrays of atoms or genes with DNA molecules; for all these entities are *things*. It is quite a different matter to 'identify' *properties* with one another, or *events* with one another, or *states of affairs* with one another, or *processes* with one another. It is clear that far too much mileage was got out of the identification of *things* in natural science by those attempting to treat the relation between mental processes and physical *processes* as an exactly analogous matter. Many of the criticisms of the 'identity' theory of mind make the same error. Those who claim that mental processes are not spatial, and hence not identical with physical processes, and those who claim that mental processes are shown to be spatial by their identity with physical processes, even though we did not realise their spatiality prior to the discovery of the identity, are equally careless. For no *process* is spatial in the sense that a physical *object* is

spatial, and this goes for *physical* processes as well as for mental processes!

It would be pleasant to dissolve the whole issue of the 'identity' theory of mind by showing that we never identify 'entities' except those in the category of things. But this is not so. We even 'identify' properties, as in 'The colour of the sky is blue', or 'The colour of rubies is red'.

(6) Lastly, there is one kind of 'reduction' upon which we have not even touched. This is the 'reduction' which proceeds by 'identifying' one class of referents with a class of 'abstract objects' associated with (instance in) but not identical to some particular physical objects. The 'reduction' of the *logical states* of a machine to its *physical states*, or of some 'mental' properties to 'patterns of functional organisation of the physical body' are of this kind. Such 'reductions', however they are to be analysed, are hardly the 'identificatory reductions' of the sort familiar from physical science.[1]

6

We have found that reductions are distinctive scientific accomplishments of very different kinds. Some are homogeneous, others inhomogeneous. Some are 'total', others merely 'partial'. Some are simply derivations, others elaborate identifications. Worse yet, some provide deeper confirmation for the reduced theory, while others serve to eliminate the reduced theory as a viable competitor for the status of scientific truth in the very act of reducing it to another accepted theory. For all their similarities, reductions are a very diverse bunch of items to come under a single name. But this should neither surprise nor disturb us so long as we do not let the convenience of the use of one 'family name' lead to obfuscation by equivocation.

Princeton University
N.J., U.S.A.

[1] For a discussion of this notion of reduction, cf. H. Putnam, 'The Mental Life of Some Machines', to appear in H. Castaneda, (ed.), *Wayne State Symposium in the Philosophy of Mind*, and J. Fodor, 'Explanations in Psychology', in M. Black, (ed.), *Philosophy in America*, London, 1965, pp. 161-79.

SPECIAL SCIENCES (OR: THE DISUNITY OF SCIENCE AS A WORKING HYPOTHESIS)*

A typical thesis of positivistic philosophy of science is that all true theories in the special sciences should reduce to physical theories in the long run. This is intended to be an empirical thesis, and part of the evidence which supports it is provided by such scientific successes as the molecular theory of heat and the physical explanation of the chemical bond. But the philosophical popularity of the reductivist program cannot be explained by reference to these achievements alone. The development of science has witnessed the proliferation of specialized disciplines at least as often as it has witnessed their reduction to physics, so the widespread enthusiasm for reduction can hardly be a mere induction over its past successes.

I think that many philosophers who accept reductivism do so primarily because they wish to endorse the generality of physics vis à vis the special sciences: roughly, the view that all events which fall under the laws of any science are physical events and hence fall under the laws of physics.[1] For such philosophers, saying that physics is basic science and saying that theories in the special sciences must reduce to physical theories have seemed to be two ways of saying the same thing, so that the latter doctrine has come to be a standard construal of the former.

In what follows, I shall argue that this is a considerable confusion. What has traditionally been called 'the unity of science' is a much stronger, and much less plausible, thesis than the generality of physics. If this is true it is important. Though reductionism is an empirical doctrine, it is intended to play a regulative role in scientific practice. Reducibility to physics is taken to be a *constraint* upon the acceptability of theories in the special sciences, with the curious consequence that the more the special sciences succeed, the more they ought to disappear. Methodological problems about psychology, in particular, arise in just this way: the assumption that the subject-matter of psychology is part of the subject-matter of physics is taken to imply that psychological theories must reduce to physical theories, and it is this latter principle

Synthese **28** (1974) 97–115. *All Rights Reserved*
Copyright © 1974 by D. Reidel Publishing Company, Dordrecht-Holland

that makes the trouble. I want to avoid the trouble by challenging the inference.

I

Reductivism is the view that all the special sciences reduce to physics. The sense of 'reduce to' is, however, proprietary. It can be characterized as follows.[2]

Let

$$(1) \qquad S_1x \rightarrow S_2x$$

be a law of the special science S. ((1) is intended to be read as something like 'all S_1 situations bring about S_2 situations'. I assume that a science is individuated largely by reference to its typical predicates, hence that if S is a special science 'S_1' and 'S_2' are not predicates of basic physics. I also assume that the 'all' which quantifies laws of the special sciences needs to be taken with a grain of salt; such laws are typically *not* exceptionless. This is a point to which I shall return at length.) A necessary and sufficient condition of the reduction of (1) to a law of physics is that the formulae (2) and (3) be laws, and a necessary and sufficient condition of the reduction of S to physics is that all its laws be so reducible.[3]

$$(2a) \qquad S_1x \leftrightarrows P_1x$$
$$(2b) \qquad S_2x \leftrightarrows P_2x$$
$$(3) \qquad P_1x \rightarrow P_2x.$$

'P_1' and 'P_2' are supposed to be predicates of physics, and (3) is supposed to be a physical law. Formulae like (2) are often called 'bridge' laws. Their characteristic feature is that they contain predicates of both the reduced and the reducing science. Bridge laws like (2) are thus contrasted with 'proper' laws like (1) and (3). The upshot of the remarks so far is that the reduction of a science requires that any formula which appears as the antecedent or consequent of one of its proper laws must appear as the reduced formula in some bridge law or other.[4]

Several points about the connective '$\rightarrow$' are in order. First, whatever other properties that connective may have, it is universally agreed that it must be transitive. This is important because it is usually assumed that the reduction of some of the special sciences proceeds via bridge laws which connect their predicates with those of intermediate reducing

52

theories. Thus, psychology is presumed to reduce to physics via, say, neurology, biochemistry, and other local stops. The present point is that this makes no difference to the logic of the situation so long as the transitivity of '→' is assumed. Bridge laws which connect the predicates of S to those of S^* will satisfy the constraints upon the reduction of S to physics so long as there are other bridge laws which, directly or indirectly, connect the predicates of S^* to physical predicates.

There are, however, quite serious open questions about the interpretations of '→' in bridge laws. What turns on these questions is the respect in which reductivism is taken to be a physicalist thesis.

To begin with, if we read '→' as 'brings about' or 'causes' in proper laws, we will have to have some other connective for bridge laws, since bringing about and causing are presumably *a*symmetric, while bridge laws express symmetric relations. Moreover, if '→' in bridge laws is interpreted as any relation other than identity, the truth of reductivism will only guaranty the truth of a weak version of physicalism, and this would fail to express the underlying ontological bias of the reductivist program.

If bridge laws are not identity statements, then formulae like (2) claim at most that, by law, x's satisfaction of a P predicate and x's satisfaction of an S predicate are causally correlated. It follows from this that it is nomologically necessary that S and P predicates apply to the same things (i.e., that S predicates apply to a subset of the things that P predicates apply to). But, of course, this is compatible with a non-physicalist ontology since it is compatible with the possibility that x's satisfying S should not itself *be* a physical event. On this interpretation, the truth of reductivism does *not* guaranty the generality of physics *vis à vis* the special sciences since there are some events (satisfactions of S predicates) which fall in the domains of a special science (S) but not in the domain of physics. (One could imagine, for example, a doctrine according to which physical and psychological predicates are both held to apply to organisms, but where it is denied that the event which consists of an organism's satisfying a psychological predicate is, in any sense, a physical event. The up-shot would be a kind of psychophysical dualism of a non-Cartesian variety; a dualism of events and/or properties rather than substances.)

Given these sorts of considerations, many philosophers have held that bridge laws like (2) ought to be taken to express contingent event identities, so that one would read (2a) in some such fashion as 'every event which consists of x's satisfying S_1 is identical to some event which consists of x's satisfying P_1 and vice versa'. On this reading, the truth of reductivism would entail that every event that falls under any scientific law is a physical event, thereby simultaneously expressing the ontological bias of reductivism and guaranteeing the generality of physics *vis à vis* the special sciences.

If the bridge laws express event identities, and if every event that falls under the proper laws of a special science falls under a bridge law, we get the truth of a doctrine that I shall call 'token physicalism'. Token physicalism is simply the claim that all the events that the sciences talk about are physical events. There are three things to notice about token physicalism.

First, it is weaker than what is usually called 'materialism'. Materialism claims *both* that token physicalism is true *and* that every event falls under the laws of some science or other. One could therefore be a token physicalist without being a materialist, though I don't see why anyone would bother.

Second, token physicalism is weaker than what might be called 'type physicalism', the doctrine, roughly, that every *property* mentioned in the laws of any science is a physical property. Token physicalism does not entail type physicalism because the contingent identity of a pair of events presumably does not guarantee the identity of the properties whose instantiation constitutes the events; not even where the event identity is nomologically necessary. On the other hand, if every event is the instantiation of a property, then type physicalism does ential token physicalism: two events will be identical when they consist of the instantiation of the same property by the same individual at the same time.

Third, token physicalism is weaker than reductivism. Since this point is, in a certain sense, the burden of the argument to follow, I shan't labour it here. But, as a first approximation, reductivism is the conjunction of token physicalism with the assumption that there are natural kind predicates in an ideally completed physics which correspond to each natural kind predicate in any ideally completed special science. It will be one of my morals that the truth of reductivism cannot be inferred from the assumption that token physicalism is true. Reductivism is a sufficient, but not a necessary, condition for token physicalism. In what follows, I shall assume a reading of reductivism which entails token physicalism. Bridge laws thus state nomologically necessary contingent event identities, and a reduction of psychology to neurology would entail that any event which consists of the instantiation of a psychological property is identical with some event which consists of the instantiation of some neurological property.

Where we have got to is this: reductivism entails the generality of physics in at least the sense that any event which falls within the universe of discourse of a special science will also fall within the universe of discourse of physics. Moreover, any prediction which follows from the laws of a special science and a statement of initial conditions will also follow from a theory which consists of physics and the bridge laws, together with the statement of initial conditions. Finally, since 'reduces to' is supposed to be an asymmetric relation, it will also turn out that physics is *the* basic science; that is, if reductivism is true, physics is the only science that is general in the sense just specified. I now want to argue that reductivism is too strong a constraint upon the unity of science, but that the relevantly weaker doctrine will preserve the desired consequences of reductivism: token physicalism, the generality of physics, and its basic position among the sciences.

II

Every science implies a taxonomy of the events in its universe of discourse. In particular, every science employs a descriptive vocabulary of theoretical and observation predicates such that events fall under the laws of the science by virtue of satisfying those predicates. Patently, not every true description of an event is a description in such a vocabulary. For example, there are a large number of events which consist of things having been transported to a distance of less than three miles from the Eiffel Tower. I take it, however, that there is no science which contains 'is transported to a distance of less than three miles from the Eiffel Tower' as part of its descriptive vocabulary. Equivalently, I take it that there is no natural law which applies to events in virtue of their being instantiations of the property *is transported to a distance of less than three miles from the Eiffel Tower* (though I suppose it is conceivable that there is some law that applies to events in virtue of their being instantiations of some distinct but co-extensive property). By way of abbreviating these facts, I shall say that the property *is transported*... does not determine a *natural kind*, and that predicates which express that property are not natural kind predicates.

If I knew what a law is, and if I believed that scientific theories consist just of bodies of laws, then I could say that P is a natural kind predicate relative to S iff S contains proper laws of the form $P_x \rightarrow \alpha_x$ or $\alpha_x \rightarrow P_x$; roughly, the natural kind predicates of a science are the ones whose terms are the bound variables in its proper laws. I am inclined to say this even in my present state of ignorance, accepting the consequence that it makes the murky notion of a natural kind viciously dependent on the equally murky notions *law* and *theory*. There is no firm footing here. If we disagree about what is a natural kind, we will probably also disagree about what is a law, and for the same reasons. I don't know how to break out of this circle, but I think that there are interesting things to say about which circle we are in.

For example, we can now characterize the respect in which reductivism is too strong a construal of the doctrine of the unity of science. If reductivism is true, then *every* natural kind is, or is co-extensive with, a physical natural kind. (Every natural kind *is* a physical natural kind if bridge laws express property identities, and every natural kind is co-extensive with a physical natural kind if bridge laws express event identities.) This follows immediately from the reductivist premise that every predicate which appears as the antecedent or consequent of a law of the special sciences must appear as one of the reduced predicates in some bridge, together with the assumption that the natural kind predicates are the ones whose terms are the bound variables in proper laws. If, in short, some physical law is related to each law of a special science in the way that (3) is related to (1), then every natural kind predicate of a special science is related to a natural kind predicate of physics in the way that (2) relates 'S_1' and 'S_2' to 'P_1' and 'P_2'.

I now want to suggest some reasons for believing that this consequence of reductivism is intolerable. These are not supposed to be knock-down reasons; they couldn't be, given that the question whether reductivism is too strong is finally an *empirical* question. (The world could turn out to be such that every natural kind corresponds to a physical natural kind, just as it could turn out to be such that the property *is transported to a distance of less than three miles from the Eiffel Tower* determines a natural kind in, say, hydrodynamics. It's just that, as things stand, it seems very unlikely that the world *will* turn out to be either of these ways.)

The reason it is unlikely that every natural kind corresponds to a physical natural kind is just that (a) interesting generalizations (e.g., counter-factual supporting generalizations) can often be made about events whose physical descriptions have nothing in common, (b) it is often the case that *whether* the physical descriptions of the events subsumed by these generalizations have anything in common is, in an obvious sense, entirely irrelevant to the truth of the generalizations, or to their interestingness, or to their degree of confirmation or, indeed, to any of their epistemologically important properties, and (c) the special sciences are very much in the business of making generalizations of this kind.

I take it that these remarks are obvious to the point of self-certification; they leap to the eye as soon as one makes the (apparently radical) move of taking the special sciences at all seriously. Suppose, for example, that Gresham's 'law' really is true. (If one doesn't like Gresham's law, then any true generalization of any conceivable future economics will probably do as well.) Gresham's law says something about what will happen in monetary exchanges under certain conditions. I am willing to believe that physics is general *in the sense that it implies that any event which consists of a monetary exchange* (hence any event which falls under Gresham's law) *has a true description in the vocabulary of physics and in virtue of which it falls under the laws of physics.* But banal considerations suggest that a description which covers all such events must be wildly disjunctive. Some monetary exchanges involve strings of wampum. Some involve dollar bills. And some involve signing one's name to a check. What are the chances that a disjunction of physical predicates which covers all these events (i.e., a disjunctive predicate which can form the right hand side of a bridge law of the form 'x is a monetary exchange $\leftrightarrow$ ...') expresses a physical natural kind? In particular, what are the chances that such a predicate forms the antecedent or consequent of some proper law of physics? The point is that monetary exchanges have interesting things in common; Gresham's law, if true, says what one of these interesting things is. But what is interesting about monetary exchanges is surely not their commonalities under *physical* description. A natural kind like a monetary exchange *could* turn out to be co-extensive with a physical natural kind; but if it did, that would be an accident on a cosmic scale.

In fact, the situation for reductivism is still worse than the discussion thus far suggests. For, reductivism claims not only that all natural kinds are co-extensive with physical natural kinds, but that the co-extensions are nomologically necessary: bridge laws are *laws*. So, if Gresham's law is true, it follows that there is a (bridge) law of nature such that 'x is a monetary exchange $\rightleftarrows x$ is P', where P is a term for a physical natural kind. But, surely, there is no such law. If there were, then P would have to cover not only all the systems of monetary exchange that there *are*, but also all the systems of monetary exchange that there *could be*; a law must succeed with the counterfactuals. What physical predicate is a candidate for 'P' in 'x is a nomologically possible monetary exchange iff P_x'?

To summarize: an immortal econophysicist might, when the whole show is over, find a predicate in physics that was, in brute fact, co-extensive with 'is a monetary exchange'. If physics is general – if the ontological biases of reductivism are true – then there must *be* such a predicate. But (a) to paraphrase a remark Donald Davidson made in a slightly different context, nothing but brute enumeration could convince us of this brute co-extensivity, and (b) there would seem to be no chance at all that the physical predicate employed in stating the coextensivity is a natural kind term, and (c) there is still less chance that the co-extension would be lawful (i.e., that it would hold not only for the nomologically possible world that turned out to be real, but for any nomologically possible world at all).

I take it that the preceding discussion strongly suggests that economics is not reducible to physics in the proprietary sense of reduction involved in claims for the unity of science. There is, I suspect, nothing special about economics in this respect; the reasons why economics is unlikely to reduce to physics are paralleled by those which suggest that psychology is unlikely to reduce to neurology.

If psychology is reducible to neurology, then for every psychological natural kind predicate there is a co-extensive neurological natural kind predicate, and the generalization which states this co-extension is a law.

Clearly, many psychologists believe something of the sort. There are departments of 'psycho-biology' or 'psychology and brain science' in universities throughout the world whose very existence is an institutionalized gamble that such lawful co-extensions can be found. Yet, as has been frequently remarked in recent discussions of materialism, there are good grounds for hedging these bets. There are no firm data for any but the grossest correspondence between types of psychological states and types of neurological states, and it is entirely possible that the nervous system of higher organisms characteristically achieves a given psychological end by a wide variety of neurological means. If so, then the attempt to pair neurological structures with psychological functions is foredoomed. Physiological psychologists of the stature of Karl Lashley have held precisely this view.

The present point is that the reductivist program in psychology is, in any event, *not* to be defended on ontological grounds. Even if (token) psychological events are (token) neurological events, it does not follow that the natural kind predicates of psychology are co-extensive with the natural kind predicates of any other discipline (including physics). That is, the assumption that every psychological event is a physical event does not guaranty that physics (or, *a fortiori*, any other discipline more general than psychology) can provide an appropriate vocabulary for psychological theories. I emphasize this point because I am convinced that the make-or-break commitment of many physiological psychologists to the reductivist program stems precisely from having confused that program with (token) physicalism.

What I have been doubting is that there are neurological natural kinds co-extensive with psychological natural kinds. What seems increasingly clear is that, even if there is such a co-extension, it cannot be lawlike. For, it seems increasingly likely that there are nomologically possible systems other than organisms (namely, automata) which satisfy natural kind predicates in psychology, and which satisfy no neurological predicates at all. Now, as Putnam has emphasized, if there are any such systems, then there are probably vast numbers, since equivalent automata can be made out of practically anything. If this observation is correct, then there can be no serious hope that the class of automata whose psychology is effectively identical to that of some organism can be described by *physical* natural kind predicates (though, of course, if token physicalims is true, that class can be picked out by some physical predicate or other). The upshot is that the classical formulation of the unity of science is at the mercy of progress in the field of computer simulation. This is, of course, simply to say that that formulation was too strong. The unity of science was intended to be an empirical hypothesis, defeasible by possible scientific findings. But no one had it in mind that it should be defeated by Newell, Shaw and Simon.

I have thus far argued that psychological reductivism (the doctrine that every psychological natural kind is, or is co-extensive with, a neurological natural kind) is not equivalent to, and cannot be inferred from, token physicalism (the doctrine that every psychological event is a neurological event). It may, however, be argued that one might as well take the doctrines to be equivalent since the only possible *evidence* one could have for token physicalism would also be evidence for reductivism: namely, the discovery of type-to-type psychophysical correlations.

A moment's consideration shows, however, that this argument is not well taken. If type-to-type psychophysical correlations would be evidence for token physicalism, so would correlations of other specifiable kinds.

We have type-to-type correlations where, for every n-tuple of events that are of the same psychological kind, there is a correlated n-tuple of events that are of the same neurological kind. Imagine a world in which such correlations are *not* forthcoming. What is found, instead, is that for every n-tuple of type identical psychological events, there is a spatio-temporally correlated n-tuple of type *distinct* neurological events. That is, every psychological event is paired with some neurological event or other, but psychological events of the same kind may be paired with neurological events of different kinds. My present point is that such pairings would provide as much support for token physicalism as type-to-type pairings do *so long as we are able to show that the type distinct neurological events paired with a given kind of psychological event are identical in respect of whatever properties are relevant to type-identification in psychology.* Suppose, for purposes of explication, that psychological events are type identified by reference to their behavioral consequences.[5] Then what is required of all the neurological events paired with a class of type homogeneous psychological events is only that they be identical in respect of their behavioral consequences. To put it briefly, type identical events do not, of course, have *all* their properties in common, and type distinct events must nevertheless be identical in *some* of their properties. The empirical confirmation of token physicalism does not depend on showing that the neurological counterparts of type identical psychological events are themselves type identical. What needs to be shown is only that they are identical in respect of those properties which determine which kind of *psychological* event a given event is.

Could we have evidence that an otherwise heterogeneous set of neurological events have these kinds of properties in common? Of course we could. The neurological theory might itself explain why an *n*-tuple of neurologically type distinct events are identical in their behavioral consequences, or, indeed, in respect of any of indefinitely many other such relational properties. And, if the neurological theory failed to do so, some science more basic than neurology might succeed.

My point in all this is, once again, not that correlations between type homogeneous psychological states and type heterogeneous neurological states would prove that token physicalism is true. It is only that such correlations might give us as much reason to be token physicalists as type-to-type correlations would. If this is correct, then the epistemological arguments from token physicalism to reductivism must be wrong.

It seems to me (to put the point quite generally) that the classical construal of the unity of science has really misconstrued the *goal* of scientific reduction. The point of reduction is *not* primarily to find some natural kind predicate of physics co-extensive with each natural kind predicate of a reduced science. It is, rather, to explicate the physical mechanisms whereby events conform to the laws of the special sciences. I have been arguing that there is no logical or epistemological reason why success in the second of these projects should require success in the first, and that the two are likely to come apart *in fact* wherever the physical mechanisms whereby events conform to a law of the special sciences are heterogeneous.

III

I take it that the discussion thus far shows that reductivism is probably too strong a construal of the unity of science; on the one hand, it is incompatible with probable results in the special sciences, and, on the

other, it is more than we need to assume if what we primarily want
is just to be good token physicalists. In what follows, I shall try to sketch
a liberalization of reductivism which seems to me to be just strong
enough in these respects. I shall then give a couple of independent reasons
for supposing that the revised doctrine may be the right one.

The problem all along has been that there is an open empirical possibility that what corresponds to the natural kind predicates of a reduced
science may be a heterogeneous and unsystematic disjunction of predicates in the reducing science, and we do not want the unity of science
to be prejudiced by this possibility. Suppose, then, that we allow that
bridge statements may be of the form

(4) $S_x \rightleftarrows P_1x \lor P_2x \lor ... \lor P_nx$,

where '$P_1 \lor P_2 \lor ... \lor P_n$' is *not* a natural kind predicate in the reducing
science. I take it that this is tantamount to allowing that at least some
'bridge laws' may, in fact, not turn out to be laws, since I take it that a
necessary condition on a universal generalization being lawlike is that
the predicates which consitute its antecedent and consequent should pick
out natural kinds. I am thus supposing that it is enough, for purposes
of the unity of science, that every law of the special sciences should be
reducible to physics by bridge statements which express true empirical
generalizations. Bearing in mind that bridge statements are to be construed as a species of identity statements, (4) will be read as something
like 'every event which consists of x's satisfying S is identical with some
event which consists of x's satisfying some or other predicate belonging
to the disjunction '$P_1 \lor P_2 \lor ... \lor P_n$'.'

Now, in cases of reduction where what corresponds to (2) is not a law,
what corresponds to (3) will not be either, and for the same reason.
Namely, the predicates appearing in the antecedent or consequent will, by
hypothesis, not be natural kind predicates. Rather, what we will have is
something that looks like (5) (see next page).

That is, the antecedent and consequent of the reduced law will each be
connected with a disjunction of predicates in the reducing science, and, if
the reduced law is exceptionless, there will be laws of the reducing science
which connect the satisfaction of each member of the disjunction associated with the antecedent to the satisfaction of some member of the
disjunction associated with the consequent. That is, if $S_1x \rightarrow S_2x$ is

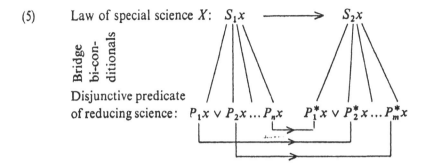

(5)   Law of special science X: $S_1x \longrightarrow S_2x$

Bridge bi-con-ditionals

Disjunctive predicate
of reducing science: $P_1x \vee P_2x \ldots P_nx$ $P_1^*x \vee P_2^*x \ldots P_m^*x$

exceptionless, then there must be some proper law of the reducing science which either states or entails that $P_1x \to P^*$ for some P^*, and similarly for $P_2 x$ through P_nx. Since there must be such laws, it follows that each disjunct of '$P_1 \vee P_2 \vee \ldots \vee P_n$' is a natural kind predicate, as is each disjunct of '$P_1^* \vee P_2^* \vee \ldots \vee P_n^*$'.

This, however, is where push comes to shove. For, it might be argued that if each disjunct of the P disjunction is lawfully connected to some disjunct of the P^* disjunction, it follows that (6) is itself a law.

(6) $P_1x \vee P_2x \vee \ldots \vee P_nx \to P_1^*x \vee P_2^*x \vee \ldots \vee P_n^*x.$

The point would be that (5) gives us $P_1x \to P_2^*x$, $P_2x \to P_m^*x$, etc., and the argument from a premise of the form $(P \supset R)$ and $(Q \supset S)$ to a conclusion of the form $(P \vee Q) \supset (R \vee S)$ is valid.

What I am inclined to say about this is that it just shows that 'it's a law that ——' defines a non-truth functional context (or, equivalently for these purposes, that not all truth functions of natural kind predicates are themselves natural kind predicates). In particular, that one may not argue from 'it's a law that P brings about R' and 'it's a law that Q brings about S' to 'it's a law that P or Q brings about R or S'. (Though, of course, the argument from those premises to 'P or Q brings about R or S' simpliciter is fine.) I think, for example, that it is a law that the irradiation of green plants by sunlight causes carbohydrate synthesis, and I think that it is a law that friction causes heat, but I do not think that it is a law that (either the irradiation of green plants by sunlight or friction) causes (either carbohydrate synthesis or heat). Correspondingly, I doubt that 'is either carbohydrate synthesis or heat' is plausibly taken to be a natural kind predicate.

It is not strictly mandatory that one should agree with all this, but one denies it at a price. In particular, if one allows the full range of truth functional arguments inside the context 'it's a law that ———', then one gives up the possibility of identifying the natural kind predicates of a science with those predicates which appear as the antecedents or the consequents of its proper laws. (Thus (6) would be a proper law of physics which fails to satisfy that condition.) One thus inherits the need for an alternative construal of the notion of a natural kind, and I don't know what that alternative might be like.

The upshot seems to be this. If we do not require that bridge statements must be laws, then either some of the generalizations to which the laws of special sciences reduce are not themselves lawlike, or some laws are not formulable in terms of natural kinds. Whichever way one takes (5), the important point is that it is weaker than standard reductivism: it does not require correspondences between the natural kinds of the reduced and the reducing science. Yet it is physicalistic on the same assumption that makes standard reductivism physicalistic (namely, that the bridge statements express true token identies). But these are precisely the properties that we wanted a revised account of the unity of science to exhibit.

I now want to give two reasons for thinking that this construal of the unity of science is right. First, it allows us to see how the laws of the special sciences could reasonably have exceptions, and, second, it allows us to see why there are special sciences at all. These points in turn.

Consider, again, the model of reduction implicit in (2) and (3). I assume that the laws of basic science are strictly exceptionless, and I assume that it is common knowledge that the laws of the special sciences are not. But now we have a painful dilemma. Since '→' expresses a relation (or relations) which must be transitive, (1) can have exceptions only if the bridge laws do. But if the bridge laws have exceptions, reductivism looses its ontological bite, since we can no longer say that every event which consists of the instantiation of an S predicate is identical with some event which consists of the instantiation of a P predicate. In short, given the reductionist model, we cannot consistently assume that the bridge laws and the basic laws are exceptionless while assuming that the special laws are not. But we cannot accept the violation of the bridge laws unless we are willing to vitiate the ontological claim that is the main point of the reductivist program.

We can get out of this (*salve* the model) in one of two ways. We can give up the claim that the special laws have exceptions or we can give up the claim that the basic laws are exceptionless. I suggest that both alternatives are undesirable. The first because it files in the face of fact. There is just no chance at all that the true, counter-factual supporting generalizations of, say, psychology, will turn out to hold in strictly each and every condition where their antecedents are satisfied. Even where the spirit is willing, the flesh is often weak. There are always going to be behavioral lapses which are physiologically explicable but which are uninteresting from the point of view of psychological theory. The second alternative is only slightly better. It may, after all, turn out that the laws of basic science have exceptions. But the question arises whether one wants the unity of science to depend upon the assumption that they do.

On the account summarized in (5), however, everything works out satisfactorily. A nomologically sufficient condition for an exception to $S_1x \rightarrow S_2x$ is that the bridge statements should identify some occurrence of the satisfaction of S_1 with an occurrence of the satisfaction of a P predicate which is not itself lawfully connected to the satisfaction of any $P*$ predicate. (I.e., suppose S_1 is connected to a P' such that there is no law which connects P' to any predicate which bridge statements associate with S_2. Then any instantiation of S_1 which is contingently identical to an instantiation of P' will be an event which constitutes an exception to $S_1x \rightarrow S_2x$.) Notice that, in this case, we need assume no exceptions to the laws of the *reducing* science since, by hypothesis, (6) *is not a law*.

In fact, strictly speaking, (6) has no status in the reduction at all. It is simply what one gets when one universally quantifies a formula whose antecedent is the physical disjunction corresponding to S_1 and whose consequent is the physical disjunction corresponding to S_2. As such, it will be true when $S_1 \rightarrow S_2$ is exceptionless and false otherwise. What does the work of expressing the physical mechanisms whereby n-tuples of events conform, or fail to conform, to $S_1 \rightarrow S_2$ is not (6) but the laws which severally relate elements of the disjunction $P_1 \vee P_2 \vee ... \vee P_n$ to elements of the disjunction $P_1^* \vee P_2^* \vee ... \vee P_n^*$. When there *is* a law which relates an event that satisfies one of the P disjuncts to an event which satisfies one of the $P*$ disjuncts, the pair of events so related conforms to $S_1 \rightarrow S_2$. When an event which satisfies a P predicate is *not* related by law to an event which satisfies a P^* predicate, that event will constitute an exception to $S_1 \to S_2$. The point is that none of the laws which effect these several connections need themselves have exceptions in order that $S_1 \to S_2$ should do so.

To put this discussion less technically: we could, if we liked, *require* the taxonomies of the special sciences to correspond to the taxonomy of physics by insisting upon distinctions between the natural kinds postulated by the former wherever they turn out to correspond to distinct natural kinds in the latter. This would *make* the laws of the special sciences exceptionless if the laws of basic science are. But it would also loose us precisely the generalizations which we want the special sciences to express. (If economics were to posit as many *kinds* of monetary systems as there are kinds of physical realizations of monetary systems, then the generalizations of economics *would* be exceptionless. But, presumbaly, only vacuously so, since there would be no generalizations left to state. Graham's law, for example, would have to be formulated as a vast, open disjunction about what happens in monetary system₁ or monetary systemₙ under conditions which would themselves defy uniform characterization. We would not be able to say what happens in monetary systems *tourt court* since, by hypothesis, 'is a monetary system' corresponds to no natural kind predicate of physics.)

In fact, what we do is precisely the reverse. We allow the generalizations of the special sciences to *have* exceptions, thus preserving the natural kinds to which the generalizations apply. But since we know that the *physical* descriptions of the natural kinds may be quite heterogeneous, and since we know that the physical mechanisms which connect the satisfaction of the antecedents of such generalizations to the satisfaction of their consequents may be equally diverse, we expect both that there will be exceptions to the generalizations and that these exceptions will be 'explained away' at the level of the reducing science. This is one of the respects in which physics really is assumed to be bedrock science; exceptions to *its* generalizations (if there are any) had better be random, because there is nowhere 'further down' to go in explaining the mechanism whereby the exceptions occur.

This brings us to why there are special sciences at all. Reducitivism as we remarked at the outset, flies in the face of the facts about the scientific institution: the existence of a vast and interleaved conglomerate of special scientific disciplines which often appear to proceed with only the most token acknowledgment of the constraint that their theories must turn out to be physics 'in the long run'. I mean that the acceptance of this constraint, *in practice,* often plays little or no role in the validation of theories. Why is this so? Presumably, the reductivist answer must be *entirely* epistemological. If only physical particles weren't so small (if only brains were on the *out*side, where one can get a look at them), *then* we would do physics instead of palentology (neurology instead of psychology; psychology instead of economics; and so on down). There is an epistemological reply; namely, that even if brains were out where they can be looked *at,* as things now stand, we wouldn't know what to look *for:* we lack the appropriate theoretical apparatus for the psychological taxonomy of neurological events.

If it turns out that the functional decomposition of the nervous system corresponds to its neurological (anatomical, biochemical, physical) decomposition, then there are only epistemological reasons for studying the former instead of the latter. But suppose there is no such correspondence? Suppose the functional organization of the nervous system crosscuts its neurological organization (so that quite different neurological structures can subserve identical psychological functions across times or across organisms). Then the existence of psychology depends not on the fact that neurons are so sadly small, but rather on the fact that neurology does not posit the natural kinds that psychology requires.

I am suggesting, roughly, that there are special sciences not because of the nature of our epistemic relation to the world, but because of the way the world is put together: not all natural kinds (not all the classes of things and events about which there are important, counterfactual supporting generalizations to make) are, or correspond to, physical natural kinds. A way of stating the classical reductionist view is that things which belong to different physical kinds *ipso facto* can have no projectible descriptions in common; that if *x* and *y* differ in those descriptions by virtue of which they fall under the proper laws of physics, they must differ in those descriptions by virtue of which they fall under any laws at all. But why should we believe that this is so? Any pair of entities, howeve: different their physical structure, must nevertheless converge in indefinitely many of their properties. Why should there not be, among those convergent properties, some whose lawful inter-relations support the generalizations of the special sciences? Why, in short, should not the natural kind predicates of the special sciences *cross-classify* the physical natural kinds?[6]

Physics develops the taxonomy of its subject-matter which best suits its purposes: the formulation of exceptionless laws which are basic in the several senses discussed above. But this is not the only taxonomy which may be required if the purposes of science in general are to be served: e.g., if we are to state such true, counterfactual supporting generalizations as there are to state. So, there are special sciences, with their specialized taxonomies, in the business of stating some of these generalizations. If science is to be unified, then all such taxonomies must apply *to the same things*. If physics is to be basic science, then each of these things had better be a physical thing. But it is not further required that the taxonomies which the special sciences employ must themselves reduce to the taxonomy of physics. It is not required, and it is probably not true.

Massachusetts Institute of Technology

NOTES

* I wish to express my gratitude to Ned Block for having read a version of this paper and for the very useful comments he made.

[1] I shall usually assume that sciences are about events, in at least the sense that it is the occurrence of events that makes the laws of a science true. But I shall be pretty free with the relation between events, states, things and properties. I shall even permit myself some latitude in construing the relation between properties and predicates. I realize that all these relations are problems, but they aren't my problem in this paper. Explanation has to *start* somewhere, too.

[2] The version of reductionism I shall be concerned with is a stronger one than many philosophers of science hold; a point worth emphasizing since my argument will be precisely that it is too strong to get away with. Still, I think that what I shall be attacking is what many people have in mind when they refer to the unity of science, and I suspect (though I shan't try to prove it) that many of the liberalized versions suffer from the same basic defect as what I take to be the classical form of the doctrine.

[3] There is an implicit assumption that a science simply *is* a formulation of a set of laws. I think this assumption is implausible, but it is usually made when the unity of science is discussed, and it is neutral so far as the main argument of this paper is concerned.

[4] I shall sometimes refer to 'the predicate which constitutes the antecedent or consequent of a law'. This is shorthand for 'the predicate such that the antecedent or consequent of a law consists of that predicate, together with its bound variables and the quantifiers which bind them'. (Truth functions of elementary predicates are, of course, themselves predicates in this usage.)

⁵ I don't think there is any chance at all that this is true. What is more likely is that type-identification for psychological states can be carried out in terms of the 'total states' of an abstract automaton which models the organism. For discussion, see Block and Fodor (1972).
⁶ As, by the way, the predicates of natural languages quite certainly do. For discussion, see Chomsky (1965).

BIBLIOGRAPHY

Block, N. and Fodor, J., 'What Psychological States Are Not', *Philosophical Review* **81** (1972) 159–181.
Chomsky, N., *Aspects of the Theory of Syntacs*, MIT Press, Cambridge, 1965.

Explanation, Reduction, and Empiricism

The main contention of the present paper is that a formal account of reduction and explanation is impossible for general theories, or noninstantial theories,[1] as they have also been called. More especially, it will be asserted and shown that wherever such theories play a decisive role both Nagel's theory of reduction[2] and the theory of explanation associated with Hempel and Oppenheim[3] cease to be in accordance with actual scientific practice and with a reasonable empiricism. It is to be admitted that these two "orthodox" accounts fairly adequately represent the relation between sentences of the 'All-ravens-are-black' type, which abound in the more pedestrian parts of the scientific enterprise.[4] But if the attempt is made to extend these accounts to such comprehensive structures of thought as the Aristotelian theory of motion, the impetus theory, Newton's celestial mechanics, Maxwell's electrodynamics, the theory of relativity, and the quantum theory, then complete failure is the result. What happens here when transition is made from a theory T' to a wider theory T (which, we shall assume, is capable of covering all the phenomena

---

[1] In what follows, the usual distinction will be drawn between *empirical generalizations*, on the one side, and *theories*, on the other. Empirical generalizations are statements, such as 'All A's are B's' (the A's and B's are not necessarily observational entities), which are tested by inspection of instances (the A's). Universal theories, such as Newton's theory of gravitation, are not tested in this manner. Roughly speaking their test consists of two steps: (1) derivation, with the help of suitable boundary conditions, of empirical generalizations; and (2) tests, in the manner indicated above, of these generalizations. One should not be misled by the fact that universal theories, too, can be (and usually are) put in the form 'All A's are B's'; for, whereas, in the case of generalizations, this form reflects the test procedure in a very direct way, such an immediate relation between the form and the test procedures does not obtain in the case of theories. Many thinkers have been seduced by the similarity of form into thinking that the test procedures will be the same in both cases.

[2] Nagel has explained his theory in [60]. I shall quote from the reprint of the article in [20], pp. 288–312.

[3] For the theory of Hempel and Oppenheim see [47]. I shall quote from the reprint in [23], pp. 319–352.

[4] For important exceptions, see fn. 90.

that have been covered by T') is something much more radical than incorporation of the unchanged theory T' (unchanged, that is, with respect to the meanings of its main descriptive terms as well as to the meanings of the terms of its observation language) into the context of T. What does happen is, rather, a complete replacement of the ontology (and perhaps even of the formalism) of T' by the ontology (and the formalism) of T and a corresponding change of the meanings of the descriptive elements of the formalism of T' (provided these elements and this formalism are still used). This replacement affects not only the theoretical terms of T' but also at least some of the observational terms which occurred in its test statements. That is, not only will description of things and processes in the domain in which so far T' had been applied be infiltrated, either with the formalism and the terms of T, or if the terms of T' are still in use, with the meanings of the terms of T, but the sentences expressing what is accessible to direct observation inside this domain will now mean something different. In short: introducing a new theory involves changes of outlook both with respect to the observable and with respect to the unobservable features of the world, and corresponding changes in the meanings of even the most "fundamental" terms of the language employed. So far this is the position which will be defended in the present paper.

This position may be said to consist of two ideas. The first idea is that the influence, upon our thinking, of a comprehensive scientific theory, or of some other general point of view, goes much deeper than is admitted by those who would regard it as a convenient scheme for the ordering of facts only. According to this first idea scientific theories are ways of looking at the world; and their adoption affects our general beliefs and expectations, and thereby also our experiences and our conception of reality. We may even say that what is regarded as "nature" at a particular time is our own product in the sense that all the features ascribed to it have first been invented by us and then used for bringing order into our surroundings. As is well known, it was Kant who most forcefully stated and investigated this all-pervasive character of theoretical assumptions. However, Kant also thought that the very generality of such assumptions and their omnipresence would forever prevent them from being refuted. As opposed to this, the second idea implicit in the position to be defended here demands that our theories be testable and that they be abandoned as soon as a test does not produce the predicted result. It is this second idea which makes sciences proceed to better and better theories and which creates the changes described in the introductory paragraphs of the present paper.

Now, it is easily seen that the mere statement of the second idea will not do. What we need is a guarantee that despite the all-pervasive character of a scientific theory as it is asserted in the first idea, it is still possible to specify facts that are inconsistent with it. Such a possibility has been denied by some philosophers. These philosophers started out by reacting against the claim that scientific theories are nothing but predictive devices; they recognized that their influence goes much deeper; however, they then doubted that it would be possible ever to get outside any such theory; and they therefore either became apriorists (Poincaré, Eddington), or they returned to instrumentalism. For these thinkers there seemed to exist only a choice between two evils—instrumentalism or apriorism.

Now, a closer look at the arguments leading up to this dilemma shows that they all proceed from a test model in which a *single* theory is confronted with the facts. As soon as this model is replaced by a model in which we make use of at least two factually adequate but mutually inconsistent theories, the first idea becomes compatible with the demand for testability which must now be interpreted as a demand for crucial tests either between two explicitly formulated theories, or between a theory and our "background knowledge." In this form, however, the test model turns out to be inconsistent with the "orthodox" theory of explanation and reduction. It is one of the aims of the present paper to exhibit this inconsistency.

It will be necessary, for this purpose, to discuss two principles which underlie the orthodox approach: (A) the principle of deducibility; and (B) the principle of meaning invariance. According to the principle of deducibility, explanation is achieved by deduction in the strict logical sense. This principle leads to the demand, which is incompatible with the test model just outlined, that all successful theories in a given domain must be mutually consistent. According to the principle of meaning invariance, an explanation must not change the meanings of the main descriptive terms of the explanandum. This principle, too, will be found to be inconsistent with empiricism.

It is interesting to note that (A) and (B) play a role both within modern empiricism and within some very influential "school philosophies." Thus it is one of the basic assumptions of Platonism that the key terms of sentences expressing knowledge (epistēmē) refer to unchangeable entities and must therefore possess a stable meaning. Similarly, the key terms of Cartesian physics—i.e., the terms 'matter,' 'space,' 'motion'—and the terms of Cartesian metaphysics—such as the terms 'god' and 'mind'—are supposed to remain unchanged in any explanation involving them. Compared with these similarities[5] between the school philosophies, on the one side, and modern empiricism, on the other, the differences are of very minor importance. These differences lie in the *terms* of which stability of meaning is required. A Platonist will direct his attention to numbers and other "ideas," and he will demand that words referring to these entities retain their (Platonic) meanings. Modern empiricism, on the other hand, regards empirical terms as fundamental and demands that their meanings remain unchanged.

Now, it will turn out, in the course of this essay, that any form of meaning invariance is bound to lead to difficulties when the task arises either to give a proper account of the growth of knowledge, and of discoveries contributing to this growth, or to establish correlations between entities which are described with the help of what we shall later call incommensurable concepts. It will also turn out that these are exactly the difficulties we encounter in trying to solve such age-old problems as the mind-body problem, the problem of the reality of the external world, and the problem of other minds. That is, it will usually turn out that a solution of these problems is deemed satisfactory only if it leaves unchanged the meanings of certain key terms and that it is exactly *this* condition, i.e., the condition of meaning invariance, which makes them insoluble. It will also be shown that the demand for meaning invariance is incompatible with empiricism. Taking all this into account, we may hope that once contemporary empiricism has been freed from the elements which it still shares with its more dogmatic opponents, it will be able to make swift progress in the solution of the above problems. It is the purpose of the present paper to develop and to defend the outlines of such a disinfected empiricism.[6]

Popper's admirable *Logic of Scientific Discovery* and his paper "The

[5] Concerning these similarities, see Popper's discussion of essentialism in [65], Ch. III and *passim*, as well as Dewey's very different account in [21], especially Ch. II.
[6] As will be shown in Sec. 2, the empiricism of the thirties was disinfected in the sense desired here. However, later on modern empiricism readopted some very undesirable principles of traditional philosophy.

P. K. Feyerabend

Aim of Science"[7] have been both the starting point and the motive force of the investigations to follow. I have also profited a great deal from discussions with Professors Bohm (Bristol-Haifa), Feigl (Minneapolis), Körner (Bristol), Maxwell (Minneapolis), Putnam (Princeton), and Tranekjaer-Rasmussen (Copenhagen). Both Professor Körner and Professor Sellars (New Haven) seem to hold similar views with respect to the character of the observation language, and reading their publications has therefore been a great help.[8]

While the present paper was in progress I had an opportunity to consult various as yet unpublished papers by Professor T. S. Kuhn (Berkeley) in which the noncumulative character of scientific progress is illustrated very forcefully by historical examples. Despite some important and perhaps unalterable differences, the area of agreement between Professor Kuhn and myself seems to be quite considerable. One most important point of agreement is the emphasis which both of us put upon the need, in the process of the refutation of a theory, for at least another theory. As far as I am aware, this point has been made previously by K. R. Popper in his lectures on scientific method which I attended in 1948 and 1952. Popper has also pointed out[9] that the alternative theory used in the process of refutation need not be explicitly stated but can be part of our "background knowledge."

Bohm's theory of levels and Putnam's considerations in the present volume seem to lead in the same direction. What I regard as a most important feature of the situation—a feature, by the way, that has been emphasized by Bohm and Vigier—is that direct refutation of a fairly complicated theory may be impossible for *empirical* reasons. That this is so will be shown with the help of an example. Finally, I would like to thank Professor Popper and Mr. J. W. N. Watkins (London) for constructive criticism that has been utilized in the final version of the paper.

1. Two Assumptions of Contemporary Empiricism.

Nagel's theory of reduction is based upon two assumptions. The first assumption concerns the relation between the secondary science, i.e., the

[7] See [68] and [66]. The basic ideas of [66] can be found in [64], which was written earlier.

[8] I am referring here to Körner's [50] and to Sellars' admirable [70].

[9] In a lecture given at Stanford University in September 1960.

74

discipline to be reduced, on the one side, and the primary science, i.e., the discipline to which reduction is made, on the other. It is asserted that this relation is the relation of deducibility. Or, to quote Nagel,

(1)　"The objective of the reduction is to show that the laws, or the general principles of the secondary science, are simply logical consequences of the assumptions of the primary science."[10]

The second assumption concerns the relation between the meanings of the primitive descriptive terms of the secondary science and the meanings of the primitive descriptive terms of the primary science. It is asserted that the former will not be affected by the process of reduction. Of course, this second assumption is an immediate consequence of (1), since a derivation is not supposed to influence the meanings of the statements derived. However, for reasons which will become clear later, it is advisable to formulate this invariance of meaning as a separate principle. This is also done by Nagel, who says: "It is of the utmost importance to observe that the expressions peculiar to a science will possess meanings that are fixed by its own procedures, and are therefore intelligible in terms of its own rules of usage, *whether or not the science has been, or will be, reduced to some other discipline.*"[11] Or, to express it in a more concise manner:

(2)　Meanings are invariant with respect to the process of reduction.

(1) and (2) admit of two different interpretations, just as does any theory of reduction and explanation: such a theory may be regarded either as a *description* of actual scientific practice, or as a *prescription* which must be followed if the scientific character of the whole enterprise is to be guaranteed. Similarly, (1) and (2) may be interpreted as *assertions* concerning actual scientific practice, or as *demands* to be satisfied by the theoretician who wants to follow the scientific method. Both of these interpretations will be scrutinized in the present paper.

Two very similar assumptions, or demands, play a decisive role in the orthodox theory of explanation, which may be regarded as an elaboration of suggestions that were first made, in a less definite form by Popper.[12] The first assumption (demand) concerns again the relation

---

[10] [20], p. 301. A more elaborate form of this condition is called the "condition of derivability" on p. 354 of [61].
[11] [20], p. 301. My italics. See also [61], p. 345, 352.
[12] [68], Sec. 12.

between the explanandum, or the laws, or the facts to be explained, on the one side, and the explanans, or the discipline which functions as the basis of explanation, on the other. It is again asserted (required) that this relation is (be) the relation of deducibility. Or, to quote Hempel and Oppenheim

(3)   "The explanandum must be a logical consequence of the explanans; in other words, the explanandum must be logically deducible from the information contained in the explanans, for otherwise the explanans would not constitute adequate grounds for the explanation."[13]

Considering what has been said in the case of reduction one would expect the assumption (demand) concerning meanings to read as follows:

(4)   Meanings are invariant with respect to the process of explanation.

However, despite the fact that (4) is a trivial consequence of (3), this assumption has never been expressed in as clear and explicit a way as (2).[14] There was even a time when a consequence of (4), viz., the assertion that *observational* meanings are invariant with respect to the process of explanation, seemed to be in doubt. It is for this reason that I have separated (2) from (1), and (4) from (3).

It is not difficult to show that, with respect to observational terms, (4) or its implications, is consistent with the earlier positivism of the Vienna Circle. Their main thesis that all descriptive terms of a scientific theory can be explicitly defined on the basis of observation terms guarantees the stability of the meanings of observational terms (unless one assumes that an explicit definition changes the meaning of the definiens, a possibility that to my knowledge has never been considered by empiricists). And as the chain of definitions leaves unchanged terms already defined, (4) turns out to be correct as well.

However, since these happy and carefree days of the *Aufbau*, logical

---

[13] [47], p. 321.

[14] An exception is Nagel who, in [61], p. 338, defines reduction as "the explanation of a theory or a set of experimental laws established in one area of inquiry by a theory usually, though not invariably, formulated for some other domain." This implies that the condition of meaning invariance formulated by him for the process of reduction is supposed to be valid in the case of explanation also. On pp. 86–87, meaning invariance for observational terms is stated quite explicitly: an experimental law "retains a meaning that can be formulated independently of [any] theory . . . [It] has . . . a life of its own, not contingent on the continued life of any particular theory that may explain the law."

empiricism has been greatly modified. The changes that took place were mainly of two kinds. On the one side, new ideas were introduced concerning the relation between observational terms and theoretical terms. On the other side, the assumptions made about the observational language itself were modified. In both cases the changes were quite drastic. For our present purpose a brief outline must suffice: The early positivists assumed that observational terms refer to subjective impressions, sensations, and perceptions of some sentient being. Physicalism for some time retained the idea that a scientific theory should be based upon experiences, and that the ultimate constituents of experience were sensations, impressions, and perceptions. Later, however, a behavioristic account was given of these perceptions to make them accessible to intersubjective testing. Such a theory was held, for some time, by Carnap and Neurath.[15] Soon afterwards the idea that it is *experiences* to which we must refer when trying to interpret our observation statements was altogether abandoned.[16] According to Popper, who has been responsible for this decisive turn, we must "*distinguish sharply between objective science on the one hand, and 'our knowledge' on the other.*" It is conceded "we can become aware of facts only by observation"; but it is denied that this implies an interpretation of observation sentences in terms of experiences, whether these experiences are explained subjectivistically or as features of objective behavior.[17] For example, we may admit that the sentence 'this is a raven' uttered by an observer who points at a bird in front of him is an observational sentence and that the observer has produced it because of the impressions, sensations, and perceptions he possesses. We may also admit that he would not have uttered the sentence had he not possessed these impressions. Yet, the sentence is not therefore about impressions; it is about a bird which is neither a sensation nor the behavior of some sentient being. Similarly, it may be admitted that the observation sentences which a scientific observer produces are prompted by his impressions. However, their content will again be determined, not by these impressions, but by the entities allegedly described. In the case of classical physics, therefore, "every basic statement must either be itself a statement about relative positions

---

[15] For this and the following, see Carnap's account in [13], especially the passages in small print on pp. 223–224.

[16] *Ibid.*, p. 223: "It is stipulated that under given circumstances any concrete statement may be regarded as a protocol statement."

[17] Popper [68], p. 98. Italics in the original.

of physical bodies . . . or it must be equivalent to some basic statement of this 'mechanistic' . . . kind."[18]

The descriptive terms of Carnap's "thing-language," too, no longer refer to experiences. They refer to properties of objects of medium size which are accessible to observation, i.e., which are such that a normal observer can quickly decide whether or not an object possesses such a property.[19] "What we have called observable predicates," says Carnap, "are predicates of the thing-language (they have to be clearly distinguished from what we have called perception terms . . . whether these are now interpreted subjectivistically, or behavioristically)."[20]

Now it is most important to realize that the characterization of observation statements implicit in the above quotations is a *causal* characterization, or if one wants to use more recent terminology, a *pragmatic* characterization:[21] an observation sentence is distinguished from other sentences of a theory, not, as was the case in earlier positivism, by its *content*; but by the *cause* of its production, or by the fact that its production conforms to certain *behavioral patterns*.[22] This being the case, the fact that a certain sentence belongs to the observation language does not allow us to infer anything about its content; more especially, it does not allow us to make any inference concerning the *kind* of entities described in it.

It is worthwhile to dwell a little longer on the features of this *pragmatic theory of observation*, as I shall call it. In the case of measuring instruments, the pragmatic theory degenerates into a triviality: nobody would ever dream of asserting that the way in which we interpret the movements of, say, the hand of a voltmeter is uniquely determined either by the character of this movement itself or by the processes inside the instrument; a person who can see and understand only these processes will be unable to infer that what is indicated is voltage, and he will be equally unable to understand what voltage is. Taken by themselves the indications of instruments do not mean anything unless we possess a theory which teaches us what situations we are to expect in

[18] Popper [68], p. 103. Popper himself does not restrict his characterization to the observation statements of classical physics.
[19] Carnap [14], p. 63, Explanation 1. Page references are to the reprint of this article in [22], pp. 47–92.
[20] *Ibid.*, p. 69.
[21] For this terminology see Morris [59], pp. 6ff.
[22] See again Explanation 1 of [14], as well as my elaboration of this explanation in [31].

the world, and which guarantees that there exists a reliable correlation between the indications of the instrument and such a particular situation. If a certain theory is replaced by a different theory with a different ontology, then we may have to revise the interpretation of *all* our measurements, however self-evident such a particular interpretation may have become in the course of time: according to the phlogiston theory, measurements of weight before and after combustion are measurements of the amount of phlogiston added or lost in the process. Today we must give a completely different interpretation of the results of these measurements. Again, Galileo's thermoscope was initially supposed to measure an intrinsic property of a heated body; however, with the discovery of the influence of atmospheric pressure, of the expansion of the substance of the thermoscope (which, of course, was known beforehand), and of other effects (nonideal character of the thermoscopic fluid), it was recognized that the property measured by the instrument was a very complicated function of such an intrinsic property, of the atmospheric pressure, of the properties of the particular enclosure used, of its shape, and so on.[23] Indeed, the point of view outlined in the beginning of the present paper gives an excellent account of the way in which results of measurement, or indications of instruments, are reinterpreted in the light of fresh theoretical insight. Nobody would dream of using the insight given by a new theory for the readjustment of some general beliefs only, leaving untouched the interpretation of the results of measurement. And nobody would dream of demanding that the meanings of observation statements *as obtained with the help of measuring instruments* remain invariant with respect to the change and progress of knowledge. Yet, precisely this is done when the measuring instrument is a human being, and the indication is the behavior of this human being, or the sensations he has, at a particular time.

It is not easy to set down in a few lines the reasons for this exceptional treatment of human observers. Nor is it possible to criticize them thoroughly and thereby fully pave the way for the acceptance of the pragmatic theory of observation. However, such a comprehensive criticism is not really necessary here. It was partly given by those very same philosophers who are responsible for the formulation of the pragmatic

---

[23] For historical references, see [18], especially the articles on the phlogiston theory (J. B. Conant) and on the early development of the concept of temperature (D. Roller).

theory[24] (which most of them dropped later on, their own excellent arguments in favor of it notwithstanding). I shall therefore content myself with giving an outline of the idea leading to the assumption that human observers are something special and cannot be treated in the same manner as physical measuring instruments.

These ideas are connected with the (very old) belief that (a) some states of the mind (sensations or abstract ideas) can be known with certainty; that (b) it is exactly this knowledge that constitutes the foundation of whatever assertion we make about the world; and that (c) meaning invariance is obtained in the following manner: if it is indeed the case that statements about, say, sensations are irrevocable once produced, then the same applies to the descriptive terms contained in them; their meaning is uniquely and irrevocably determined by the structure of the statements in which they occur as well as by the circumstances which lead to the certain production of these statements. (Similar considerations apply if we are dealing, not with sensations, but with the 'clear and distinct' appearance of ideas.)

The theories behind meaning invariance are, of course, a little more complicated than I have just indicated, and they should perhaps be outlined in greater detail in order that their force be duly realized. Nevertheless, their most fundamental assumptions—viz., (a), (b), and (c)—can be eliminated on the basis of some very simple and almost trivial considerations. These considerations, which cannot be found in the above-mentioned writings of the original defenders of the pragmatic theory, proceed from the remark that in the argument leading up to (c) the distinction is obliterated between (psychological and sociological) *facts* and (linguistic) *conventions*.[25] It is assumed that the urge we feel under certain circumstances to say 'I am in pain' and the peculiar character of this urge (it is different from the urge we feel when we say 'I am hungry') *already determine* the meaning of the main descriptive term of the sentence uttered, viz., the meaning of the term 'pain' or 'hunger').

Now, quite apart from leading to some very undesirable paradoxes[26] this procedure assumes that a *fact*, viz., the existence of either an urge

[24] Cf. Carnap [11] and [12].

[25] For a very clear presentation of this distinction, see Popper [65], Ch. V.

[26] For a more detailed discussion of these paradoxes, see my paper [31], especially Secs. 4 and 5.

to produce a sentence of a certain kind or the existence of psychological phenomenon, can without further ado transfer meaning upon a sentence, viz., the sentence 'I am in pain.' It is therefore unacceptable to any philosopher who takes seriously the distinction between facts and conventions. Conversely, the attempt to uphold this distinction leads at once to the separation, characteristic of the pragmatic theory, of the observational character of a statement from its meaning: according to the pragmatic theory, the fact that a statement belongs to the observational domain has no bearing upon its meaning. Even if its production is accompanied by very forceful sensations and related to them in a manner that makes substitution by a different sentence psychologically very difficult or perhaps even impossible, even then we are still free to interpret the sentence in whatever way we like. It is very important to point out that this freedom of interpretation obtains also in psychology, where our sentences are indeed about subjective events. Whatever restrictions of interpretation we accept are determined by the language we use, or by the theories or general points of view whose development has led to the formulation of this language.[27]

To repeat: strict adherence to the distinction between nature and convention at once eliminates the third of the three assumptions mentioned above and thereby introduces a very fundamental element of the pragmatic theory, namely, its emphasis upon the separation between observability and meaning. However, we cannot retain the first assumption either. The reason is that the sciences are the result of a decision to use only testable statements for the expression of laws and singular facts. This being the case we cannot admit any irrefutable statement,

---

[27] A pointed criticism of the idea that the interpretation of a statement is uniquely determined by the sensations that accompany its production has been given by Wittgenstein in his [74]. This book also emphasizes the dependence of interpretations upon the incorporation of the corresponding sentence into a language. It does not seem to me, however, that Wittgenstein possesses a clear idea of what we have called the pragmatic theory of observation. He fails to recognize that languages are not ends in themselves but are means for expressing theories and that they can and should be abandoned as soon as the corresponding theory has been refuted. Quite to the contrary—he dwells upon the difficulties one will encounter when trying to change a language in a very fundamental way, and he thereby insinuates that it may be altogether impossible to carry out decisive changes. The reason for this pessimism seems to be identical with the one I briefly discussed in the introductory part of this paper: it is assumed that the all-pervasive character of a language makes it impossible to specify grounds for abandoning it. For an application of this pessimism to more concrete problems, see Hanson [43], especially Chs. III and V. For a criticism, see my review of Hanson in [35].

however elevated and noble its source may seem to be.[28] Indeed, a whole theory may at some time turn out to be unsatisfactory, and the need may arise to replace it by a completely different idiom based upon a different point of view. Clearly, then, the interpretation of the observation sentences will have to follow suit, for again there is no way of conferring an interpretation upon them except by incorporating them into a new and better theory.

The pragmatic theory of observation thus turns out to be a presupposition of the feasibility of the point of view which has been outlined in my introductory remarks (and a consequence of the distinction between nature and convention). This position, this point of view, and especially the idea that our theories determine our entire conception of reality now emerge as a combination of (a) the demand to apply the terminology and the ontology of a given theory everywhere inside the domain of its validity with (b) the pragmatic theory of observation. It is in this form that I shall defend my position in the present paper.

The freedom of interpretation admitted by the pragmatic theory did not exist in the earlier positivism. Here sensations were thought to be the objects of observation. According to it, whether or not a statement is a sense-datum statement and, therefore, part of the observation language could be determined by logical analysis. Conversely, the assertion that a certain statement belongs to the observation language there implied an assertion about the kind of entities described (e.g., sense data). The ontology of the observational domain was therefore fixed independently of theorizing. This being the case, the demand for a unified ontology (which is still retained) could be met only by adopting the one or the other of the following two procedures: it could be met by either denying a descriptive function to the sentences of the theory and by declaring that these sentences are nothing but part of a complicated prediction machine (*instrumentalism*), or by conferring upon these sentences an interpretation that completely depends upon their connection with the observational language as well as upon the (fixed) interpretation of the latter (*reductionism*). It is important to realize that it is the clash between realism, on the one side, and the combination of the theory of sense data with the demand for a unified ontology, on the

---

[28] In [33] I discuss some of the consequences of the use of irrefutable statements of observation and thereby provide some reasons for their elimination from the body of the sciences and of knowledge in general.

other, which necessitates this transition to either instrumentalism or reductionism.

Now one of the most surprising features of the development of contemporary empiricism is that the very articulate formulation of the pragmatic account of observation was not at once followed by an equally articulate formulation of a realistic interpretation of scientific theories. After all, realism had been abandoned mainly because the theory of sense data had made it incompatible with the demand for a unified ontology. The arrival of the pragmatic theory of observation removed this incompatibility and thereby opened the way for a hypothetical realism of the kind outlined earlier. Yet, in spite of this possibility, the actual historical development was in a completely different direction. The pragmatic theory was retained for a while (and is still retained, in footnotes, by some empiricists[29]), but it was soon combined either with instrumentalism or with reductionism. As the reader can verify for himself, such a combination in effect amounts to abandoning the pragmatic theory, a more complicated language with a more complicated ontology now taking the place of the sense-data language of the earlier point of view. How close the most recent offspring of this development is to the old sense-data ideology may be seen in a recent paper by Professor Carnap.

In this paper Carnap analyzes scientific theories with the help of his well-known double-language model consisting of an observational language, L_O and a theoretical language, L_T, the latter containing a postulate system, T. The languages are connected to each other by correspondence rules, i.e., by sentences containing observational terms and theoretical terms. With respect to such a system, Carnap asserts that "there is no independent interpretation for L_T. The system T is itself an uninterpreted postulate system. The terms of $[L_T]$ obtain only an indirect and incomplete interpretation by the fact that some of them are connected by correspondence rules with observational terms, and the remaining terms of $[L_T]$ are connected with the first-ones by the postulates of T."[30]

This procedure quite obviously presupposes that the meaning of the observational terms is fixed independently of their connection with theoretical systems. If the pragmatic theory of observation were still retained in this essay of Carnap's, then the interpretation of an observational statement would have to be independent of the behavioral pattern exhibited in the observational situation as well. It is not clear how, then, the observation sentence could be given any meaning at all. Now, Carnap is very emphatic about the fact that incorporation into a theoretical context is not sufficient for providing an interpretation, since no theoretical context possesses an "independent interpretation."[31] We must, therefore, suspect that, for Carnap, incorporation of a sentence into a complicated behavioral pattern has implications for its meaning, i.e., we must suspect that Carnap has silently dropped the pragmatic theory. This is indeed the case. He asserts that "a complete interpretation of L_0" is given since "L_0 is used by a certain language community as a means of communication,"[32] adding in a later passage[33] that if people use a term in such a fashion that for some sentences containing the term "any possible observational result can never be absolutely conclusive evidence, but at best evidence yielding a high probability, then the appropriate place for [the term] in a dual language system . . . is in L_T rather than in L_0 . . ." These two passages together seem to imply that the meaning of an observational statement is already fixed by the way in which the sentence expressing it is handled in the *immediate* observational situation (note the emphasis upon absolute confirmability for observational sentences!), i.e., they seem to imply the rejection of the pragmatic theory.

As I said above, this tacit withdrawal from the pragmatic theory of observation is one of the most surprising features of modern empiricism. It is responsible for the fact that this philosophy, despite the apparent progress that has been made since the thirties, is still in accordance with the assumption that observational meanings are invariant with respect to the process of explanation and perhaps even with full meaning invariance (if we only consider that the behavioristic criterion of observability will be satisfied by any language that has been used for a long time, a long history and observational plausibility brought about by it are the best preconditions for the petrification of meanings; this applies to Platonism as well as to modern empiricism).

This finishes a somewhat lengthy digression which started immediately after the pronouncement of (4). I will make only two points before

---

[31] For a detailed criticism of this assertion, cf. my [27] and [39].
[32] Carnap [15], p. 40.
[33] *Ibid.*, p. 69.

returning to the main argument of the present paper: first, that the un-witting and partial return to the ideology of sense data is responsible for many of the 'inner contradictions' which are so characteristic of contem-porary empiricism as well as for the pronounced similarity of this phi-losophy to the "school philosophies" it has attacked; second, that (4) has been accepted, not only by philosophers, but also by many physicists who believe in the so-called Copenhagen interpretation of microphysics. It is one of Niels Bohr's most fundamental ideas that "however far the new phenomena" found on the microlevel "transcend the scope of classi-cal physical explanation, the account of all evidence must be expressed in classical terms."[34] I shall not discuss, in the present section, the argu-ments which Bohr has developed in favor of this idea. Let me only say that this idea immediately leads to the invariance of the meanings of the descriptive terms of the observation language, the classical signs now playing the role of the observational vocabulary.

To sum up: two ideas which are common to both the modern empiri-cist's theory of reduction and to his theory of explanation are:

(A)   reduction or explanation is (or should be) by derivation;
(B)   the meanings of (observational) terms are invariant with respect to both reduction and explanation.

In the sections to follow it will be my task to scrutinize these two basic principles. I shall begin with (A).

2. *Criticism of Reduction or Explanation by Derivation.*

The task of science, so it is assumed by those who hold the theory about to be criticized, is the explanation, and the prediction, of known singular facts and regularities with the help of more general theories. In what follows we shall assume T' to be the totality of facts and regular-ities to be explained, D' the domain in which T' makes correct predic-tions, and T (domain D' ⊂ D) the theory which functions as the basis of explanation.[35] Considering (3) we shall have to demand that T be either strong enough to contain T' as a logical consequence, or at least

[34] [6], pp. 209ff. For a more detailed account of Bohr's philosophy of science, see [32].
[35] In what follows it will not be necessary explicitly to distinguish between " 'T' " and "T," and this distinction will therefore not be made. Also terms such as 'con-sistent,' 'incompatible,' and 'follows from' will be applied to pairs of theories, (T,T'), and they will then mean that T *taken together with the conditions of validity of* T', *or the boundary conditions characterizing* D', is compatible with, consistent with, or sufficient to derive, T'.

P. K. Feyerabend

compatible with T' (inside D', that is). Only theories which satisfy one or the other of the two demands just stated are admissible as explanatia. Or, taking the demand for explanation for granted,

(5)   only such theories are admissible (for explanation and prediction) in a given domain which either *contain* the theories already used in this domain, or are at least consistent with them.

It is in this form that (A) will be discussed in the present section and in the sections to follow.

As has just been shown, condition (5) is an immediate consequence of the logical empiricist's theory of explanation and reduction, and it is therefore adopted—at least by implication—by all those who defend that theory. However, its correctness has been taken for granted by a much wider circle of thinkers, and it has also been adopted independently of the problem of explanation. Thus, in his essay "Studies in the Logic of Confirmation" C. G. Hempel demands that "every logically consistent observation report" be "logically compatible with the class of all the hypotheses which it confirms," and more especially, he has emphasized that observation reports do "not confirm any hypotheses which contradict each other."[86] If we adopt this principle, then a theory T (see the notation introduced at the beginning of the present section) will be confirmed by the observations confirming a more narrow theory T' only if it is compatible with T'. Combining this with the principle that a theory is admissible only if it is confirmed to some degree by the evidence available, we at once arrive at (5).

Outside philosophy, (5) has been taken for granted by many physicists. Thus, in his *Waermelehre*, Ernst Mach makes the following remark: "Considering that there is, in a purely mechanical system of absolutely elastic atoms no real analogue for the *increase* of entropy, one can hardly suppress the idea that a violation of the second law . . . should be possible if such a mechanical system were the *real* basis of thermodynamic processes."[87] And he insinuates that, for this reason, the mechanical hypothesis must not be taken too seriously.[88] More recently, Max Born has based his arguments against the possibility of a return to

[86] [45], p. 105, condition (8.3). It was J. W. N. Watkins who drew my attention to this property of Hempel's theory.
[87] [53], p. 364.
[88] For a much more explicit statement of what appears in [53] only as an insinuation, see [54].

determinism upon (5) and the assumption, which we shall here take for granted,[39] that the theory of wave mechanics is incompatible with determinism. "If any future theory should be deterministic," says he, "it cannot be a modification of the present one, but must be entirely different. How this should be possible without sacrificing a whole treasure of well-established results I leave the determinist to worry about."[40]

The use of (5) is not restricted to such general remarks, however. A decisive part of the quantum theory itself, viz., the so-called quantum theory of measurement, is the immediate result of the postulate that the behavior of macroscopic objects, such as measuring instruments, must obey some classical laws precisely and not only approximately. For example, macroscopic objects must always dwell in a well-defined classical state, and this despite the fact that their microscopic constituents exhibit a very different behavior. It is this postulate which leads to the introduction of abrupt jumps in addition to the continuous changes that occur in accordance with Schrödinger's equation.[41] An account of measurement which very clearly exhibits this feature has been given by Landau and Lifshitz.[42] These authors point out that "the classical nature of the apparatus means that . . . the reading of the apparatus . . . has some definite value." "This," they continue, "enables us to say that the state of the system apparatus + electron after the measurement will in actual fact be described, not by the entire sum $[\Sigma A_n(q)\Phi_n(\mathfrak{s})$ where q is the coordinate of the electron, $\mathfrak{s}$ the apparatus coordinate] but by only the one term which corresponds to the 'reading' g_n of the apparatus, $A_n(q)\Phi_n(\mathfrak{s})$." Moreover, most of the arguments against suggestions such as those put forth by Bohm, de Broglie, and Vigier make more or less explicit use of (5).[43] A discussion of this condition is therefore very

[39] Born believes that this assumption has been established by von Neumann's proof. In this he is mistaken; see [29]. However, there exist different and quite plausible arguments for the incompatibility of determinism and wave mechanics, and it is for this reason that I take the assumption for granted. An outline of these plausible arguments is given in [37]. It should be noted that von Neumann himself did not share Born's inductivism. See [62], p. 327.

[40] [7], p. 109. In his treatment of the relation between Kepler's laws and Newton's theory, which, he thinks, applies to all pairs of theories which overlap in a certain domain and are adequate in this domain, Born explicitly accepts (5). For an analysis of Born's inductivism, see Popper [67].

[41] See [30].

[42] [52], p. 22. See also von Neumann's treatment of the Compton effect in [62], pp. 211–215.

[43] Cf. [32], [36], [38].

topical and leads right into the center of contemporary arguments about microphysics.

This discussion will be conducted in three steps. It will first be argued that most of the cases which have been used as shining examples of scientific explanation *do not* satisfy (5) and that it is not possible to adapt them to the deductive scheme. It will then be shown that (5) cannot be defended on empirical grounds and that it leads to very unreasonable consequences. Finally, it will turn out that once we have left the domain of emp`irical generalizations, (5) *should not* be satisfied either. In connection with this last, methodological step, the elements of a positive methodology for theories will be developed, and the historical, psychological, and semantical aspects of such a methodology will be discussed. Altogether the three steps will show that (A) is in disagreement both with actual scientific practice and with reasonable methodological demands. I start now with the discussion of the *actual inadequacy* of (5).

3. The First Example.

A favorite example of both reduction and explanation is the reduction of what Nagel calls the Galilean science to the physics of Newton,[44] or the explanation of the laws of the Galilean physics on the basis of the laws of the physics of Newton. By the Galilean science (or the Galilean physics) is meant, in this connection, the body of theory dealing with the motion of material objects (falling stone, penduli, balls on an inclined plane) near the surface of the earth. A basic assumption here is that the vertical accelerations involved are constant over any finite (vertical) interval. Using T' to express the laws of this theory, and T to express the laws of Newton's celestial mechanics, we may formulate Nagel's assertion to the effect that the one is reducible to the other (or explainable on the basis of the other) by saying that

(6) $T \& d \vdash T'$

where d expresses, in terms of T, the conditions valid inside D'. In the case under discussion d will include description of the earth and its surroundings (supposed to be free from air; we shall also abstract from

[44] [20], p. 291. I am aware that, from a historical point of view, the discussion to follow is not adequate. However, I am here interested in the systematic aspect, and I have therefore allowed myself what could only be regarded as great liberties if the main interest were historical.

all those phenomena which are due to the rotation of the earth and whose inclusion would strengthen, rather than weaken our case), and reference will be made to the fact that the variation H of the height above ground level in the processes described is very small if compared with the radius R of the earth.

As is well known (6) cannot be correct: as long as H/R has some finite value, *however small*, T' will not follow (logically) from T and d. What will follow will rather be a law, T'', which, while being experimentally indistinguishable from T' (on the basis of the experiments which formed the inductive evidence for T' in the first place), is yet inconsistent with T'. If, on the other hand, we want to derive T' precisely, then we must replace d by a statement which is patently false, as it would have to describe the conditions in the close neighborhood of the earth as leading to a vertical acceleration that is constant over a finite interval of vertical distance. It is therefore impossible, *for quantitative reasons*, to establish a deductive relationship between T and T', or even to make T and T' compatible. This shows that the present example is not in agreement with (5) and is, therefore, also incompatible with (A), (1), and (3).

Now in this situation, we may adopt one or the other of the following two procedures. We may either declare that the Galilean science can neither be reduced to, nor explained in, terms of Newton's physics;[45] or we may admit that reduction and explanation are possible, but deny that deducibility, or even consistency (on the basis of suitable boundary conditions), is a necessary condition of either. It is clear that the question as to which of these two procedures is to be adopted is of subordinate importance (after all, it is purely a matter of terminology that is to be settled here!) if compared with the question whether newly invented theories should be consistent with, or contain, those of their predecessors with whom they overlap in empirical content. We shall therefore defer settlement of the terminological problem raised above and concentrate on the question of consistency, or derivability. And we shall use the terms 'explanation' and 'reduction' either in a vague and general sense, awaiting further explication, or in the manner suggested by Nagel and by Hempel and Oppenheim. The usage adopted should always be clear from the context.

[45] This suggestion was made to me by Professor Viktor Kraft.

P. K. Feyerabend

The objection which has just been developed—so it is frequently pointed out—cannot be said to endanger the correct theory of explanation, since everybody would admit that explanation may be by approximation only. This is a curious remark indeed! It criticizes us for taking seriously, and objecting to, a criterion which has either been universally stated as a necessary condition of explanation, or which plays a central role in some theories of confirmation, viz., condition (3). Now dropping (3) means altogether giving up the orthodox theory, for (3) formed the very core of this theory.[46] On the other hand, the remark that we explain "by approximation'" is much too vague and general to be regarded as the statement of an alternative theory. As a matter of fact, it will turn out that the idea of approximation cannot any more be incorporated into a formal theory, since it contains elements which are essentially subjective. However, before dealing with this aspect of explanation we shall inquire a little more closely into the reasons for the failure of (3). Such an inquiry will lead to the result not only that (3) is false, but it is also very unreasonable to assume that it could be true.

4. Reasons for the Failure of (5) and (3).

The basic argument is really very simple, and it is very surprising that it has not been used earlier. It is based upon the fact that *one and the same set of observational data is compatible with very different and mutually inconsistent theories.* This is possible for two reasons: first, because theories, which are universal, always go beyond any set of observations that might be available at any particular time; second, because the truth of an observation statement can always be asserted within a certain margin of error only.[47] The first reason allows for theories to differ in domains where experimental results are not yet available. The second reason allows for such differences even in those domains where observations have been made, provided the differences are restricted to the margin of error connected with the observations.[48] Both reasons taken together sometimes allow considerable freedom in the construction of our theories.

[46] This has been emphasized, in private communication, by Professors Kraft (Vienna) and Rynin (Berkeley).
[47] As J. W. N. Watkins has pointed out to me, this invalidates Hempel's conditions 9.1 and 9.2 (in [45]). An attempt to bring logical order into the relation between observation statements and the more precise statements derived from a theory has been made by Professor S. Körner [50], p. 140.
[48] Even this condition is too strong, as will be shown below.

90

Now, it is very important to realize that this freedom which experience grants the theoretician is nearly always restricted by conditions of an altogether different character. These additional conditions are neither universally valid, nor objective. They are connected partly with the tradition in which the scientist works, with the beliefs and the prejudices which are characteristic of that tradition; and they are partly connected with his own personal idiosyncrasies. The formal apparatus available and the structure of the language he speaks will also strongly influence the activity of the scientist. Whorff's assertion to the effect that the properties of the Hopi language are not very favorable for the development of a physics like the one with which we are acquainted may very well be correct.[49] Of course, it must not be overlooked[50] that man is capable not only of applying, but also of inventing, languages. Still, the influence of the language from which he starts should never be underestimated. Another factor which strongly influences theorizing is metaphysical beliefs. The Neoplatonism of Copernicus was at least a contributing factor in his acceptance of the system of Aristarchus.[51] Also, the contemporary issue between the followers of Niels Bohr and the realists, being still undecidable on the basis of contemporary experimentation, is mainly metaphysical in character.[52] That the choice of theories may be influenced even by aesthetic motives can be seen from Galileo's reluctance to accept Kepler's ellipses.[53]

Taking all this into account we see that the theory which is suggested by a scientist will also depend, apart from the *facts* at his disposal, on the *tradition* in which he participates, on the mathematical instruments he accidentally knows, on his preferences, on his aesthetic prejudices, on the suggestions of his friends, and on other elements which are rooted, not in facts, but in the mind of the theoretician and which are therefore subjective. This being the case it is* to be expected that theoreticians working in different traditions, in different countries, will arrive at theories which, although in agreement with all the known facts, are yet mutually inconsistent. Indeed, any consistency over a long period

[49] See [73].
[50] As is done by Bohr, Heisenberg, and von Weizsaecker in their philosophical writings as well as by some Wittgensteinians. For the point of view of these physicists, see [34] and [38], as well as the end of Sec. 7 of the present paper.
[51] See T. S. Kuhn [51], pp. 128ff.
[52] See [36].
[53] See E. Panofsky [63].

of time would have to be regarded not, as is suggested by (3), (A), and (5), as a methodological virtue, but as an alarming sign that no new ideas are being produced and that the activity of theorizing has come to an end. Only the inductivistic doctrine that theories are uniquely determined by the facts could have persuaded people that lack of ideas is praiseworthy and that its consequences are an essential feature of the development of our knowledge.[54]

At this point it is worth mentioning what will be explained in great detail later: that the freedom of theorizing granted by the indeterminateness of facts is of great methodological importance. It will turn out that many test procedures presuppose the existence of a class of mutually incompatible, but factually adequate, theories. Any attempt to reduce this class to a single theory would result in a decisive decrease of the empirical content of this remaining theory and would therefore be undesirable from the point of view of empiricism. The freedom granted by the indeterminateness of facts is therefore not only psychologically important (it allows scientists of different temperament to follow their different inclinations and thereby gives them satisfaction which goes beyond the satisfaction derived from the exclusive consideration of facts); it is also needed for methodological reasons.

The gist of the argument developed so far is that because of the latitude which experience allows the theoretician, and because of the different way in which this latitude will be exercised by thinkers of different tradition, temperament, and interests, it is to be expected that two different theories, and especially two theories of a different degree of generality, will be inconsistent with each other even in those cases where both are confirmed by the set. In this argument it was assumed that the experimental evidence which inside D' confirms T and T' is the same in both cases. Although this may be so in the specific example discussed, it is certainly not true in general. Experimental evidence does

---

[54] This is true mainly of those more crude theories of induction which are held, by implication, by many physicists. It would seem to me that discussion and criticism of these theories is a much more effective way of advancing scientific knowledge than invention of highly technical theories of confirmation which are of no interest to the scientist because they cannot be applied to a single noninstantial theory. Unfortunately, many philosophers of science consider it below their dignity to discuss such crude but effective theories, and they prefer the construction of sophisticated theories which are totally ineffective and useless. The brief discussion of Hempel's paper seems to show that similar objections must be raised against some ideas held by contemporary empiricists.

not consist of facts pure and simple, but of facts analyzed, modeled, and manufactured according to some theory. The first indication of this manufactured character of the evidence is seen in the corrections which we apply to the readings of our measuring instruments, and in the selection which is made among those readings. Both the corrections and the selection made depend upon the theories held, and they may be different for the theoretical complex containing T, and for the theoretical complex containing T'. Usually T will be more general, more sophisticated, than T', and it will also be invented a considerable time after T'. New experimental techniques may have been introduced in the meantime. Hence, the 'facts,' within D', which count as evidence for T will be different from the 'facts,' within D', which counted as evidence for T' when the latter theory was first introduced. An example is the very different manner in which the apparent brightness of stars was determined in the seventeenth century and is determined now. This is another important reason why T usually will not satisfy (5) with respect to T': not only are T and T' connected with different theoretical ideas leading to different predictions even in the domain where they overlap and are both confirmed, but the better experimental techniques and the improved theories of measurement will usually provide evidence for T which is different from the evidence for T' even within the domain of common validity. In short: introducing T very often leads to recasting the evidence for T'. The demand that T should satisfy (5) with respect to T' would in this case imply the demand that new and refined measurements not be used, which is clearly inconsistent with empiricism.

Against the argument in the last paragraph it might be pointed out that results of measurement which are capable of improvement, and which therefore change, do not belong to the observational domain, but must be formulated with the help of singular statements of the theoretical language.[55] Observational statements proper are such qualitative statements as "pointer A coincides with mark B," or "A is greater than B"—and these statements will not change, or be eliminated, whatever the development of the theory, or of the methods of measurement. This point will be dealt with, and refuted, in Section 7 of this paper.

A further indication of the "manufactured" character of the experi-

[55] For this move, cf. Carnap [15], p. 40.

P. K. Feyerabend

mental evidence is seen in the fact that observable results, and indeed anything conveyed with the help of a language, are always expressed in some theory or other. Because this fact will also be of importance in connection with my criticism of (B), and because it leads to a further criticism of (A), I shall discuss at length the example I have chosen for its elucidation.

5. Second Example: The Problem of Motion.[56]

From its very beginning, rational cosmology, this creation of the Ionian "physiologists," was faced with the problem of change and motion (in the general sense in which it includes locomotion, qualitative alteration, quantitative augmentation and diminution, as well as generation and corruption). The problem arose in two forms. The first was the possibility of change and motion. This form of the problem had to be solved by the invention of a cosmology which allowed for change, i.e., which was not such that the occurrence of change was (unwittingly) excluded from it by the very nature of the assumptions upon which it was based. The second form of the problem which arose, once the first had been solved in a satisfactory manner, was the cause of change. As was shown by Parmenides, the early monistic theories of Thales, Anaximander, and others could not solve the first form of the problem. For Parmenides himself, this did not refute monism; it refuted the existence of change.

The majority of thinkers went a different path, however. They regarded monism as refuted and started with pluralistic theories. In the case of the atomic theory, which was one of these pluralistic theories, this relation between Parmenides' arguments and pluralism is very clear. Leucippus, who "had associated with Parmenides in philosophy,"[57] "thought he had a theory which was in harmony with the senses, and did not do away with coming into being and passing away, nor motion, nor with the multiplicity of things."[58] This is how the atomic theory arose, as an attempt to solve problems created by the empirical inadequacy of the early monism of the Ionians.

However, the theory which was most influential in the Middle Ages

---

[56] For a more detailed account of the theories mentioned in this section, see M. Clagett [17]. Concerning the first part of the present section, see J. Burnet [10], as well as Clagett [16] and Popper [67].

[57] Aristotle [2], A, 8 324b35.

[58] Theophrastus quoted from Burnet [10], p. 333.

and which also tried to solve what I have above called the second form of the problem was Aristotle's theory of motion as the actualization of potentiality. According to Aristotle

(7)   "motion is a process arising from the continuous action of a source of motion, or a 'motor,' and a 'thing moving.' "[59]

This principle, according to which any motion (and not only accelerated motion) is due to the action of some kind of force, can be easily supported by such common observations as a cart drawn by a horse and a chair pushed around by an angry husband. It gets into difficulties when one considers the motion of things thrown: stones continue to move despite the fact that contact with the motor apparently ceases when they leave the hand. Various theories have been suggested to eliminate this difficulty. From the point of view of later developments, the most important one of these theories is the impetus theory. The impetus theory retains (7) and the general background of the Aristotelian theory of motion. Its distinction lies in the specific assumptions it makes concerning the causes that are responsible for the motion of the projectile. According to the impetus theory, the motor (for example the hand) transfers upon the projectile an inner moving force which is responsible for its continuation of path, and which is continually decreased by the resisting air and by the gravity of the projectile. A stone in empty space would therefore either remain at rest or move (along a straight line[60]) with constant speed, depending on whether its impetus is zero or possesses a finite value.

At this point a few words must be said about the characterization of locomotion. The question as to its proper characterization was a matter of dispute. To us it seems quite natural to characterize motion by space transversed, and, as a matter of fact, one of the suggested characterizations did just this: it defined motion kinematically by reference to space transversed. This apparently very simple characterization needs further specification if an account is to be given of nonuniform movements where the distinction becomes relevant between average velocity and instantaneous velocity. Compared with the actual space transversed by a given body, the instantaneous velocity is a rather abstract notion since

[59] Clagett [17], p. 425.
[60] The parentheses I have added because of the absence from the earlier forms of the impetus theory of an explicit consideration of direction.

it refers to the space that would be transversed if the velocity were to retain constancy over a finite interval of time.

Another characterization of motion is the dynamical. It defines motion in terms of the forces which bring it about in accordance with (7). Adopting the impetus theory the motion of a stone thrown would have to be characterized by its inherent impetus, which pushes it along until it is exhausted by the opposing forces of friction and gravity.

Which characterization is the better one to take? From an operationalistic point of view (and we shall adopt this point of view, since we want to follow the empiricist as far as possible), the dynamical characterization is definitely to be preferred: while it is fairly easy to observe the impetus enclosed in a moving body by bringing it to a stop in an appropriate medium (such as soft wax) and then noting the effect of such a maneuver, it is much more difficult, if not nearly impossible, to arrange matters in such a way that from a given moment on, a nonuniformly moving object assumes a constant speed with a value identical with the value of the instantaneous velocity of the object at that moment and then to watch the effect of this procedure.

With the use of the dynamical characterization, the "inertial law" pronounced above reads as follows:

(8)   The impetus of a body in empty space which is not under the influence of any outer force remains constant.

Now, in the case of inertial motions (8) gives correct predictions about the behavior of material objects. According to (3), explanation of this fact will involve derivation of (8) from a theory and suitable initial conditions. Disregarding the demand for explanation, we can also say, on the basis of (5), that any theory of motion that is more general than (8) will be adequate only if it contains (8) which, after all, is a very basic law. According to (2), the meanings of the key terms of (8) will be unaffected by such a derivation. Assuming Newton's mechanics to be the primary theory, we shall therefore have to demand that (8) be derivable from it *salva significatione*. Can this demand be satisfied?

At first sight it would seem that it is much easier to derive (8) from Newton's theory than it is to establish the correctness of (6): as opposed to Galileo's law (8) is not in quantitative disagreement with anything asserted by Newton's theory. Even better: (8) seems to be identical with Newton's first law so that the process of derivation seems to degenerate into a triviality.[61]

In the remainder of the present section, it will be shown that this is not so and that it is impossible to establish a deductive relationship between (8) and Newton's theory. Later on this will be the starting point of our criticism of (B).

Let me repeat, before beginning the argument, that (8), *taken by itself*, cannot be attacked on empirical grounds. Indeed, we have indicated a primitive method of measurement of impetus, and the attempt to confirm (8) by using this method will certainly show that within the domain of error connected with such crude measurements, (8) is perfectly all right. It is, therefore, quite in order to ask for the explanation, or the reduction, of (8), and the failure to arrive at a satisfactory solution of this task cannot be blamed upon the empirical inadequacy of (8).

We now turn to an analysis of the main terms of (8). According to Nagel the meaning of these terms is to be regarded as "fixed" by the procedures and assumption of the impetus theory, and any one of them is "therefore intelligible in terms of its own rules of usage."[62] What are these meanings, and what are the rules which establish them?

Take the term 'impetus.' According to the theory of which (8) is a part, the impetus is the force responsible for the movement of the object that has ceased to be in direct contact, by push, or by pull, with the material mover. If this force did not act, i.e., if the impetus were destroyed, then the object would cease to move and fall to the ground (or simply remain where it is, in case the movement were on a frictionless horizontal plane). A moving object which is situated in empty space and which is influenced neither by gravity nor by friction is not outside the reach of any force. It is pushed along by the impetus, which may be pictured as a kind of inner principle of motion (similar, perhaps, to the vital force of an organism which is the inner principle of *its* motion).

We now turn to Newton's celestial mechanics and the description, in terms of this theory, of the movement of an object in empty space. (Newton's theory still retains the notion of absolute space and allows therefore for such a description to be formed.) Quantitatively, the same

---

[61] There existed theories, among them the theory of *mail* by Abu'l-Barakat, where quantitative disagreement with Newton's laws was to be expected: in these theories, the impetus decreased with time in the same manner in which a hot poker that is removed from the fire gradually loses the heat stored in it. Cf. Clagett [17], p. 513.

[62] See Nagel [20], p. 301.

movement results. But can we discover in the description of this movement, or in the explanation given for it, anything resembling the impetus of (8)? It has been suggested that the momentum of the moving object is the perfect analogue of the impetus. It is correct that the measure of this magnitude (viz., mv) is identical with the measure that has been suggested for the impetus.[63] However, it would be very mistaken if we were, on that account, to identify impetus and momentum. For whereas the impetus is supposed to be something that pushes the body along,[64] the momentum is the result rather than the cause of its motion. Moreover, the inertial motion of classical mechanics is a motion which is supposed to occur by itself, and without the influence of any causes. After all, it is this feature which according to most historians, radical empiricists included, constitutes one of the main differences between the Aristotelian theory and the celestial mechanics of the seventeenth, eighteenth, and nineteenth centuries: in the Aristotelian theory, the natural state in which an object remains without the assistance of any causes is the state of rest. A body at rest (in its natural place, we should add) is not under the influence of any forces. In the Newtonian physics it is the state of being at rest or in uniform motion which is regarded as the natural state. This means, of course, the explicit denial of a force such as the impetus is supposed to represent.

Now this denial need not mean that the concept of such a force cannot be formed within Newton's mechanics. After all, we deny the existence of unicorns and use in this denial the very concept of a unicorn. Is it then perhaps possible to define a concept such as impetus in terms of the theoretical primitives of Newton's theory? The surprising fact is that any attempt to arrive at such a definition leads to disappointment (which shows, by the way, that theories such as Newton's are expressed in a language that is much more tightly knit than is the language of everyday life). I have already pointed out that the momentum, which would give us the correct mathematical value, is not what we want. What we want is a force that acts upon the isolated object and is responsible for its motion. The concept of such a force can of course be formed within Newton's theory. But considering (a) that the movement under review (the inertial movement) occurs with constant ve-

[63] See Clagett [17], p. 523.
[64] For an elaborate discussion of the difference between momentum and impetus, see Anneliese Maier [58]. For what follows, see also M. Bunge [9], Ch. 4.4.

locity, and (b) Newton's second law, we obtain in all relevant cases zero for the value of this force which is not the measure we want. A positive measure is obtained only if it is assumed that the movement occurs in a resisting medium (which is, of course, the original Aristotelian assumption), an assumption which is inconsistent with another feature of the case considered, i.e., with the fact that the inertial movement is supposed by Newton's theory to occur in empty space. I conclude from this that the concept of impetus, as fixed by the usage established in the impetus theory, cannot be defined in a reasonable way within Newton's theory. And this is not further surprising. For this usage involves laws, such as (7), which are inconsistent with the Newtonian physics.

In the last argument, the assumption that the concept force is the same in both theories played an essential role. This assumption was used in the transition from the assertion, made by the impetus theory, that inertial motions occur under the influence of forces to the calculation of the magnitude of these forces on the basis of Newton's second law. Its legitimacy may be derived from the fact that both the impetus theory and Newton's theory apply the concept force under similar circumstances (paradigm-case argument!). Still, meaning and application are not the same thing, and it might well be objected that the transition performed is not legitimate, since the different contexts of the impetus theory, on the one hand, and of Newton's theory, on the other, confer different meanings upon one and the same word 'force.' This being the case, our last argument is based upon a quaternio terminorum and is, therefore, invalid. In order to meet this objection, we may repeat our argument using the word 'cause' instead of the word 'force' (the latter has a somewhat more specific meaning). But if someone again retorts that 'cause' has a different meaning in Newton's theory from what it has in the impetus theory, then all I can say is that a consistent continuation of that kind of objection will in the end establish what I wanted to show in a more simple manner, viz., the impossibility of defining the notion of an impetus in terms of the descriptive terms of Newton's theory. To sum up: the concept impetus is not "explicable in terms of the theoretical primitives of the primary science."[65] And this is exactly as it should be, considering the inconsistency between some very basic principles of these two theories.

[65] Nagel [20], p. 302.

However, explication in terms of the primitives of the primary science is not the only method which was considered by Nagel in his discussion of the process of reduction. Another way to achieve reduction, which he mentions immediately after the above quotation, "is to adopt a material, or physical hypothesis according to which the occurrence of the properties designated by some expression in the premises of the primary science is a sufficient, or a necessary and sufficient condition for the occurrence of the properties designated by the expressions of the secondary discipline." Both procedures are in accordance with (4), or with (2), or at least Nagel thinks that they are: ". . . in this case" he says, referring to the procedure just outlined, "the meaning of the expressions of the secondary science *as fixed by the established usage of the latter,* is not declared to be analytically related to the meanings of the corresponding expressions of the primary science."[66] Let us now see what this second method achieves in the present case.

To start with, this method amounts to introducing a hypothesis of the form

(9)   impetus = momentum

where each side retains the meaning it possesses in its respective discipline. The hypothesis then simply asserts that wherever momentum is present, impetus will also be present (see the above quotation of Nagel's), and it also asserts that the measure will be the same in both cases. Now this hypothesis, although acceptable within the impetus theory (after all, this theory permits the incorporation of the concept of momentum), is incompatible with Newton's theory. It is therefore not possible to achieve reduction and explanation by the second method.

To sum up: a law such as (8) which, as I have argued, is empirically adequate, and in quantitative agreement with Newton's first law, is yet incapable of reduction to Newton's theory and therefore incapable of explanation in terms of the latter. Whereas the reasons we have so far found for irreducibility were of a quantitative nature, this time we met a qualitative reason, as it were, i.e., the incommensurable character of the conceptual apparatus of (8), on the one side, with that of Newton's theory, on the other.

Taking together the quantitative as well as the qualitative argument, we are now presented with the following situation: there exist pairs of

[66] Ibid. My italics.

theories, T and T', which overlap in a domain D' and which are incompatible (though experimentally indistinguishable) in this domain. Outside D', T has been confirmed, and it is also more coherent, more general, and less ad hoc than T'. The conceptual apparatus of T and T' is such that it is possible neither to define the primitive descriptive terms of T' on the basis of the primitive descriptive terms of T nor to establish correct empirical relations involving both these terms (correct, that is, from the point of view of T). This being the case, explanation of T' on the basis of T or reduction of T' to T is clearly impossible if both explanation and reduction are to satisfy (A) and (B). Altogether, *the use of T will necessitate the elimination both of the conceptual apparatus of T' and of the laws of T'.* The conceptual apparatus will have to be eliminated because its use involves principles, such as (7) in the example above, which are inconsistent with the principles of T; and the laws will have to be eliminated because they are inconsistent with what follows from T for events inside D'. (This would apply to the example above if the theory of mail had been used instead of the impetus theory.) This being the case the demand for explanation and reduction clearly cannot arise if this demand is interpreted as the demand for the explanation, or reduction, of T', rather than of a set of laws that is in some respect similar to T' but in other respects (meanings of fundamental terms included) very different from it. For such a demand would imply the demand to derive, from correct premises, what is false, and to incorporate what is incommensurable.

The effect of the transition from T' and T is rather to be described in the manner indicated in the introductory remarks of the present paper: where I said: What happens when transition is made from a restricted theory T' to a wider theory T (which is capable of covering all the phenomena which have been covered by T') is something much more radical than incorporation of the unchanged theory T' into the wider context of T. What happens is rather a *complete replacement* of the ontology of T' by the ontology of T, and a corresponding change in the meanings of all descriptive terms of T' (provided these terms are still employed). Let me add here that the not-too-well-known example of the impetus theory versus Newton's mechanical theory is not the only instance where this assertion holds. As I shall show a little later, more recent theories also correspond to it. Indeed, it will turn out that the principle correctly describes the relation between the elements of any

P. K. Feyerabend

pair of noninstantial theories satisfying the conditions which I have just enumerated.

This finishes step one of the argument against the assumption that reduction and explanation are by derivation. What I have shown (and shall show in later sections) is that some very important cases which have been, or could be used as examples of reduction (and explanation) are not in agreement with the condition of derivability. It will be left to the reader to verify that this holds in almost all cases of explanation by theories: assumption (A) does not give a correct account of actual scientific practice. It has also been shown that in this respect the thesis formulated in the beginning of this paper is much more adequate.

Now, as against this result it may be pointed out, with complete justification, that scientific method, as well as the rules for reduction and explanation connected with it, is not supposed to describe what scientists are actually doing. Rather, it is supposed to provide us with normative rules which should be followed, and to which actual scientific practice will correspond only more or less closely.[67] It is very important nowadays to defend such a normative interpretation of scientific method and to uphold reasonable demands even if actual scientific practice should proceed along completely different lines. It is important because many contemporary philosophers of science seem to see their task in a very different light. For them actual scientific practice is the material from which they start, and a methodology is considered reasonable only to the extent to which it mirrors such practice.[68] Looking at disciplines such as medicine, they discover (whether rightly or wrongly—this I do not want to discuss at the present moment) that what is called "explanation" here is not always the inverse of prediction, and they infer from this that the orthodox model which demands such an inverse relationship to hold between explanation and prediction[69] is unduly restrictive.

Two elements must be distinguished in this "discovery."[70] The first is of a purely linguistic character. It is connected with the problem as to what meaning should be given to the word 'explanation.' Clearly this

[67] This point has been made, most forcefully, by K. R. Popper [68], Sec. 10.
[68] If I understand him correctly, this is also the point of view held by my colleague, Professor T. S. Kuhn.
[69] Cf. Hempel and Oppenheim [47], p. 323.
[70] The "discovery" is due among others to Professor Barker. See his contribution to [24], my criticism in the same volume, and his reply.

102

element is without serious interest. It may be that the word 'explanation' sounds beautiful to some ears—but who would think it reasonable to start a war for, or against, its elimination? The second element, however, which usually remains hidden beneath the linguistic analysis, is much more serious. For what the suggested procedure amounts to is increased leniency with respect to questions of test: a certain medical hypothesis (which, let us say, is expressed by saying that a patient died because of tuberculosis) is accepted, and retained, despite the fact that independent tests (independent, that is, of the *past* histories of this case and of other cases) are not available, and its further use is defended by reference to the fact that it is in accordance with the "logic of medicine." Expressed in more pedestrian terms, this maneuver propagates the acceptance of unsatisfactory hypotheses on the grounds that this is what everybody is doing. It is conformism covered up with high-sounding language.[71] It is clear, however, that if this conformism had been propagated successfully in the Middle Ages, modern science with its so very different "logic" would never have come into existence. Modern science is the result of a conscious criticism of the theses propagated and the methods employed by the great majority of scholastic philosophers. For the thinker who demands that a subject be judged "according to its own standards," such criticism is of course impossible; he will be strongly inclined to reject any interference and to "leave everything as it

[71] It should be pointed out that almost all theses which terminate (if at all) the long-winded inquiries of linguistic philosophers possess this two-faced character. On the one side, they seem to be about the meanings of terms only, and therefore rather harmless and uninteresting (although there are enough enthusiastic buyers even for such products). However, on closer inspection, it often turns out that beneath these linguistic trappings there are hidden, and thereby removed from criticism, some highly questionable theories or methodological rules. Only consider the example discussed in the text: it may well be the case that in some disciplines the word 'explanation' is used in a manner that does not lead to the demand for additional predictions. This linguistic result may be expressed by saying that prediction is not essential to explanation (viz., to 'explanation' in the new sense that is characteristic for the disciplines under review). Now from this last result it is then very often inferred that the search for additional predictions is unnecessary and that all is well. Clearly this methodological consequence can be derived from the linguistic premise only if it is further assumed that all is well once an explanation has been given, and this regardless of the sense in which the word 'explanation' is being used in the discipline in question. This assumption, which, I submit, is the silent premise of many contemporary linguistic arguments about explanation, amounts to asserting that all is well as long as the word 'explanation' occurs somewhere in the description of the procedure to be analyzed. This is, of course, pure word magic. Curiously enough, it is the linguistic philosopher who is swayed by such word magic. Which only shows how little language is understood by some of its most verbose champions.

P. K. Feyerabend

is."[72] It is somewhat puzzling to find that such demands are nowadays advertised under the title of philosophy of science.

Against such conformism it is of paramount importance to insist upon the normative character of scientific method. Adopting this point of view, one cannot regard the arguments of the last few sections as ultimately decisive. They are satisfactory insofar as they show that the "orthodox" are wrong when asserting that (A), (B), and (5) reflect actual scientific practice. But they do not dispose of these principles if they are interpreted as demands to be followed by the scientist (although, of course, they provide ample material for such disproof). I therefore proceed now to a methodological criticism of the demands of the orthodox. The first move in this criticism will be the examination of an argument which has sometimes been used to defend (5).

6. Methodological Considerations.

The argument runs as follows: (α) a good theory is a summary of facts; (β) the predictive success of T' (I will continue to use the notation introduced in Section 2) has shown T' to be good theory inside D'; hence (τ), if T, too, is to be successful inside D', then it must either give us all the facts contained in T', i.e., it must give us T', or at least it must be compatible with T'.

It is easily seen that this very popular argument[73] will not do. We can show this by considering its premises. Premise (α) is acceptable if it is not taken in too strict a sense (for example, if it is not interpreted as implying an ontology of mutually independent 'facts' as has been suggested by Mach and the early Wittgenstein). Interpreted in such a loose manner (α) simply says that a good theory not only will be able to answer many questions, but will also answer them correctly. Now if this is to be the interpretation of (α), then (β) cannot possibly be correct: in (β) the predictive success of T' is taken to indicate that T' will give a correct account of all the facts inside its domain. However, one must remember that because of the general character of statements expressing laws and theories, their predictive success can be established

[72] Wittgenstein [74], Sec. 124.
[73] A sloppy version of this argument occurs frequently in arguments by physicists. It ought to be mentioned, by the way, that Hempel's condition 8 leads to the very same result, viz., to the demand that new theories be consistent with their confirmed predecessors. A justification for discussing "crude" arguments of the kind outlined in the text above is given in fn. 54.

104

only with respect to part of their content. Only part of a theory can at any time be known to be in agreement with observation. From this limited knowledge nothing can be inferred (logically!) with respect to the remainder.[74]

We have also to consider the margin of error involved in every single test. Hence, from a purely logical point of view, new theories will be restricted only to the extent to which their predecessors have been tested and confirmed.[75] Only to this extent will it be necessary for them to agree with their predecessors. In domains where tests have not yet been carried out, or where only very crude tests have been made, we have complete freedom on how to proceed, and this quite independently of which theories were originally used here for the purpose of prediction. Clearly this last condition, which is in agreement with empiricism, is much less restrictive than either (3) or (5).

One might hope to arrive at more restrictive conditions by adding inductive argument to logical reasoning. True, from a logical point of view we can only say that part of T' has been found to be in agreement with observation and that T need agree only with that part and not, as is demanded in (5), with the whole of T'. However, if inductive reasoning is used as well, then we shall perhaps have to admit that this partial confirmation has established T', and that therefore the whole of T' should be covered by T. Does this help us to strengthen the condition mentioned at the end of the last paragraph and to demonstrate (5) after all?

It is clear that inductive reasoning cannot establish (5) either. For let us assume that T agrees with T' only where T' has been confirmed and is different from T' in all other instances without having as yet been refuted. In this case T will satisfy our own condition of the last paragraph, and it will not satisfy any stronger condition (except accidentally). Can inductive reasoning prompt us to eliminate T? It is not easily seen how this could be the case, since T shares all its confirming instances with T'. Hence, if T' is established by these instances, then so is T— unless we use formal considerations (which I shall discuss later). Again

---

[74] This point is indebted to Hume. That Hume's arguments are still not understood by many thinkers and are therefore still in need of repetition has been emphasized by Popper [68], Reichenbach [69], Goodman [42], and others.

[75] As was mentioned in Sec. 4, it is hardly ever the case that two theories which have been discussed in very different historical periods will be based upon exactly the same observations. The condition is therefore still too strict.

we arrive at the result that from the point of view of fact there is not much to choose between T and T', and that (5) cannot be defended on empirical grounds.

It is worthwhile to inquire a little more closely into the effects which adoption of (5), and, incidentally, also of Hempel's condition 8,[76] would have upon the development of scientific knowledge. Such adoption would lead to the elimination of a theory, not because it is inconsistent with the facts, but because it is inconsistent with another, and as yet unrefuted, theory whose confirming instances it shares. This is a strange procedure to be adopted by thinkers who, above anything, claim to be empiricists! However, the situation becomes even worse when we inquire why the one theory is retained and the other rejected. The answer (which is, of course, not the answer given by the empiricist) can only be that the theory which is retained was there first. This shows that in practice the allegedly empirical procedure (5) leads to the preservation of the old theories and to the rejection of the new theories even before these new theories have been confronted with the facts. That is, it leads to the same result as transcendental deduction, intuitive argumentation, and other forms of a priori reasoning, the only difference being that now it is in the name of experience that such results are obtained. This is not the only instance where, on closer scrutiny, a rather close relation emerges between some versions of modern empiricism and the 'school philosophies' it attacks.

We have now to consider the argument that formal criteria may provide a principle of choice between T and T' that is independent of fact. Such formal criteria can indeed be given.[77] However, while usually a more general and coherent theory is preferred to a less general collection of laws, and this on account of being less ad hoc, (5) tends to reverse this procedure. This is due to the fact that general theories of a high degree of coherence usually violate (5). Again this principle is seen to be incompatible with reasonable methodology.

Two things have been shown so far. First, the invalidity of an argument used for establishing (5). Second, the undesirability, from an empirical point of view, of some consequences of this argument. However, all this has little weight when compared with the following most important consideration.

[76] See [45], p. 105.
[77] See Popper [68], Ch. VI.

Within contemporary empiricism, discussions of test and of empirical content are usually carried out in the following manner: it is inquired how a theory is related to its empirical consequences and what these consequences are. True, in the derivation of these consequences reference will have to be made to principles or theorems which are borrowed from other disciplines and which then occur in the correspondence rules. However, these principles and these theorems play a subordinate role when compared with the theory under review; and it is, of course, also assumed that they are mutually consistent and consistent with the theory. One may therefore say that, for the orthodox procedure, the natural unit to which discussions of empirical content and of test methods are referred is always a single theory taken together with those of its consequences that belong to the observation language.

This manner of discussion does not allow us to give an adequate account of crucial experiments which involve more than one theory, none of which are expendable or of psychological importance only. A very good example of the structure of such crucial tests is provided by the more recent development of thermodynamics. As is well known, the Brownian particle is a perpetual motion machine of the second kind, and its existence refutes the (phenomenological) second law. However, could this fact have been discovered in a direct manner, i.e., by a direct investigation of the observational consequences of thermodynamics? Consider what such a refutation would have required! The proof that the Brownian particle is a perpetual motion machine of the second kind would have required (a) measurement of the exact motion of the particle in order to ascertain the changes of its kinetic energy plus the energy spent on the overcoming of the resistance of the fluid, and (b) precise measurements of temperature and heat transfer in the surrounding medium in order to ascertain that any loss occurring here was indeed compensated by the increase of the energy of the moving particle and the work done against the fluid as mentioned in (a). Such measurements, however, are beyond experimental possibilities.[78] Hence, a direct refutation of the second law, i.e., a refutation based upon an investigation of the testable consequences of thermodynamics alone, would have had to wait for one of those rare, not repeatable, and therefore, prima facie suspicious, large fluctuations in which the transferred heat is indeed

[78] Concerning the extreme difficulties of following the motion of the Brownian particle in all its details, see R. Fuerth [41].

107

accessible to measurement. This means that such a refutation would have never taken place, and, as is well known, the actual refutation of the second law was brought about in a very different manner. It was brought about via the kinetic theory and Einstein's utilization of it in the calculation of the statistical properties of the Brownian motion. In the course of this procedure the phenomenological theory (T') was incorporated into the wider context of statistical physics (T) in such a manner that (5) was violated; and then a crucial experiment was staged (Perrin's investigations).

Now it seems to me that the more general our knowledge gets the more important it will be to carry out tests in the manner indicated, i.e., not by comparing a single theory with experience, but by staging crucial experiments between theories which, although in accordance with all the known facts, are mutually inconsistent and give widely different answers in unexplored domains. This suggests that outside the domain of empirical generalizations the methodological unit to which we refer when discussing questions of test and empirical content consists of a whole set of partly overlapping, factually adequate, but mutually inconsistent theories. To the extent to which utilization of such a set provides additional tests which, for empirical reasons, could not have been carried out in a direct manner, the use of a set of this kind is demanded by empiricism. For the basic principle of empiricism is, after all, to increase the empirical content of whatever knowledge we claim to possess.[79]

On the other hand, the fact that (5) does not allow for the formation of such sets now proves this principle to be inconsistent with empiricism. By excluding valuable tests it decreases the empirical content of the theories that are permited to remain (and which, as indicated above, will usually be the theories which were there first). This last result of a consistent application of (5) is of very topical interest: it may well be, as has been pointed out by Bohm and Vigier,[80] that the refutation of the quantum-mechanical uncertainties presupposes just an incorporation of the present theory into a wider context, which is not any more in

[79] With respect to this last demand, it might be objected that, given a certain theory, such an extension of content is not possible without changing the theory. This argument would be correct if it could be granted that the interpretation of a physical theory is wholly empirical. As I have shown elsewhere (see [39]), this is not the case. The demand to increase the empirical part of this interpretation is therefore a sensible demand.

[80] See the discussion remarks of these two physicists in [49], as well as those of Bohm [4].

accordance with the idea of complementarity and which therefore suggests new and decisive experiments. And it may also be that the insistence on part of the majority of contemporary physicists upon (5) will, if successful, forever protect these uncertainties from refutation. This is how modern empiricism may finally lead to a situation where a certain point of view petrifies into dogma by being, in the name of experience, completely removed from any conceivable criticism.

To sum up the arguments of the present section: it has been shown that neither (5) nor (A) can be defended on the basis of experience. Quite on the contrary, a strict empiricism will admit theories which are factually adequate and yet mutually inconsistent.[81] An analysis of the character of tests in the domain of theories has revealed, moreover, that the existence of sets of partly overlapping, mutually inconsistent, and yet empirically adequate theories is not only possible, but also required. I shall now conclude the present section by discussing a little more in detail the logical and psychological consequences of the use of such a set.

Increase of testability will not be the only result. The use of a set of theories with the properties indicated above will also improve our understanding of each of its members by making it very clear what is denied by the theory that happens to be accepted in the end. Thus, it seems to me that our understanding of Newton's somewhat obscure notion of absolute space and of its merits is greatly improved when we compare it with the relational ideas of Berkeley, Huyghens, Leibnitz, and Mach, and when we consider the failure of the latter ideas to give a satisfactory account of the phenomenon of inertial forces. Also, the study of general relativity will lead to a deeper understanding of this notion than could be obtained from a study of the *Principia* alone.[82] This is not meant to

[81] It is interesting to study in detail the dilemma of an inductivistic philosophy of science. To start with, a radical empiricist demands close adherence to the facts and is ultimately suspicious of any generalization; a summary of facts is all he will admit on empirical grounds. However, generalizations do play an important role in the sciences; scientific knowledge is, after all, a collection of theories. Hence, methods must be found to justify these theories on the basis of experience, and a "logic" must be constructed which allows us, again on the basis of experience, to confer some kind of certainty upon them. A "logic" of this kind is introduced, and theories are established with its help. But now the demand that the future theories should be consistent with the theories thus established turns out to be far too strict. The way out of this dilemma can only consist in abandoning the idea that theories can be "established" by experience and in admitting that insofar as they go beyond the facts we have no means whatever (except, perhaps, psychological ones) to guarantee their trustworthiness.

[82] This, by the way, is one of the reasons why an axiomatic exposition of physical

be understood in a psychological sense only. For just as the meaning of a term is not an intrinsic property of it but is dependent upon the way in which the term has been incorporated into a theory, in the very same manner the content of a whole theory (and thereby again the meaning of the descriptive terms which it contains) depends upon the way in which it is incorporated into both the set of its empirical consequences and the set of all the alternatives which are being discussed at a given time:[83] once the contextual theory of meaning has been adopted, there is no reason to confine its application to a single theory, or a single language, especially as the boundaries of such a theory or of such a language are almost never well defined. The considerations above have shown, moreover, that the unit involved in the test of a specific theory is not this theory taken together with its own consequences; they have shown

principles, such as Newton's, is inferior by far to a dialectical exposition where many ideas are considered, and the pros and cons discussed, until finally one theory is pronounced the most satisfactory one. Of course, if one holds that, concerning theories, the only relation of interest is the relation between a single theory and "the facts," and if one also believes that these facts single out a certain theory more or less uniquely, then one will be inclined to regard discussion of alternatives as a matter of history, or of psychology, and one will even wish to hide, with some embarrassment, the situation at the time when the clear message of the facts had not yet been grasped. However, as soon as it is recognized that the refutation (and thereby also the confirmation) of a theory necessitates its incorporation into a family of mutually inconsistent alternatives, in the very same moment, the discussion of these alternatives becomes of paramount importance for methodology and should be included in the presentation of the theory that is accepted in the end. For the same reason, adherence to either the distinction between context of discovery (where alternatives are considered, but given a psychological function only) and context of justification (where they are not mentioned any more), or strict adherence to the axiomatic approach must be regarded as an arbitrary and very misleading restriction of methodological discussion: much of what has been called "psychological," or "historical," in past discussions of method is a very relevant part of the theory of test procedures.

Considering all this, the increased attention paid to the historical aspects of a subject, and the attempts to break down the distinction between the synthetic and the analytic must be welcomed as steps in the right direction. However, even here there are drawbacks. Only very few of the enthusiastic proponents of an increased study of the history of a subject realize the methodological importance of their investigations. The justification they give for their interest is either sentimental or psychological ("it gives me ideas"), or based upon some very implausible (because Hegelian) notions concerning the "growth" of knowledge. What these thinkers need in order not to fall victims to all sorts of quasi-philosophies is a methodological backbone, and I hope that the theory of test which has been sketched above in its merest outlines will provide such a backbone.

[88] In the twentieth century, the contextual theory of meaning has been defended most forcefully by Wittgenstein; see [74] as well as my summary in [28]. However, it seems that Wittgenstein is inclined to restrict this theory to the inside of his language games: Platonism of concepts is replaced by Platonism of (theories or) games. For a brief criticism of this attitude see [35].

that this unit is a whole class of mutually incompatible and factually adequate theories. Hence, both consistency and methodological considerations suggest such a class as the context from which meanings are to be made clear.[84]

Also, the use of such a class rather than of a single theory is a most potent antidote against dogmatism. Psychologically speaking, dogmatism arises, among other things, from the inability to imagine alternatives to the point of view in which one believes. This inability may be due to the fact that such alternatives have been absent for a considerable time and that, therefore, certain ways of thinking have been left undeveloped; it may also be due to the conscious elimination of such alternatives. However that may be, persistence of a single point of view will lead to the gradual establishment of well-circumscribed methods of observation and measurement; it will lead to codification of the ways in which these results are interpreted; it will lead to a standardized terminology and to other developments of a similarly conservative kind. This being the case, the gradual acceptance of the theory by an ever-increasing number of people must finally bring about a transformation even of the most common idiom that is taught in very early youth. In the end, all the key terms will be fixed in an unambiguous manner, and the idea (which may have led to such a procedure in the first place) that they are copies of unchanging entities and that change of meaning, if it should happen, is due to human mistake—this idea will now be very plausible. Such plausibility reinforces all the maneuvers which may be used for the preservation of the theory (elimination of opponents included[85]).

The conceptual apparatus of the theory having penetrated nearly all means of communication, such methods as transcendental deduction and analysis of usage, which are further means of solidifying the theory, will be very successful. Altogether it will seem that at last an absolute and irrevocable truth has been arrived at. Disagreement with facts may of course occur, but, being now convinced of the truth of the existing

[84] Textbooks and historical presentations very often create the impression either that such classes never existed and that physicists (at least the "great" ones) at once arrived at the one and good theory or that their existence must not be taken too seriously. This is quite understandable. After all, historians have been just as much under the influence of inductivistic ideas as the physicists and the philosophers.

[85] Today, of course, the "elimination" takes the more "refined" form of a refusal to publish (or to read) what is not in agreement with the accepted doctrine. However, this "liberalism" applies to physical theories only. It does not seem to apply to political theories.

point of view, its proponents will try to save them with the help of ad hoc hypotheses. Experimental results that cannot be accommodated, even with the greatest ingenuity, will be put aside for later consideration. The result will be absolute truth, but, at the same time, it will decrease in empirical content to such an extent that all that remains will be no more than a verbal machinery which enables us to accompany any kind of event with noises (or written symbols) which are considered true statements by the theory.[86]

The picture painted above is by no means exaggerated. The way in which, for example, the theory of witchcraft and daemonic influence crept into the most common way of thinking, and could be preserved for quite a considerable time, offers a vivid illustration of each point mentioned in the last paragraph. Moreover, the story of its overthrow furnishes another illustration of our thesis that comprehensive theories cannot be eliminated by a direct confrontation with "the facts."

Now let us compare such a dogmatic procedure with the effects of the use of a class of theories rather than a single theory. First of all such a procedure will encourage the building of a great variety of measuring instruments. There will be no one way of interpreting the results, and the theoretician will be trained to switch quickly from one interpretation to another.[87] Intuitive appeal will lose its paralyzing effect, transcendental deduction which, after all, presupposes uniformity of usage, will be impossible; and the question of agreement with the facts will assume a very prominent position. Experimental results which are inconsistent with one theory may be consistent with a different theory; this elimi-

[86] The fact that all the features of absolute knowledge can be manufactured by exercising an absolute conformism I regard as the most important objection to any claim to finality: it would seem to show that whereas hypothetical knowledge, being the result of repeated corrections in the light of criticism, is at least in partial contact with the world, absolute knowledge is entirely man made and can therefore not raise any claim to factual content. Even the stability of a testable hypothesis cannot be regarded as a sign of its truth because it may be due to the fact that, owing to some particular astigmatism on our part, we have as yet overlooked some very decisive tests. There is no sign by which factual truth may be recognized.

It is interesting to note that the development toward dogmatism as it has been described in the text occurred both in the "school philosophies" and in empiricism, and, in the latter case, notably in physics: there is no indication that empiricism provides special protection against dogmatic petrification. The only difference is that its defense will be in terms of "experience" rather than in terms of "intuition" or "revelation." For this see Dewey [21], esp. Ch. II.

[87] As Professor Agassi has pointed out to me, this method was consciously used by Faraday in order to escape the influence of prejudice. Concerning its role in modern discussions about the microlevel, see [37].

nates the motives for *ad hoc* hypotheses, or at least reduces them considerably. Nor will it be necessary to use instrumentalism as a means of getting out of trouble, since a coherent account may be provided by an alternative to the theory considered. The likelihood that empirical results will be left lying around will also be smaller; if they do not fit one theory, they will fit another. It is not at all superfluous to mention the tremendous development of human capabilities encouraged by such a procedure and the antidotes it contains against the wish to set up, and to obey, all-powerful regimes, be they political, religious, or scientific. Taking all this into account, we are inclined to say that whereas *unanimity of opinion may be fitting for a church, or for the willing followers of a tyrant, or some other kind of 'great man,' variety of opinion is a methodological necessity for the sciences and, a fortiori, for philosophy.* Neither (A), nor (B), nor (5) allows for such variety. It follows that, to the extent to which both principles (and the philosophy behind them) delimit variety and demand future theories to be consistent with theories already in existence, they contain a theological element (which lies, of course, in the worship of "facts" so characteristic for nearly all empiricism).

The paralyzing effect of familiarity and intuitive appeal upon social reconstruction and the progress of knowledge has been understood and described in a most excellent manner by Bertolt Brecht. It is true that Brecht was mainly concerned with the function of the theater in the process of making familiar, and thereby creating the impression of the unchangeability of, social relations. But in his analysis, which was to establish the need for a new kind of theater, he hit upon a very important fact of a much more general character, namely, the paralyzing effect of familiarity and of the methods used to bring about such familiarity. In the theater these methods consist in trying to represent what is accidental and changeable (a particular social situation, for example) as essential to man or nature, and therefore unchangeable. "The theatre with which we are confronted shows the structure of society (represented on the stage) as being removed from the influence of society (which is situated in the audience). Oedipus, who has sinned against some principles which were supported by the society of his time, is executed; the gods take care of that—they cannot be criticised" ([8], p. 146). "What we need," Brecht continues, "is a theatre which does not only allow for those sensations, insights, and impulses that are permitted by

P. K. Feyerabend

the field of human relations where the action takes place; what we need is a theatre which uses and creates thoughts and feelings that play a role in the change of the field itself" ([8], p. 147). *It is exactly the same thing that is needed in the domain of epistemology.* What is needed is a method which does not—in the name of either "universal principles," "revelation," or "experience"—put fetters on the scientist's imagination but which enables him to use alternatives to the point of view which is the one commonly accepted. What is needed is a method that also enables him to take a critical attitude with respect to any element of this point of view, be it a law, or a so-called empirical fact. I am afraid that only very few scientists have ever been aware of the need for such a method, and that most of them are to be compared to the transfixed audience of one of the familiar pieces of the "classical" repertoire which has been described so vividly by Brecht—textbooks and scientific journals replacing the images projected by the actors.

At all times, the existence, within a certain tradition, of a variety of opinions (or a variety of theories) has been regarded as proof of the unsoundness of the method adopted by the members of this tradition. It was assumed, as being nearly self-evident, that the proper method must lead to the truth, that the truth is one, and that the proper method must therefore result in the establishment of a single theory and the perennial elimination of all alternatives. Conversely, the existence of various points of view and of a community where discussion of alternatives was regarded as fundamental was always regarded as a sign of confusion. Curiously enough, this attitude is found in thinkers who otherwise have very little in common. This can be seen from an examination of various criticisms of the pre-Socratic philosophers that have been proffered in the course of history.

As has been shown by Popper [67], these early philosophers were not only the inventors of a theoretical science (as opposed to a science which is content with assembling empirical generalizations as was the physics, the mathematics, and the astronomy of the Egyptians), but they also invented the method characteristic for this kind of science, i.e., the method of test within a class of mutually inconsistent, partly overlapping, and to that extent empirically adequate theories. All this was thoroughly misunderstood. Thus the sophists are reported to have ridiculed the Ionians by pointing out that their motto seemed to be "To every philosopher his own principle." Plato made full use of this popular sentiment (*Sophist* 242ff), and so did the church fathers later on: "As of canonical authors," writes St. Augustine (quoted from [16], p. 132), "God forbid that they should differ . . . [But] let one look amongst all the multitude of philosophers' writings, and if he finds two that tell both one tale in all respects, it may be registered for a rarity." Soon after the rise of Baconian empiricism with its so very different message, the variety of the theories discussed by the pre-Socratics was used as an example of where one gets when leaving the solid ground, not of revelation, but of experience. The following quotation is very characteristic: "As to the particular tenets of Thales, and his successors of the *Ionian* school, the sum of what we learn from the imperfect accounts we have of them is that each overthrew what his predecessor had advanced; and met with the same treatment himself from his successor . . . So early did the passion for systems begin." [88] In fact, nearly all inventors of new methods in philosophy and in the sciences have been inspired by the hope that they would be able to put an end to the quarrel of the schools and to establish the one true body of knowledge (this inspiration is present even today in some of the defenders of the Copenhagen interpretation of the quantum theory). What emerges from our own considerations is that *once dispute has been made empirically testable* (and this was already the case with the early Ionians—see again Popper's article [67]), *it becomes an essential element of the development of knowledge;* and it also emerges that cessation of dispute is not to be regarded as a sign that now we have finally arrived at the truth, but rather as a sign of fatigue in those originally participating in the dispute (just as the more recent return to religious beliefs is a sign of fatigue and of despair in the capabilities of reason).

According to Popper this procedure of testing a theory by comparing it with experience, as seen in the light of alternatives, is identical with the scientific method. Professor Matson, who also emphasizes that "the key" to the method of the pre-Socratics lies in the fact that "they were not dogmatic about their predecessors, but rather . . . criticized them most acutely," [89] differs from Popper insofar as he regards this method as relevant either to part of the sciences only (he regards it as relevant to contemporary cosmology), or even as a nondogmatic alternative to

[88] McLaurin [55], p. 28.
[89] [57], p. 445.

empiricism (for Popper the test procedure outlined *is* the empirical method). That the pre-Socratics were inventors of *method* as well as *theories* is emphasized in Professor Matson's admirable paper on Anaximander [56]. As far as I can make out, Popper's views on the matter were already developed at the time of the first publication of his *Open Society*. Both these thinkers represent a minority view which, in my opinion, is of the greatest importance for the understanding of the history of early Greek philosophy.

This finishes my criticism of (A) and (5). (A) has been shown to be in disagreement not only with actual scientific practice but also with the principles of a sound empiricism. The account of theorizing given in the introduction has been shown to be superior to the hierarchy of axioms and theorems which seems to be the favorite model of contemporary empiricism. The use of a set of mutually inconsistent and partially overlapping theories has been found to be of fundamental importance for methodology. The desideratum mentioned in connection with what has here been called the second idea has thereby been fulfilled. Serious doubt has been thrown upon the correctness and the desirability of (B). I now turn to the refutation of (B).

7. *Criticism of the Assumption of Meaning Invariance.*

In Section 5 it was shown that the "inertial law" (8) of the impetus theory is incommensurable with Newtonian physics in the sense that the main concept of the former, viz., the concept of impetus, can neither be defined on the basis of the primitive descriptive terms of the latter, nor related to them via a correct empirical statement. The reason for this incommensurability was also exhibited: although (8), *taken by itself*, is in quantitative agreement both with experience and with Newton's theory, the "rules of usage" to which we must refer in order to explain the meanings of its main descriptive terms contain the law (7) and, more especially, the law that constant forces bring about constant velocities. Both of these laws are inconsistent with Newton's theory. Seen from the point of view of this theory, any concept of a force whose content is dependent upon the two laws just mentioned will possess zero magnitude, or zero denotation, and will therefore be incapable of expressing features of actually existing situations. Conversely, it will be capable of being used in such a manner only if all connections with Newton's theory have first been severed. It is clear that this example refutes (B) if we interpret that thesis as the description of how science actually proceeds.

We may generalize this result in the following fashion: consider two theories, T' and T, which are both empirically adequate inside D', but which differ widely outside D'. In this case the demand may arise to explain T' on the basis of T, i.e., to derive T' from T and suitable initial conditions (for D'). Assuming T and T' to be in quantitative agreement inside D', such derivation will still be impossible if T' is part of a theoretical context whose "rules of usage" involve laws inconsistent with T.[90]

It is my contention that the conditions just enumerated apply to many pairs of theories which have been used as instances of explanation and reduction. Many (if not all) of such pairs on closer inspection turn out to consist of elements which are incommensurable and therefore incapable of mutual reduction and explanation. However, the above conditions admit of still wider application and then lead to very important consequences with regard to the structure and development both of our knowledge and of the language used for the expression of it. After all, the principles of the context of which T' is a part need not be explicitly formulated, and as a matter of fact they rarely are. To bring about the situation described above (sets of mutually incommensurable concepts), it is sufficient that they govern the use of the main terms of T'. In such a case T' is formulated in an idiom some of whose implicit rules of usage are inconsistent with T (or with some consequences of T in the domain where T' is successful). Such inconsistency will not be obvious at a glance; it will take considerable time before the incommensurability of T and T' can be demonstrated. However, as soon as this demonstration has been carried out, in the very same moment, the idiom of T' must be given up and must be replaced by the idiom of T. Of course, one need not go through the laborious and very uninteresting task of analyzing the context of which T' is part.[91] All that is needed is the adoption of the terminology and the 'grammar' of the most detailed and most

[90] Since this difficulty can arise even in the domain of empirical generalizations, the orthodox account may be inappropriate for them as well.

[91] There are many philosophers (including my friends in the Minnesota Center) who would admit that the importance of linguistic analysis is very limited. However, they would still hold that its application is necessary in order to find out to what extent the advent of a new theory modifies the customary idiom. The considerations above would show that even this is granting too much and that one travels best without any linguistic ballast.

successful theory throughout the domain of its application.[92] This automatically takes care of whatever incommensurabilities may arise, and it does so without any linguistic detective work (which therefore turns out to be entirely unnecessary for the progress of knowledge).

What has just been said applies most emphatically to the relation between (theories formulated in) some commonly understood language and more abstract theories. That is, I assert that languages such as the "everyday language," this notorious abstraction of contemporary linguistic philosophy, frequently contain (not explicitly formulated, that is, but implicit in the way in which its terms are used) principles which are inconsistent with newly introduced theories, and that they must therefore be either abandoned and replaced by the language of the new and better theories even in the most common situations, or they must be completely separated from these theories (which would lead to a situation where it is possible to believe in various kinds of "truth"): it is far from correct to assume that the everyday languages are so widely conceived, so tolerant, indefinite, and vague that they will be compatible with any scientific theory, that science can at most fill in details, and that a scientific theory will never run against the principles implicitly contained in them. *The very opposite is the case.* As will be shown later, even everyday languages, like languages of highly theoretical systems, have been introduced in order to give expression to some theory or point of view, and they therefore contain a well-developed and sometimes very abstract ontology. It is very surprising that the champions of the "ordinary language" should have such a low opinion of its descriptive power.

However, before turning to this part of the argument, I shall briefly discuss another example where the questionable principles of T' have been explicitly formulated, or can at least be easily unearthed.

The example which is dealt with by Nagel is the relation between phenomenological thermodynamics and the kinetic theory. Employing his own theory of reduction and, more especially, the condition I have quoted in the text adjacent to my footnote 11, Nagel claims that the

[92] One hears frequently that a complete replacement of the grammar and the terminology of the "old language" is impossible because this old language will be needed for introducing the new language and will, therefore, infect at least part of the new language. This is curious reasoning indeed if we consider that children learn languages without the help of a previously known idiom. Is it really asserted that what is possible for a small child will be impossible for a philosopher, a linguistic philosopher at that?

terms of the statements which have been derived from the kinetic theory (with the help of correlating hypotheses similar to (9)) will have the meanings they originally possessed within the phenomenological theory, and he repeatedly emphasizes that these meanings are fixed by "its own procedures" (i.e., by the procedures of the phenomenological theory) "whether or not [this theory] has been, or will be, reduced to some other discipline."[93]

As in the case of the impetus theory, we shall begin our study of the correctness of this assertion with an examination of these "procedures" and "usages"; more especially, we shall start with an examination of the usage of the term 'temperature,' "as fixed by the established procedures" of thermodynamics.

Within thermodynamics proper,[94] temperature ratios are defined by reference to reversible processes of operating between two levels, L' and L'', each of these levels being characterized by one and the same temperature throughout. The definition, viz.,

$$(10) \quad T':T'' = Q':Q''$$

identifies (after a certain arbitrary choice of units) the ratio of the temperature with the ratio between the amount of heat absorbed at the higher level and the amount of heat rejected at the lower level. Closer inspection of the "established usage" of the temperature thus defined shows that it is supposed to be

(11)   independent of the material of the substance chosen for the cycle, and unique.

This property can be inferred from the extension of the concept of temperature thus defined to radiation fields and from the fact that the constants of the main laws in this domain are universal, rather than dependent upon either the thermometric substance or the substance of the system investigated.

Now, it can be shown by an argument not to be presented here that (10) and (11) taken together imply the second law of thermodynamics in its strict (phenomenological) form: the concept of temperature as "fixed by the established usages" of thermodynamics is such that its application to concrete situations entails the strict (i.e., nonstatistical) second law.

[93] Nagel [20], p. 301.
[94] See Fermi [22], Sec. 9.

P. K. Feyerabend

Now whatever procedure is adopted, the kinetic theory does *not* give us such a concept. First of all, there does not exist any dynamical concept that possesses the required property.[95] The statistical account, on the other hand, allows for fluctuations of heat back and forth between two levels of temperature and, therefore, again contradicts one of the laws implicit in the "established usage" of the thermodynamic temperature. The relation between the thermodynamic concept of temperature and what can be defined in the kinetic theory, therefore, can be seen to conform to the pattern that has been described at the beginning of the present section: we are again dealing with two incommensurable concepts. The same applies to the relation between the purely thermodynamic entropy and its statistical counterpart; whereas the latter admits of very general application, the former can be measured by infinitely slow reversible processes only. Taking all this into consideration we must admit that it is impossible to relate the kinetic theory and the phenomenological theory in the manner described by Nagel, or to explain all the laws of the phenomenological theory in the manner demanded by Hempel and Oppenheim on the basis of the statistical theory. Again replacement rather than incorporation, or derivation (with the help, perhaps, of premises containing statistical as well as phenomenological concepts), is seen to be the process that characterizes the transition from a less general theory to a more general one.

It ought to be pointed out that the discussion is very idealized. The reason is that a purely kinetic account of the phenomena of heat does not yet seem to exist. What exists is a curious mixture of phenomenological and statistical elements, and it is this mixture which has received the name 'statistical thermodynamics.' However, even if this is admitted, it remains that the concept of temperature as it is used in this new and mixed theory is different from the original, purely phenomenological concept. To our point of view, according to which terms change their meanings with the progress of science, Nagel raises the following objection: "The redefinition of expressions with the development of inquiry [so it is noted], is a recurrent feature in the history of science. Accordingly, though it must be admitted that in an earlier use the word 'temperature' had a meaning specified exclusively by the rules and procedures of thermometry and classical thermodynamics, it is *now* so used

---

[95] I shall not discuss, in the present paper, the somewhat different situation with respect to the first law.

120

that temperature is 'identical by definition' with molecular energy. The deduction of Boyle-Charles' law does not therefore require the introduction of a further postulate, whether in the form of a coordinating definition or a special empirical hypothesis, but simply makes use of this definitional identity. This objection illustrates the unwitting double talk into which it is so easy to fall. It is certainly possible to redefine the word 'temperature' so that it becomes synonymous with 'mean kinetic energy.' But it is equally certain that on this redefined usage the word has a different meaning from the one associated with it in the classical science of heat, and therefore a meaning different from the one associated with the word in the statement of the Boyle-Charles law. However, if thermodynamics is to be reduced to mechanics, it is temperature in the sense of the term in the classical science of heat which must be asserted to be proportional to the mean kinetic energy of gas molecules. Accordingly, if the word 'temperature' is redefined as suggested by the objection, the hypothesis must be invoked that the state of bodies described as 'temperature' (in the classical thermodynamic sense) is also characterized by 'temperature' in the redefined sense of the term. This hypothesis, however, will then be one that does not hold as a matter of definition . . . Unless this hypothesis is adopted, it is not the Boyle-Charles law which can be derived from the assumptions of the kinetic theory of gases. What is derivable without the hypothesis is a sentence similar in syntactical structure to the standard formulation of the law, but possessing a sense that is unmistakably different from what the law asserts."[96] So far Nagel.

Commencing my criticism, I shall at once admit the correctness of the last assertion. After all, it has been my contention all through this paper that extension of knowledge leads to a decisive modification of the previous theories both as regards the quantitative assertions made and as regards the meanings of the main descriptive terms used. Applying this to the present case I shall therefore at once admit that incorporation into the context of the statistical theory is bound to change the meanings of the main descriptive terms of the phenomenological theory. The difference between Nagel and myself lies in the following. For me, such a change to new meanings and new quantitative assertions is a natural occurrence which is also desirable for methodological reasons

[96] [61], pp. 357–358.

(the last point will be established later in the present section). For Nagel such a change is an indication that reduction has not been achieved, for reduction in Nagel's sense is supposed to leave untouched the meanings of the main descriptive terms of the discipline to be reduced (cf. his "if thermodynamics is to be reduced to mechanics, it is temperature in the sense of the term in the classical science of heat which must be asserted to be proportional to the mean kinetic energy of gas-molecules"). "Accordingly," he continues, quite obviously assuming that reduction in his sense can be carried through, "if the word 'temperature' is redefined as suggested by the objection, the hypothesis must be invoked that the state of bodies described as 'temperature' (in the classical thermodynamic sense) is also characterized by 'temperature' in the redefined sense of the term. This hypothesis . . . will then be one that does not hold as a matter of definition." It will also be a false hypothesis because the conditions for the definition of the phenomenological temperature are never satisfied in nature (see the arguments above in the text and compare also the arguments in connection with formula (9)), which is only another sign of the fact that reduction, in the sense of Nagel, of the phenomenological theory to the statistical theory is not possible (obviously the additional premises used in the reduction are not supposed to be false). Once more arguments of meaning have led to quite unnecessary complications.

Further examples exhibiting the same features can be easily provided. Thus in classical, prerelativistic physics the concept of mass (and, for that matter, the concept of length and the concept of time duration) was absolute in the sense that the mass of a system was not influenced (except, perhaps, causally) by its motion in the coordinate system chosen. Within relativity, however, mass has become a relational concept whose specification is incomplete without indication of the coordinate system to which the spatiotemporal descriptions are all to be referred. Of course, the values obtained on measurement of the classical mass and of the relativistic mass will agree in the domain D', in which the classical concepts were first found to be useful. This does not mean that what is measured is the same in both cases: what is measured in the classical case is an *intrinsic property* of the system under consideration; what is measured in the case of relativity is a *relation* between the system and certain characteristics of D'. It is also impossible to define the exact classical concepts in relativistic terms or to relate them with the help of an empirical generalization. Any such procedure would imply the false assertion that the velocity of light is infinitely large. It is therefore again necessary to abandon completely the classical conceptual scheme once the theory of relativity has been introduced; and this means that it is imperative to use relativity in the theoretical considerations put forth for the explanation of a certain phenomenon as well as in the observation language in which tests for these considerations are to be formulated; after all, the empirically untenable consequences of the attempts above to give a reduction of classical terms to relativistic terms emerges whether or not the elements of the definition belong to the observation language.

Many more examples can be added to those discussed in the present paper (viz., the impetus theory, phenomenological thermodynamics, and the classical conception of mass). All these examples show that the postulate of meaning invariance is incompatible with actual scientific practice. That is, it has been shown that in most cases it is impossible to relate successive scientific theories in such a manner that the key terms they provide for the description of a domain D', where they overlap and are empirically adequate, either possess the same meanings or can at least be connected by empirical generalizations. It is also clear that the methodological arguments against meaning invariance will be the same as the arguments against the derivability condition and the consistency condition. After all, the demand for meaning invariance implies the demand that the laws of later theories be compatible with the principles of the context of which the earlier theories are part, and this demand is, therefore, seen to be a special case of condition (5). Using our earlier arguments against (5) we may now infer the untenability, on methodological grounds, of meaning invariance as well. And as our argument is quite general we may also infer that it is undesirable that the "ordinary" usage of terms be preserved in the course of the progress of knowledge. Wherever such preservation is observed, we shall feel inclined to think that the suggested new theories are not as revolutionary as they perhaps ought to be, and we shall have the suspicion that some ad hoc procedures have perhaps been adopted. Violation of ordinary usage, and of other "established" usages, on the other hand, is a sign that real progress has been made, and it is to be welcomed by anybody interested in such progress (provided of course that this violation is connected with the suggestion of a new point of view or a new theory and is not just the result of linguistic arbitrariness).

P. K. Feyerabend

Our argument against meaning invariance is simple and clear. It proceeds from the fact that usually some of the principles involved in the determination of the meanings of older theories or points of view are inconsistent with the new, and better, theories. It points out that it is natural to resolve this contradiction by eliminating the troublesome and unsatisfactory older principles and to replace them by principles, or theorems, of the new and better theory. And it concludes by showing that such a procedure will also lead to the elimination of the old meanings and thereby to the violation of meaning invariance.

The most important method used for escaping the force of this clear and simple argument is the transition to instrumentalism. Instrumentalism maintains that the new theory must not be interpreted as a series of statements but that it is rather to be understood as a predictive machine whose elements are tools rather than statements and therefore cannot be incompatible with any principle already in existence. This very popular move (popular, that is, because used also by scientists) admittedly cuts the ground from beneath our argument and makes it inapplicable. However, it has never been explained why a new and satisfactory theory should be interpreted as an instrument, whereas the principles behind the established usage, which can easily be shown to be empirically inadequate, are not so interpreted. After all, the only advantage of the latter is that they are familiar—an advantage which is a psychological and historical accident and which should therefore not have any influence upon questions of interpretation and of reality. One may try to answer this criticism by ascribing an instrumental function to all principles, old or new, and not only to those contained in the most recent theory. Such a procedure means acceptance of a sense-data account of knowledge. Having shown elsewhere[97] that such an account is impossible, I can now say that this consequence of a universal instrumentalism is tantamount to its refutation. Result: neither a restricted nor a universal instrumentalism can be carried through in a satisfactory manner. This disposes of the instrumentalistic move.

While instrumentalism possesses at least a semblance of plausibility, the arguments to be discussed now are devoid even of this feature. Indeed, I am very hesitant to apply the word 'arguments' to these expressions of confused thinking, their wide acceptance and asserted self-evi-

[97] [38].

124

dence notwithstanding. Consider for example the following question (which is supposed to be a criticism of our suggestion that after the acceptance of the kinetic theory the word 'temperature' will be in need of reinterpretation):[98] "If the meaning of 'temperature' is [now] the same as that of 'mean kinetic energy of molecular motion,' what are we talking about when milk is said to have a temperature of 10° Cels? Surely not the kinetic energy of the molecular constituents of the liquid, for the uninstructed layman is able to understand what is here said without possessing any notion about the molecular composition of the milk."

Now it may be quite correct that the "uninstructed layman"[99] does not think of molecules when speaking about the temperature of his milk and that he has not the slightest notion of the molecular constitution of the liquid either. However, what has the reference to him got to do with our argument according to which a person who has already accepted and understood the theory of the molecular constitution of gases, liquids, and solids cannot at the same time demand that the premolecular concept of temperature be retained? It is not at all denied by our argument that the "uninstructed layman" may possess a concept of temperature that is very different from the one connected with the molecular theory (after all, some "uninstructed laymen," intelligent clergymen included, still believe in ghosts and in the devil). What is denied is that anybody can consistently continue using this more primitive concept and at the same time believe in the molecular theory. Again, this does not mean that a person may not, on different occasions, use concepts which belong to different and incommensurable frameworks. The only thing that is forbidden for him is the use of both kinds of concepts *in the same argument*; for example, he may not use the one kind of concept in his observation language and the other kind in his theoretical language. Any such combination—and this is the gist of our considerations in the pres-

[98] The argument in connection with this question can be found in Nagel [20], p. 293. It is not clear to me whether or not Nagel would be prepared to support the argument.

[99] By the way, who is this uninstructed layman? From the purpose for which he is being employed in many arguments, it would seem to emerge that he is not supposed to know much science, or much politics, or much religion, or much of anything. This means that in these times of mass communication and mass education he must be very careful not to read the wrong parts of his newspaper, he must be careful not to leave his television set on for too long a time, and he must also not allow himself to converse too much with his friends, his children, etc. That is, he must be either a savage or an idiot. I really wonder what are the motives which lead to a philosophy where the most interesting language is the language of savages or of idiots.

125

ent section—would introduce principles which are mutually inconsistent and thereby destroy the argument in which it is supposed to occur. It is evident that this position is not at all endangered by the objection implied by the question above.

However, quite apart from being so obviously irrelevant to our thesis, the objection reflects an attitude that must appear quite incredible to anybody who possesses even the slightest acquaintance with the history of knowledge. The question insinuates that the layman's ability to handle the word 'temperature' according to the rules prescribed for it in some simple idiom indicates his understanding of the thermal properties of bodies. It insinuates that the existence of an idiom allows us to infer the truth of the principles which underlie this idiom. Or, to be more specific, it insinuates that *what is being used is, on that account alone, already exhibited as adequate, useful, and perhaps irreplaceable.* After all, the reference to the layman's understanding of the word 'temperature' is not made without purpose. It is made with the purpose of preserving the common meaning of this word since, it is alleged, this common meaning can be understood and is not in need of replacement. The discussion of a specific example will at once show the detrimental effect of any such procedure.

The example chosen now brings us to the second part of the present section where the relation is investigated, not between explicitly formulated theories, but between a theory and the implicit principles that govern the usage of the descriptive terms of some idiom. As has been said a little earlier, it is our conviction that "everyday languages," far from being so widely and generally conceived that they can be made compatible with any scientific theory, contain principles that may be inconsistent with some very basic laws. It was also pointed out that these principles are rarely expressed in an explicit manner (except, perhaps, in those cases where there is an attempt to defend the corresponding idiom against replacement or change) but that they are implicit in the rules that govern the use of its main descriptive terms. And our point was that, once these principles are found to be empirically inadequate, they must be given up and with them the concepts that are obtained by using terms in accordance with them. Conversely, the attempt to retain these concepts will lead to the conservation of false laws and to a situation where every connection between concepts or facts is severed.

The example which I have chosen to show this involves the pair 'up-down.' There existed a time when this pair was used in an absolute fashion, i.e., without reference to a specified center, such as the center of the earth. That it was used in such a manner can be easily seen from the "vulgar" remark that the antipodes would "fall off" the earth if the earth were spherical,[100] as well as from the more sophisticated attempts of Thales, Xenophanes, and others to find support for the earth as a whole, assuming that it would otherwise fall "down."[101] These attempts, as well as that remark about the antipodes employ two assumptions: first, that any material object is under the influence of a force; second, that this force acts in a privileged direction in space and must therefore be regarded as anisotropic. It is this privileged direction to which the pair 'up-down' refers. The second assumption is not explicitly made; it can only be derived from the way in which the pair 'up-down' is used in arguments such as those mentioned above.[102] We have here an example of a cosmological assumption (anisotropic character of space) implicit in the common idiom.

This example refutes the thesis which has been defended by some philosophers that "everyday languages" are fairly free from hypothetical elements and therefore ideally suited as observational languages.[103] It refutes the thesis by showing that even the most harmless part of a common idiom may rest upon very far-reaching hypotheses and must therefore be regarded as hypothetical to a very high degree.

Another remark concerns the changes of meaning needed once the Newtonian (or perhaps even the Aristotelian) explanation of the fall of heavy bodies is adopted. Newtonian space is isotropic and homogeneous. Hence, accepting this theory, one cannot anymore use the pair 'up-down' in the previous fashion and at the same time assume that one is describing actual features of physical situations. More especially, one cannot retain the absolute use of this pair for the description of observable features, since such features are quite obviously assumed to exist. Any per-

[100] For a discussion of this remark and of a related "vulgar" remark concerning the shape and arrangement of the terrestrial waters, see Pliny, *Natural History*, II, 161–166, quoted in Cohen and Drabkin [19], pp. 159–161.

[101] For a description and criticism of these attempts, see [1], 294a12ff; also quoted in [19], pp. 143–148.

[102] For the atomist's conception of space, which, at least since Epicurus, seems to be influenced by the popular ideas discussed above, cf. M. Jammer [48], p. 11.

[103] This thesis was introduced by Professor Herbert Feigl in discussions with me. For my own position, see also Philipp Frank [40].

son accepting Newton's physics and the conception of space it contains must, therefore, give a new meaning even to such a familiar pair of terms as is the pair 'up-down,' and he must now interpret it as a relation between the direction of a motion and a center that has been fixed in advance. And as Newton's theory is preferable, on empirical grounds, to the older and "absolute" cosmology, it follows that the relational usage of the pair 'up-down' will be preferable, too. Conversely, the attempt to retain the old usage amounts to retaining the old cosmology, and this despite the discoveries which have shown it to be obsolete.

To this argument it may be, and has been, objected that the "vulgar" usage of the pair 'up-down' was never supposed to be so general as to be applicable to the universe as a whole. This may be the case (although I do not see any reason for assuming that "ordinary" people are so very cautious as to apply the pair to the surface of the earth only; all the passages referred to in the above quotations contradict this assumption and so does the fact that at all times *real* ordinary people—and not only their Oxford substitutes—were very much interested in celestial phenomena[104]). However, even such a restriction would not invalidate our argument. It would rather show that the pair was used for singling out an absolute direction near the surface of the earth and that it did not assume such a direction to exist throughout the universe. It is clear that even this modest position is incompatible with the ideas implicit in the Newtonian point of view, which does not allow for local anisotropies either.

Consider now, after this example, the following argument in favor of the thesis that what is being used is, on that account alone, already exhibited as adequate, useful, and perhaps irreplaceable. The argument is the late Professor Austin's, and it has been repeated by G. J. Warnock.[105] "Language," writes Warnock, "is to be used for a vast number of highly important purposes; and it is at the very least unlikely that it should contain either much more, or much less, than those purposes require. If so, the existence of a number of different ways of speaking is very likely indeed to be an indication that there is a number of different things to be said . . . Where the topic at issue really is one that does

---

[104] The reason why Oxford philosophers so rarely discuss the influence of astronomy upon everyday languages may perhaps be found in the weather of their favorite discussion place. However, this reason unfortunately does not explain their ignorance in physics, theology, mythology, biology, and even linguistics.

[105] [71], pp. 150–151.

constantly concern most people in some practical way—as for example perception, the ascription of responsibility, or the assessment of human character or conduct—then it is certain that everyday language is as it is for some extremely good reasons; its verbal variety is certain to provide clues to important distinctions."[106]

If I understand this passage correctly, it means that the existence of certain distinctions in a language may be taken as an indication of similar distinctions in the nature of things, situations, and the like. And the reason for this is that people who are in constant contact with things and situations will soon develop the correct linguistic means for describing their properties. In short: human beings are good inductive machines in domains of concentrated interest, and their inductive ability will be the better the greater their concern, or the greater the practical value of the topic treated. Consequently, languages containing distinctions of practical interest are very likely to be adequate and irreplaceable.

There are many objections against this train of reasoning. First of all, it would seem to be somewhat arbitrary to restrict interests to those which can be derived from the immediate necessities of the physical life of the human race. From history we learn that the motives emerging from abstract considerations such as those found in a myth, or in a theo-astronomical system, are at least as strong as the more pedestrian motives connected with the immediate fulfillment of material needs (after all, people have died for their convictions!). Now if a language can be trusted because of the commitment of those who use it and if commitment is found to range over a much wider area than has first been imagined, if it is found to range over physics, astronomy (think of Giordano Bruno!), and biology, then the result will be that the principle we are discussing at the present moment (viz., the principle that what is being used for a purpose is on that account alone already useful and irreplaceable) must be applied to any language and any theory that has ever been developed and seriously tested. However—and this is the second point—there exist many theories and languages which have been found to be inadequate, and this despite their usefulness and despite the zeal of those who had developed them. This applies to the language of

[106] Astronomy is again omitted. It would seem to me that problems of astronomy had a much greater influence upon the formation of our language than problems of perception, which are of a very ephemeral nature and also are very technical. The skies and the stars (which, after all, were assumed to be gods) were everyone's concern.

129

P. K. Feyerabend

the Aristotelian physics, which had to be introduced into medieval thinking under very great difficulties and whose influence went much further than is sometimes realized; it applies to the language of the physics of Newton (mechanicism); and it applies to many other languages. Of course, this is the result one would expect: success under even very severe tests does not guarantee infallibility; no amount of commitment and no amount of success can guarantee the perennial reliability of inductions.

The principle which we have been discussing just now does not occur only in philosophy. Bohr's contention [107] that the account of all quantum mechanical evidence must forever "be expressed in classical terms" has been defended in a very similar manner. According to Bohr, we need our classical concepts not only if we want to give a summary of facts; without these concepts the facts to be summarized could not even be stated. As Kant before him, Bohr observes that our experimental statements are always formulated with the help of certain theoretical terms and that the elimination of these terms would lead, not to the "foundations of knowledge" as a positivist would have it, but to complete chaos. "Any experience," he asserts, "makes its appearance within the frame of our customary points of view and forms of perception" [108]— and at the present moment the forms of perception are those of classical physics.

Now does it follow, as is asserted by Bohr, that we can never go beyond the classical framework and that therefore all our future microscopic theories will have to use the notion of complementarity as a fundamental notion?

It is quite obvious that the actual use of classical concepts for the description of experiments within contemporary physics can never justify such an assumption, even if these concepts happen to have been very successful in the past (Hume's problem). For a theory may be found whose conceptual apparatus, when applied to the domain of validity of classical physics, would be just as comprehensive and useful as the classical apparatus without coinciding with it. Such a situation is by no means uncommon. The behavior of the planets, of the sun, and of the satellites can be described both by the Newtonian concepts and by the concepts

[107] See p. 43 above.
[108] [5], p. 1. For a more detailed account of what follows, see [32], [34], [36], [37], [38].

130

of general relativity. The order introduced into our experiences by Newton's theory is retained and improved upon by relativity. This means that the concepts of relativity are sufficiently rich for the formulation of all the facts which were stated before with the help of Newtonian physics. Yet the two sets of concepts are completely different and bear no logical relation to each other.

Other examples of the same kind can be provided very easily. What we are dealing with here is, of course, again the old problem of induction. No number of examples of usefulness of an idiom is ever sufficient to show that the idiom will have to be retained forever. And if it is objected, as it has been in the case of the quantum theory, that the language of classical physics is the only actual language in existence for the description of experiments,[109] then the reply must be that man is not only capable of using theories and languages but that he is also capable of inventing them.[110] How else could it have been possible, to mention only one example, to replace the Aristotelian physics and the Aristotelian cosmology with the new physics of Galileo and Newton? The only conceptual apparatus then available was the Aristotelian theory of change with its opposition of actual and potential properties, the four causes, and the like. Within this conceptual scheme, which was also used for the description of experimental results, Galileo's (or rather Descartes') law of inertia does not make sense, nor can it be formulated. Should, then, Galileo have followed Heisenberg's advice and have tried to get along with the Aristotelian concepts as well as possible, since his "actual situation . . . [was] such that [he did] use the Aristotelian concepts"[111] and since "there is no use discussing what could be done if we were other beings than we are"? By no means. What was needed was not improvement or delimitation of the Aristotelian concepts; what was needed was an entirely new theory. This concludes our argument against the principle that a useful language is to be regarded as adequate and irreplaceable and, thereby, fully restores the force of our attack against meaning invariance, as well as reinforces the positive suggestions made in connection with this attack and especially the idea that conceptual changes may occur anywhere in the system that is employed at a certain time for the explanation of the properties of the world we live in.

[109] See Heisenberg [44], p. 56, and von Weizsaecker [72], p. 110.
[110] See also fn. 92.
[111] This is a paraphrase of a passage in Heisenberg [44], p. 56.

P. K. Feyerabend

As I indicated in the introductory discussion, this transition from a point of view which demands that certain 'basic' terms retain their meaning, come what may, to a more liberal point of view which allows for changes anywhere in the system employed is bound to influence profoundly our attitude with respect to many philosophical problems and will also facilitate their solution. Let me take the mind-body problem as an example. It seems to me that the difficulties of this problem are to be sought precisely in the fact that meaning invariance is regarded as a necessary condition of its satisfactory solution. That is, it is assumed, or even demanded, that the meanings of at least some terms of the problem must remain constant throughout the discussion of the problem and further that these terms must retain their meanings in the solution as well.

Of course, different schools will apply the demand for meaning invariance to different concepts. A Platonist will demand that terms such as 'mind' and 'matter' remain unchanged, whereas an empiricist will require that some observational terms, such as the term 'pain,' or the more abstract term 'sensation,' retain their (common) meaning. Now a closer analysis of these key terms will, I think, reveal that they are incommensurable in exactly the sense in which this term has been defined at the beginning of the present section. This being the case, it is of course completely impossible either to reduce them to each other, or to relate them to each other with the help of an empirical hypothesis, or to find entities which belong to the extension of both kinds of terms. That is, the conditions under which the mind-body problem has been set up as well as the particular character of its key terms are such that a solution is forever impossible: a solution of the problem would require relating what is incommensurable without allowing for a modification of meanings which would eliminate this incommensurability.

All these difficulties disappear if we are prepared to admit that, in the course of the progress of knowledge, we may have to abandon altogether a certain point of view and the meanings connected with it—for example, if we are prepared to admit that the mental connotation of mental terms may be spurious and in need of replacement by a physical connotation according to which mental events, such as pains, states of awareness, and thoughts, are complex physical states of either the brain or the central nervous system, or perhaps the whole organism. I personally happen to

132

favor this idea that at some time sensations will turn out to be fairly complex central states which therefore possess a definite location inside the human body (which need not coincide with the place where the sensation is *felt* to be). I also hope that it will be possible to carry out a similar analysis of all so-called mental states.

Now whatever the merit of this belief of mine, it cannot be refuted by reference to the fact that what we "mean" by a sensation, or by a thought, is nothing that could have a location,[112] an internal structure, or physical ingredients. For if my belief is correct, and if it is indeed possible to develop a "materialistic" theory of human beings, then we shall of course be forced to abandon the "mental" connotations of the mental terms, and we shall have to replace them by physical connotations. According to the point of view which I am defending in the present paper, the only legitimate way of criticizing such a procedure would be to criticize this new materialistic theory by either showing that it is not in agreement with experimental findings or pointing out that it possesses some undesirable formal features (for example, by pointing out that it is *ad hoc*). Linguistic counter arguments have, I hope, been shown to be completely irrelevant.

The considerations in these last paragraphs are of course very sketchy. Still I hope that they give the reader an indication of the tremendous changes implied by the renunciation of the principle of meaning invariance as well as of the nefarious influence this principle has had upon traditional philosophy (modern empiricism included).

8. *Summary and Conclusion.*

Two basic assumptions of the orthodox theory of reduction and explanation have been found to be in disagreement with actual scientific practice and with reasonable methodology. The first assumption was that the explanandum is *derivable* from the explanans. The second assumption was that *meanings are invariant* with respect to the process of reduction and explanation. We may sum up the results of our investigation in the following manner:

Let us assume that T and T' are two theories satisfying the conditions outlined at the beginning of Section 3. Then, from the point of view of scientific method, T will be most satisfactory if it is (a) *inconsistent*

---

[112] See Wittgenstein's investigation of the "grammar" of mental terms as presented in [74].

P. K. Feyerabend

with T' in the domain where they both overlap;[113] and if it is (β) *incommensurable* with T'.

Now it is clear that a theory which is satisfactory according to the criterion just pronounced will not be capable of functioning as an explanans in any explanation or reduction that satisfies the principles put forth by Hempel and Oppenheim or Nagel. Paradoxically speaking: *Hempel-Oppenheim explanations cannot use satisfactory theories as explanantia. And satisfactory theories cannot function as explanantia in Hempel-Oppenheim explanations.* How is the theory of explanation and reduction to be changed in order to eliminate this very undesirable paradox?

It seems to me that the changes that are necessary will make it impossible to retain a formal theory of explanation, because these changes will introduce pragmatic or "subjective" considerations into the theory of explanation. This being the case, it seems perhaps advisable to eliminate altogether considerations of explanation from the domain of scientific method and to concentrate upon those rules which enable us to compare two theories with respect to their formal character and their predictive success and which guarantee the constant modification of our theories in the direction of greater generality, coherence, and comprehensiveness. I shall now give a more detailed outline of the reasons which have prompted me to adopt this pragmatic point of view.

Consider again T and T' as described above. Under these circumstances, the set of laws T'' following from T inside D' will either be inconsistent with T' or incommensurable with it. In what sense, then, can T be said to explain T'? This question has been answered by Popper for the case of the inconsistency of T' and T''. "Newton's theory," he says, "unifies Galileo's and Kepler's. But far from being a mere conjunction of these two theories—which play the part of *explicanda* for Newton—*it corrects them while explaining them.* The original explanatory task was the deduction of the earlier results. It is solved, not by deducing them, but by deducing something better in their place: new results which, under the special conditions of the older results, come numerically very close to these older results, and at the same time correct them. Thus the empirical success of the old theory may be said to corroborate the new theory; and in addition, the corrections may be tested in their

[113] This condition has been discussed with great clarity in [66]. It was this discussion (as well as dissatisfaction with [60]) that was the starting point of the present analysis of the problem of explanation.

turn . . . What is brought out strongly by [this] . . . situation . . . is the fact that the new theory cannot possibly be ad hoc . . . Far from repeating its explicandum, the new theory contradicts it and corrects it. In this way, even the evidence of the explicandum itself becomes independent evidence for the new theory."[114]

In a letter to me, J. W. N. Watkins has suggested that this theory may be summarized as follows: Explanation consists of two steps. The first step is derivation, from T, of those laws which obtain under the conditions characterizing D'. The second step is comparison of T'' and T' and realization that both are empirically adequate, i.e., fall within the domain of uncertainty of the observational results. Or, to express it in a more concise manner: T explains T' satisfactorily only if T is true and there exists a consequence T'' of T for the conditions of validity of T' such that T'' and T' are at least equally strong and also experimentally indistinguishable.

The first question that arises in connection with Dr. Watkins' formulation is this: experimentally indistinguishable on the basis of which observations? T' and T'' may be indistinguishable by the crude methods used at the time when T was first suggested, but they may well be distinguishable on the basis of later and more refined methods. Reference to a certain observational method will therefore have to be included in the clause of experimental indistinguishability. The notion of explanation will be relative to this observational material. It will not make sense any longer to ask whether or not T explains T'. The proper question will be whether T explains T' given the observational material, or the observational methods O. Using this new mode of speech we are forced to deny that Kepler's laws are explained by Newton's theory relative to the present observations—and this is perfectly in order; for these present observations in fact refute Kepler's laws and thereby eliminate the demand for explanation. It seems to me that this theory can well deal with all the problems that arise when T and T' are commensurable, but inconsistent inside D'. It does not seem to me that it can deal with the case where T' and T are incommensurable. The reason is as follows.

As soon as reference to certain observational material has been included in the characterization of what counts as a satisfactory explanation, in the very same moment the question arises as to how this observational

[114] Popper [66], p. 33.

135

material is to be presented. If it is correct, as has been argued all the way through the present paper, that the meanings of observational terms depend on the theory on behalf of which the observations have been made, then the observational material referred to in this modified sketch of explanation must be presented in terms of this theory also. Now incommensurable theories may not possess any comparable consequences, observational or otherwise. Hence, there may not exist any possibility of finding a characterization of the observations which are supposed to confirm two incommensurable theories. How, then, is the above account of explanation to be modified to cover the case of incommensurable theories also? [115]

It seems to me that the only possible way lies in closest adherence to the pragmatic theory of observation. According to this theory, it will be remembered, we must carefully distinguish between the causes of the production of a certain observational sentence, or the features of the process of production, on the one side, and the meaning of the sentence produced in this manner on the other. More especially, a sentient being must distinguish between the fact that he possesses a certain sensation, or disposition to verbal behavior, and the interpretation of the sentence being uttered in the presence of this sensation, or terminating this verbal behavior. Now our theories, apart from being pictures of the world, are also instruments of prediction. And they are good instruments if the information they provide, taken together with information about initial conditions characterizing a certain observational domain D_o, would enable a robot, who has no sense organs, but who has this information built into himself (or herself), to react in this domain in exactly the same manner as sentient beings who, without knowledge of the theory, have been trained to find their way about D_o and who are able to answer, 'on the basis of observation,' many questions concerning their surroundings.[116] This is the criterion of predictive success, and it is seen not at all to involve reference to the meanings of the reactions carried out either by the robot or by the sentient beings (which latter need not be humans, but can also be other robots). All it involves is agreement of behavior.

Now this criterion involves "subjective" elements. Agreement is de-

---

[115] As Professor Feigl has pointed out to me, this difficulty also arises in the case of crucial experiments.

[116] Of course, the motivations of the robot and of the sentient being must also be the same.

manded between the behavior of (nonsentient, but theory-fed) robots and that of sentient beings, and it is thereby assumed that the latter possesses a privileged position. Considering that perceptions are influenced by belief in theories and that behavior, too, is influenced by belief in theories, this criterion would seem to be somewhat arbitrary. It is easily seen, however, that it cannot be replaced by a less arbitrary and more "objective" criterion. What would such an objective criterion be? It would be a criterion which is either based upon behavior that is not connected with any theoretical element—and this is impossible (cf. my criticism of the theory of sense data above)—or it would be behavior that is tied up with an irrefutable and firmly established theory—which is equally impossible. We have to conclude, therefore, that a formal and "objective" account of explanation cannot be given.

REFERENCES

1. Aristotle. *De Coelo.*
2. Aristotle. *De Generatione et Corruptione.*
3. Barker, S. "The Role of Simplicity in Explanation," in *Current Issues in the Philosophy of Science*, H. Feigl and G. Maxwell, eds. New York: Holt, Rinehart, and Winston, 1961. Pp. 265–274.
4. Bohm, D. *Causality and Chance in Modern Physics.* London: Routledge and Kegan Paul, 1957.
5. Bohr, N. *Atomic Theory and the Description of Nature.* Cambridge: Cambridge University Press, 1932.
6. Bohr, N. "Discussions with Einstein," in *Albert Einstein, Philosopher-Scientist*, P. A. Schilpp, ed. Evanston, Ill.: Library of Living Philosophers, 1948. Pp. 201–241.
7. Born, M. *Natural Philosophy of Cause and Chance.* Oxford: Oxford University Press, 1948.
8. Brecht, B. *Schriften zum Theater.* Berlin and Frankfurt/Main: Suhrkamp Verlag, 1957.
9. Bunge, M. *Causality.* Cambridge, Mass.: Harvard University Press, 1959.
10. Burnet, J. *Early Greek Philosophy.* London: Adam and Charles Black, 1930.
11. Carnap, R. "Die Physikalische Sprache als Universalsprache der Wissenschaft," *Erkenntnis*, 2:432–465 (1932).
12. Carnap, R. "Psychologie in Physikalischer Sprache," *Erkenntnis*, 3:107–142 (1933).
13. Carnap, R. "Über Protokollsaetze," *Erkenntnis*, 3:215–228 (1933).
14. Carnap, R. "Testability and Meaning," *Philosophy of Science*, 3:419–471 (1936) and 4:1–40 (1937).
15. Carnap, R. "The Methodological Character of Theoretical Concepts," in *Minnesota Studies in the Philosophy of Science*, Vol. I, H. Feigl and M. Scriven, eds. Minneapolis: University of Minnesota Press, 1956. Pp. 38–76.
16. Clagett, M. *Greek Science in Antiquity.* London: Abelard-Schuman, 1957.
17. Clagett, M. *The Science of Mechanics in the Middle Ages.* Madison: University of Wisconsin Press, 1959.
18. Conant, J. B. *Case Histories in the Experimental Sciences*, Vol. I. Cambridge, Mass.: Harvard University Press, 1957.

P. K. Feyerabend

19. Cohen, M. R., and I. E. Drabkin, eds. *A Source Book in Greek Science*. New York: McGraw-Hill, 1948.
20. Danto, A., and S. Morgenbesser, eds. *Philosophy of Science*. New York: Meridian Books, 1960.
21. Dewey, John. *The Quest for Certainty*. New York: Capricorn Books, 1960.
22. Fermi, E. *Thermodynamics*. New York: Dover Publications, 1956.
23. Feigl, H., and M. Brodbeck, eds. *Readings in the Philosophy of Science*. New York: Appleton-Century-Crofts, 1953.
24. Feigl, H., and G. Maxwell, eds. *Current Issues in the Philosophy of Science*. New York: Holt, Rinehart, and Winston, 1961.
25. Feigl, H., and M. Scriven, eds. *Minnesota Studies in the Philosophy of Science*, Vol. I. Minneapolis: University of Minnesota Press, 1956.
26. Feigl, H., M. Scriven, and G. Maxwell, eds. *Minnesota Studies in the Philosophy of Science*, Vol. II. Minneapolis: University of Minnesota Press, 1958.
27. Feyerabend, P. K. "Carnap's Theorie der Interpretation Theoretischer Systeme," *Theoria*, 21:55–62 (1955).
28. Feyerabend, P. K. "Wittgenstein's 'Philosophical Investigations,'" *Philosophical Review*, 54:449–483 (1955).
29. Feyerabend, P. K. "Eine Bemerkung zum Neumannschen Beweis," *Zeitschrift für Physik*, 145:421–423 (1956).
30. Feyerabend, P. K. "On the Quantum Theory of Measurement," in *Observation and Interpretation*, S. Körner, ed. London: Butterworth, 1957. Pp. 121–130.
31. Feyerabend, P. K. "An Attempt at a Realistic Interpretation of Experience," *Proceedings of the Aristotelian Society*, New Series, 58:143–170 (1958).
32. Feyerabend, P. K. "Complementarity," *Proceedings of the Aristotelian Society*, Supplementary Vol., 32:75–104 (1958).
33. Feyerabend, P. K. "Das Problem der Existenz Theoretischer Entitaeten," *Probleme der Wissenschaftstheorie*. Vienna: Springer, 1960. Pp. 35–72.
34. Feyerabend, P. K. "O Interpretacji Relacyj Nieokreslonosci," *Studia Filozoficzne*, 19:23–78 (1960).
35. Feyerabend, P. K. "Patterns of Discovery," *Philosophical Review*, 59:247–252 (1960).
36. Feyerabend, P. K. "Professor Bohm's Philosophy of Nature," *British Journal for the Philosophy of Science*, 10:321–338 (1960).
37. Feyerabend, P. K. "Bohr's Interpretation of the Quantum Theory," in *Current Issues in the Philosophy of Science*, H. Feigl and G. Maxwell, eds. New York: Holt, Rinehart, and Winston, 1961. Pp. 371–390.
38. Feyerabend, P. K. "On the Interpretation of Microphysical Theories," to appear in *Minnesota Studies in the Philosophy of Science*, Vol. IV, H. Feigl and G. Maxwell, eds.
39. Feyerabend, P. K. "On the Interpretation of Scientific Theories," to appear in *Proceedings of the XIIth International Congress of Philosophy*, Milan.
40. Frank, P. *Relativity, a Richer Truth*. Boston: Beacon Press. 1950.
41. Fuerth, R. "Über einige Beziehungen Zwischen Klassischer Statistik und Quantenmechanik," *Zeitschrift für Physik*, 81:143–162 (1933).
42. Goodman, N. *Fact, Fiction, and Forecast*. Cambridge, Mass: Harvard University Press, 1955.
43. Hanson, N. R. *Patterns of Discovery*. Cambridge: Cambridge University Press, 1958.
44. Heisenberg, W. *Physics and Philosophy*. New York: Harper and Brothers, 1958.
45. Hempel, C. G. "Studies in the Logic of Confirmation," *Mind*, 54:1–26, 97–121 (1945).
46. Hempel, C. G. "A Logical Appraisal of Operationism," in *Validation of Scientific Theories*, P. Frank, ed. Boston: Beacon Press, 1954. Pp. 52–67.

47. Hempel, C. G., and P. Oppenheim. "Studies in the Logic of Explanation," Philosophy of Science, 15:135–175 (1948).
48. Jammer, M. Concepts of Space. Cambridge, Mass.: Harvard University Press, 1957.
49. Körner, S., ed. Observation and Interpretation. London: Butterworth, 1957.
50. Körner, S. Conceptual Thinking. New York: Dover Publications, 1960.
51. Kuhn, T. S. The Copernican Revolution. New York: Random House, 1959.
52. Landau, L. D., and E. M. Lifschitz. Quantum Mechanics. Reading, Mass.: Addison-Wesley, 1958.
53. Mach, E. Waermelehre. Leipzig: Johann Ambrosius Barth, 1897.
54. Mach, E. Zwei Aufsaetze. Leipzig: Johann Ambrosius Barth, 1912.
55. McLaurin, C. An Account of Sir Isaac Newton's Philosophical Discoveries. London: Buchanan's Head, 1750.
56. Matson, W. I. "The Naturalism of Anaximander," Review of Metaphysics, 6: 387–395 (1953).
57. Matson, W. I. "Cornford and the Birth of Metaphysics," Review of Metaphysics, 8:443–454 (1955).
58. Maier, A. Die Vorlaeufer Galilei's im 14, Jahrhundert. Rome: Edizioni di Storiae Litteratura, 1949.
59. Morris, E. "Foundation of the Theory of Signs," International Encyclopaedia of Unified Science, Sec. II/7. Chicago: University of Chicago Press, 1942.
60. Nagel, E. "The Meaning of Reduction in the Natural Sciences," in Science and Civilization, R. C. Stauffer, ed. Madison: University of Wisconsin Press, 1949. Pp. 99–145.
61. Nagel, E. The Structure of Science. New York: Harcourt, Brace, and Company, 1961.
62. Neumann, J. von. Mathematical Foundations of Quantum Mechanics. Princeton: Princeton University Press, 1957.
63. Panofsky, E. "Galileo as a Critic of the Arts," Isis, 47:3–15 (1956).
64. Popper, K. R. "Naturgesetze und Theoretische Systeme," in Gesetz und Wirklichkeit, S. Moser, ed. Innsbruck: Hochschulverlag, 1948. Pp. 65–84.
65. Popper, K. R. The Open Society and Its Enemies. Princeton: Princeton University Press, 1950.
66. Popper, K. R. "The Aim of Science," Ratio, 1:24–35 (1957).
67. Popper, K. R. "Back to the Pre-Socratics," Proceedings of the Aristotelian Society, New Series, 54:1–24 (1959).
68. Popper, K. R. The Logic of Scientific Discovery. New York: Basic Books, 1959.
69. Reichenbach, H. Experience and Prediction. Chicago: University of Chicago Press, 1948.
70. Sellars, W. "The Language of Theories," in Current Issues in the Philosophy of Science, H. Feigl and G. Maxwell, eds. New York: Holt, Rinehart, and Winston, 1961. Pp. 57–77.
71. Warnock, J. British Philosophy in 1900. Oxford: Oxford University Press, 1956.
72. Weizsaecker, C. F. von. Zum Weltbild der Physik. Leipzig: Verlag Hirzel, 1954.
73. Whorff, B. L. Language, Thought, and Reality: Selected Writings, John B. Carroll, ed. Cambridge, Mass.: Technology Press of Massachusetts Institute of Technology, 1956.
74. Wittgenstein, L. Philosophical Investigations. Oxford: Basil Blackwell, 1953.

The Nature and Necessity of
Scientific Revolutions

These remarks permit us at last to consider the problems that provide this essay with its title. What are scientific revolutions, and what is their function in scientific development? Much of the answer to these questions has been anticipated in earlier sections. In particular, the preceding discussion has indicated that scientific revolutions are here taken to be those non-cumulative developmental episodes in which an older paradigm is replaced in whole or in part by an incompatible new one. There is more to be said, however, and an essential part of it can be introduced by asking one further question. Why should a change of paradigm be called a revolution? In the face of the vast and essential differences between political and scientific development, what parallelism can justify the metaphor that finds revolutions in both?

One aspect of the parallelism must already be apparent. Political revolutions are inaugurated by a growing sense, often restricted to a segment of the political community, that existing institutions have ceased adequately to meet the problems posed by an environment that they have in part created. In much the same way, scientific revolutions are inaugurated by a growing sense, again often restricted to a narrow subdivision of the scientific community, that an existing paradigm has ceased to function adequately in the exploration of an aspect of nature to which that paradigm itself had previously led the way. In both political and scientific development the sense of malfunction that can lead to crisis is prerequisite to revolution. Furthermore, though it admittedly strains the metaphor, that parallelism holds not only for the major paradigm changes, like those attributable to Copernicus and Lavoisier, but also for the far smaller ones associated with the assimilation of a new sort of phenomenon, like oxygen or X-rays. Scientific revolutions, as we noted at the end of Section V, need seem revolutionary only to those whose paradigms are affected by them. To outsiders they may, like the Balkan revolutions of the early twentieth century, seem normal parts of the developmental process. Astronomers, for example, could accept X-rays as a mere addition to knowledge, for their paradigms were unaffected by the existence of the new radiation. But for men like Kelvin, Crookes, and Roentgen, whose research dealt with radiation theory or with cathode ray tubes, the emergence of X-rays necessarily violated one paradigm as it created another. That is why these rays could be discovered only through something's first going wrong with normal research.

This genetic aspect of the parallel between political and scientific development should no longer be open to doubt. The parallel has, however, a second and more profound aspect upon which the significance of the first depends. Political revolutions aim to change political institutions in ways that those institutions themselves prohibit. Their success therefore necessitates the partial relinquishment of one set of institutions in favor of another, and in the interim, society is not fully governed by institutions at all. Initially it is crisis alone that attenuates the role of political institutions as we have already seen it attenuate the role of paradigms. In increasing numbers individuals become increasingly estranged from political life and behave more and more eccentrically within it. Then, as the crisis deepens, many of these individuals commit themselves to some concrete proposal for the reconstruction of society in a new institutional framework. At that point the society is divided into competing camps or parties, one seeking to defend the old institutional constellation, the others seeking to institute some new one. And, once that polarization has occurred, *political recourse fails.* Because they differ about the institutional matrix within which political change is to be achieved and evaluated, because they acknowledge no supra-institutional framework for the adjudication of revolutionary difference, the parties to a revolutionary conflict must finally resort to the techniques of mass persuasion, often including force. Though revolutions have had a vital role in the evolution of political institutions, that role depends upon

their being partially extrapolitical or extrainstitutional events.
The remainder of this essay aims to demonstrate that the
historical study of paradigm change reveals very similar characteristics in the evolution of the sciences. Like the choice between competing political institutions, that between competing
paradigms proves to be a choice between incompatible modes
of community life. Because it has that character, the choice is
not and cannot be determined merely by the evaluative procedures characteristic of normal science, for these depend in
part upon a particular paradigm, and that paradigm is at issue.
When paradigms enter, as they must, into a debate about paradigm choice, their role is necessarily circular. Each group uses
its own paradigm to argue in that paradigm's defense.

The resulting circularity does not, of course, make the arguments wrong or even ineffectual. The man who premises a paradigm when arguing in its defense can nonetheless provide a
clear exhibit of what scientific practice will be like for those
who adopt the new view of nature. That exhibit can be immensely persuasive, often compellingly so. Yet, whatever its
force, the status of the circular argument is only that of persuasion. It cannot be made logically or even probabilistically
compelling for those who refuse to step into the circle. The
premises and values shared by the two parties to a debate over
paradigms are not sufficiently extensive for that. As in political
revolutions, so in paradigm choice—there is no standard higher
than the assent of the relevant community. To discover how
scientific revolutions are effected, we shall therefore have to
examine not only the impact of nature and of logic, but also the
techniques of persuasive argumentation effective within the
quite special groups that constitute the community of scientists.

To discover why this issue of paradigm choice can never be
unequivocally settled by logic and experiment alone, we must
shortly examine the nature of the differences that separate the
proponents of a traditional paradigm from their revolutionary
successors. That examination is the principal object of this section and the next. We have, however, already noted numerous
examples of such differences, and no one will doubt that history

can supply many others. What is more likely to be doubted than their existence—and what must therefore be considered first—is that such examples provide essential information about the nature of science. Granting that paradigm rejection has been a historic fact, does it illuminate more than human credulity and confusion? Are there intrinsic reasons why the assimilation of either a new sort of phenomenon or a new scientific theory must demand the rejection of an older paradigm?

First notice that if there are such reasons, they do not derive from the logical structure of scientific knowledge. In principle, a new phenomenon might emerge without reflecting destructively upon any part of past scientific practice. Though discovering life on the moon would today be destructive of existing paradigms (these tell us things about the moon that seem incompatible with life's existence there), discovering life in some less well-known part of the galaxy would not. By the same token, a new theory does not have to conflict with any of its predecessors. It might deal exclusively with phenomena not previously known, as the quantum theory deals (but, significantly, not exclusively) with subatomic phenomena unknown before the twentieth century. Or again, the new theory might be simply a higher level theory than those known before, one that linked together a whole group of lower level theories without substantially changing any. Today, the theory of energy conservation provides just such links between dynamics, chemistry, electricity, optics, thermal theory, and so on. Still other compatible relationships between old and new theories can be conceived. Any and all of them might be exemplified by the historical process through which science has developed. If they were, scientific development would be genuinely cumulative. New sorts of phenomena would simply disclose order in an aspect of nature where none had been seen before. In the evolution of science new knowledge would replace ignorance rather than replace knowledge of another and incompatible sort.

Of course, science (or some other enterprise, perhaps less effective) might have developed in that fully cumulative manner. Many people have believed that it did so, and most still

seem to suppose that cumulation is at least the ideal that historical development would display if only it had not so often been
distorted by human idiosyncrasy. There are important reasons
for that belief. In Section X we shall discover how closely the
view of science-as-cumulation is entangled with a dominant
epistemology that takes knowledge to be a construction placed
directly upon raw sense data by the mind. And in Section XI we
shall examine the strong support provided to the same historio-
graphic schema by the techniques of effective science pedagogy.
Nevertheless, despite the immense plausibility of that ideal
image, there is increasing reason to wonder whether it can possibly be an image of *science.* After the pre-paradigm period the
assimilation of all new theories and of almost all new sorts of
phenomena has in fact demanded the destruction of a prior
paradigm and a consequent conflict between competing schools
of scientific thought. Cumulative acquisition of unanticipated
novelties proves to be an almost non-existent exception to the
rule of scientific development. The man who takes historic fact
seriously must suspect that science does not tend toward the
ideal that our image of its cumulativeness has suggested. Perhaps it is another sort of enterprise.

If, however, resistant facts can carry us that far, then a second
look at the ground we have already covered may suggest that
cumulative acquisition of novelty is not only rare in fact but im-
probable in principle. Normal research, which *is* cumulative,
owes its success to the ability of scientists regularly to select
problems that can be solved with conceptual and instrumental
techniques close to those already in existence. (That is why an
excessive concern with useful problems, regardless of their relation to existing knowledge and technique, can so easily inhibit
scientific development.) The man who is striving to solve a
problem defined by existing knowledge and technique is not,
however, just looking around. He knows what he wants to
achieve, and he designs his instruments and directs his thoughts
accordingly. Unanticipated novelty, the new discovery, can
emerge only to the extent that his anticipations about nature
and his instruments prove wrong. Often the importance of the

resulting discovery will itself be proportional to the extent and stubbornness of the anomaly that foreshadowed it. Obviously, then, there must be a conflict between the paradigm that discloses anomaly and the one that later renders the anomaly lawlike. The examples of discovery through paradigm destruction examined in Section VI did not confront us with mere historical accident. There is no other effective way in which discoveries might be generated.

The same argument applies even more clearly to the invention of new theories. There are, in principle, only three types of phenomena about which a new theory might be developed. The first consists of phenomena already well explained by existing paradigms, and these seldom provide either motive or point of departure for theory construction. When they do, as with the three famous anticipations discussed at the end of Section VII, the theories that result are seldom accepted, because nature provides no ground for discrimination. A second class of phenomena consists of those whose nature is indicated by existing paradigms but whose details can be understood only through further theory articulation. These are the phenomena to which scientists direct their research much of the time, but that research aims at the articulation of existing paradigms rather than at the invention of new ones. Only when these attempts at articulation fail do scientists encounter the third type of phenomena, the recognized anomalies whose characteristic feature is their stubborn refusal to be assimilated to existing paradigms. This type alone gives rise to new theories. Paradigms provide all phenomena except anomalies with a theory-determined place in the scientist's field of vision.

But if new theories are called forth to resolve anomalies in the relation of an existing theory to nature, then the successful new theory must somewhere permit predictions that are different from those derived from its predecessor. That difference could not occur if the two were logically compatible. In the process of being assimilated, the second must displace the first. Even a theory like energy conservation, which today seems a logical superstructure that relates to nature only through independently established theories, did not develop historically without paradigm destruction. Instead, it emerged from a crisis in which an essential ingredient was the incompatibility between Newtonian dynamics and some recently formulated consequences of the caloric theory of heat. Only after the caloric theory had been rejected could energy conservation become part of science.[1] And only after it had been part of science for some time could it come to seem a theory of a logically higher type, one not in conflict with its predecessors. It is hard to see how new theories could arise without these destructive changes in beliefs about nature. Though logical inclusiveness remains a permissible view of the relation between successive scientific theories, it is a historical implausibility.

A century ago it would, I think, have been possible to let the case for the necessity of revolutions rest at this point. But today, unfortunately, that cannot be done because the view of the subject developed above cannot be maintained if the most prevalent contemporary interpretation of the nature and function of scientific theory is accepted. That interpretation, closely associated with early logical positivism and not categorically rejected by its successors, would restrict the range and meaning of an accepted theory so that it could not possibly conflict with any later theory that made predictions about some of the same natural phenomena. The best-known and the strongest case for this restricted conception of a scientific theory emerges in discussions of the relation between contemporary Einsteinian dynamics and the older dynamical equations that descend from Newton's *Principia*. From the viewpoint of this essay these two theories are fundamentally incompatible in the sense illustrated by the relation of Copernican to Ptolemaic astronomy: Einstein's theory can be accepted only with the recognition that Newton's was wrong. Today this remains a minority view.[2] We must therefore examine the most prevalent objections to it.

[1] Silvanus P. Thompson, *Life of William Thomson Baron Kelvin of Largs* (London, 1910), I, 266–81.

[2] See, for example, the remarks by P. P. Wiener in *Philosophy of Science,* XXV (1958), 298.

The gist of these objections can be developed as follows. Relativistic dynamics cannot have shown Newtonian dynamics to be wrong, for Newtonian dynamics is still used with great success by most engineers and, in selected applications, by many physicists. Furthermore, the propriety of this use of the older theory can be proved from the very theory that has, in other applications, replaced it. Einstein's theory can be used to show that predictions from Newton's equations will be as good as our measuring instruments in all applications that satisfy a small number of restrictive conditions. For example, if Newtonian theory is to provide a good approximate solution, the relative velocities of the bodies considered must be small compared with the velocity of light. Subject to this condition and a few others, Newtonian theory seems to be derivable from Einsteinian, of which it is therefore a special case.

But, the objection continues, no theory can possibly conflict with one of its special cases. If Einsteinian science seems to make Newtonian dynamics wrong, that is only because some Newtonians were so incautious as to claim that Newtonian theory yielded entirely precise results or that it was valid at very high relative velocities. Since they could not have had any evidence for such claims, they betrayed the standards of science when they made them. In so far as Newtonian theory was ever a truly scientific theory supported by valid evidence, it still is. Only extravagant claims for the theory—claims that were never properly parts of science—can have been shown by Einstein to be wrong. Purged of these merely human extravagances, Newtonian theory has never been challenged and cannot be.

Some variant of this argument is quite sufficient to make any theory ever used by a significant group of competent scientists immune to attack. The much-maligned phlogiston theory, for example, gave order to a large number of physical and chemical phenomena. It explained why bodies burned—they were rich in phlogiston—and why metals had so many more properties in common than did their ores. The metals were all compounded from different elementary earths combined with phlogiston, and the latter, common to all metals, produced common prop-

erties. In addition, the phlogiston theory accounted for a number of reactions in which acids were formed by the combustion
of substances like carbon and sulphur. Also, it explained the
decrease of volume when combustion occurs in a confined volume of air—the phlogiston released by combustion "spoils" the
elasticity of the air that absorbed it, just as fire "spoils" the
elasticity of a steel spring.[3] If these were the only phenomena
that the phlogiston theorists had claimed for their theory, that
theory could never have been challenged. A similar argument
will suffice for any theory that has ever been successfully applied to any range of phenomena at all.

But to save theories in this way, their range of application
must be restricted to those phenomena and to that precision of
observation with which the experimental evidence in hand already deals.[4] Carried just a step further (and the step can
scarcely be avoided once the first is taken), such a limitation
prohibits the scientist from claiming to speak "scientifically"
about any phenomenon not already observed. Even in its present form the restriction forbids the scientist to rely upon a theory in his own research whenever that research enters an area
or seeks a degree of precision for which past practice with the
theory offers no precedent. These prohibitions are logically un-
exceptionable. But the result of accepting them would be the
end of the research through which science may develop further.

By now that point too is virtually a tautology. Without commitment to a paradigm there could be no normal science. Furthermore, that commitment must extend to areas and to degrees
of precision for which there is no full precedent. If it did not,
the paradigm could provide no puzzles that had not already
been solved. Besides, it is not only normal science that depends
upon commitment to a paradigm. If existing theory binds the

[3] James B. Conant, *Overthrow of the Phlogiston Theory* (Cambridge, 1950),
pp. 13–16; and J. R. Partington, *A Short History of Chemistry* (2d ed.; London,
1951), pp. 85–88. The fullest and most sympathetic account of the phlogiston
theory's achievements is by H. Metzger, *Newton, Stahl, Boerhaave et la doctrine
chimique* (Paris, 1930), Part II.

[4] Compare the conclusions reached through a very different sort of analysis
by R. B. Braithwaite, *Scientific Explanation* (Cambridge, 1953), pp. 50–87,
esp. p. 76.

scientist only with respect to existing applications, then there can be no surprises, anomalies, or crises. But these are just the signposts that point the way to extraordinary science. If positivistic restrictions on the range of a theory's legitimate applicability are taken literally, the mechanism that tells the scientific community what problems may lead to fundamental change must cease to function. And when that occurs, the community will inevitably return to something much like its pre-paradigm state, a condition in which all members practice science but in which their gross product scarcely resembles science at all. Is it really any wonder that the price of significant scientific advance is a commitment that runs the risk of being wrong?

More important, there is a revealing logical lacuna in the positivist's argument, one that will reintroduce us immediately to the nature of revolutionary change. Can Newtonian dynamics really be *derived* from relativistic dynamics? What would such a derivation look like? Imagine a set of statements, E_1, E_2, ..., E_n, which together embody the laws of relativity theory. These statements contain variables and parameters representing spatial position, time, rest mass, etc. From them, together with the apparatus of logic and mathematics, is deducible a whole set of further statements including some that can be checked by observation. To prove the adequacy of Newtonian dynamics as a special case, we must add to the E_i's additional statements, like $(v/c)^2 << 1$, restricting the range of the parameters and variables. This enlarged set of statements is then manipulated to yield a new set, N_1, N_2, ..., N_m, which is identical in form with Newton's laws of motion, the law of gravity, and so on. Apparently Newtonian dynamics has been derived from Einsteinian, subject to a few limiting conditions.

Yet the derivation is spurious, at least to this point. Though the N_i's are a special case of the laws of relativistic mechanics, they are not Newton's Laws. Or at least they are not unless those laws are reinterpreted in a way that would have been impossible until after Einstein's work. The variables and parameters that in the Einsteinian E_i's represented spatial position, time, mass, etc., still occur in the N_i's; and they there still represent Einsteinian space, time, and mass. But the physical referents of these Einsteinian concepts are by no means identical with those of the Newtonian concepts that bear the same name. (Newtonian mass is conserved; Einsteinian is convertible with energy. Only at low relative velocities may the two be measured in the same way, and even then they must not be conceived to be the same.) Unless we change the definitions of the variables in the N_i's, the statements we have derived are not Newtonian. If we do change them, we cannot properly be said to have *derived* Newton's Laws, at least not in any sense of "derive" now generally recognized. Our argument has, of course, explained why Newton's Laws ever seemed to work. In doing so it has justified, say, an automobile driver in acting as though he lived in a Newtonian universe. An argument of the same type is used to justify teaching earth-centered astronomy to surveyors. But the argument has still not done what it purported to do. It has not, that is, shown Newton's Laws to be a limiting case of Einstein's. For in the passage to the limit it is not only the forms of the laws that have changed. Simultaneously we have had to alter the fundamental structural elements of which the universe to which they apply is composed.

This need to change the meaning of established and familiar concepts is central to the revolutionary impact of Einstein's theory. Though subtler than the changes from geocentrism to heliocentrism, from phlogiston to oxygen, or from corpuscles to waves, the resulting conceptual transformation is no less decisively destructive of a previously established paradigm. We may even come to see it as a prototype for revolutionary reorientations in the sciences. Just because it did not involve the introduction of additional objects or concepts, the transition from Newtonian to Einsteinian mechanics illustrates with particular clarity the scientific revolution as a displacement of the conceptual network through which scientists view the world.

These remarks should suffice to show what might, in another philosophical climate, have been taken for granted. At least for scientists, most of the apparent differences between a discarded scientific theory and its successor are real. Though an out-of-date theory can always be viewed as a special case of its up-to-date successor, it must be transformed for the purpose. And the transformation is one that can be undertaken only with the advantages of hindsight, the explicit guidance of the more recent theory. Furthermore, even if that transformation were a legitimate device to employ in interpreting the older theory, the result of its application would be a theory so restricted that it could only restate what was already known. Because of its economy, that restatement would have utility, but it could not suffice for the guidance of research.

Let us, therefore, now take it for granted that the differences between successive paradigms are both necessary and irreconcilable. Can we then say more explicitly what sorts of differences these are? The most apparent type has already been illustrated repeatedly. Successive paradigms tell us different things about the population of the universe and about that population's behavior. They differ, that is, about such questions as the existence of subatomic particles, the materiality of light, and the conservation of heat or of energy. These are the substantive differences between successive paradigms, and they require no further illustration. But paradigms differ in more than substance, for they are directed not only to nature but also back upon the science that produced them. They are the source of the methods, problem-field, and standards of solution accepted by any mature scientific community at any given time. As a result, the reception of a new paradigm often necessitates a redefinition of the corresponding science. Some old problems may be relegated to another science or declared entirely "unscientific." Others that were previously non-existent or trivial may, with a new paradigm, become the very archetypes of significant scientific achievement. And as the problems change, so, often, does the standard that distinguishes a real scientific solution from a mere metaphysical speculation, word game, or mathematical play. The normal-scientific tradition that emerges from a scientific revolution is not only incompatible but often actually incommensurable with that which has gone before.

The impact of Newton's work upon the normal seventeenth-

century tradition of scientific practice provides a striking example of these subtler effects of paradigm shift. Before Newton
was born the "new science" of the century had at last succeeded
in rejecting Aristotelian and scholastic explanations expressed in
terms of the essences of material bodies. To say that a stone fell
because its "nature" drove it toward the center of the universe
had been made to look a mere tautological word-play, something it had not previously been. Henceforth the entire flux of
sensory appearances, including color, taste, and even weight,
was to be explained in terms of the size, shape, position, and
motion of the elementary corpuscles of base matter. The attribution of other qualities to the elementary atoms was a resort to
the occult and therefore out of bounds for science. Molière
caught the new spirit precisely when he ridiculed the doctor
who explained opium's efficacy as a soporific by attributing to it
a dormitive potency. During the last half of the seventeenth
century many scientists preferred to say that the round shape of
the opium particles enabled them to sooth the nerves about
which they moved.[5]

In an earlier period explanations in terms of occult qualities
had been an integral part of productive scientific work.
Nevertheless, the seventeenth century's new commitment to
mechanico-corpuscular explanation proved immensely fruitful
for a number of sciences, ridding them of problems that had de-
fied generally accepted solution and suggesting others to replace
them. In dynamics, for example, Newton's three laws of motion
are less a product of novel experiments than of the attempt to
reinterpret well-known observations in terms of the motions and
interactions of primary neutral corpuscles. Consider just one
concrete illustration. Since neutral corpuscles could act on each
other only by contact, the mechanico-corpuscular ·view of
nature directed scientific attention to a brand-new subject of
study, the alteration of particulate motions by collisions. Descartes announced the problem and provided its first putative

[5] For corpuscularism in general, see Marie Boas, "The Establishment of the
Mechanical Philosophy," *Osiris*, X (1952), 412–541. For the effect of particle-
shape on taste, see *ibid.*, p. 483.

solution. Huyghens, Wren, and Wallis carried it still further, partly by experimenting with colliding pendulum bobs, but mostly by applying previously well-known characteristics of motion to the new problem. And Newton embedded their results in his laws of motion. The equal "action" and "reaction" of the third law are the changes in quantity of motion experienced by the two parties to a collision. The same change of motion supplies the definition of dynamical force implicit in the second law. In this case, as in many others during the seventeenth century, the corpuscular paradigm bred both a new problem and a large part of that problem's solution.[6]

Yet, though much of Newton's work was directed to problems and embodied standards derived from the mechanico-corpuscular world view, the effect of the paradigm that resulted from his work was a further and partially destructive change in the problems and standards legitimate for science. Gravity, interpreted as an innate attraction between every pair of particles of matter, was an occult quality in the same sense as the scholastics' "tendency to fall" had been. Therefore, while the standards of corpuscularism remained in effect, the search for a mechanical explanation of gravity was one of the most challenging problems for those who accepted the *Principia* as paradigm. Newton devoted much attention to it and so did many of his eighteenth-century successors. The only apparent option was to reject Newton's theory for its failure to explain gravity, and that alternative, too, was widely adopted. Yet neither of these views ultimately triumphed. Unable either to practice science without the *Principia* or to make that work conform to the corpuscular standards of the seventeenth century, scientists gradually accepted the view that gravity was indeed innate. By the mid-eighteenth century that interpretation had been almost universally accepted, and the result was a genuine reversion (which is not the same as a retrogression) to a scholastic standard. Innate attractions and repulsions joined size, shape, posi-

[6] R Dugas, *La mécanique au XVIIᵉ siècle* (Neuchatel, 1954), pp. 177–85, 284–98, 345–56.

tion, and motion as physically irreducible primary properties of matter.[7]

The resulting change in the standards and problem-field of physical science was once again consequential. By the 1740's, for example, electricians could speak of the attractive "virtue" of the electric fluid without thereby inviting the ridicule that had greeted Molière's doctor a century before. As they did so, electrical phenomena increasingly displayed an order different from the one they had shown when viewed as the effects of a mechanical effluvium that could act only by contact. In particular, when electrical action-at-a-distance became a subject for study in its own right, the phenomenon we now call charging by induction could be recognized as one of its effects. Previously, when seen at all, it had been attributed to the direct action of electrical "atmospheres" or to the leakages inevitable in any electrical laboratory. The new view of inductive effects was, in turn, the key to Franklin's analysis of the Leyden jar and thus to the emergence of a new and Newtonian paradigm for electricity. Nor were dynamics and electricity the only scientific fields affected by the legitimization of the search for forces innate to matter. The large body of eighteenth-century literature on chemical affinities and replacement series also derives from this supramechanical aspect of Newtonianism. Chemists who believed in these differential attractions between the various chemical species set up previously unimagined experiments and searched for new sorts of reactions. Without the data and the chemical concepts developed in that process, the later work of Lavoisier and, more particularly, of Dalton would be incomprehensible.[8] Changes in the standards governing permissible problems, concepts, and explanations can transform a science. In the next section I shall even suggest a sense in which they transform the world.

[7] I. B. Cohen, *Franklin and Newton: An Inquiry into Speculative Newtonian Experimental Science and Franklin's Work in Electricity as an Example Thereof* (Philadelphia, 1956), chaps. vi–vii.

[8] For electricity, see *ibid*, chaps. viii–ix. For chemistry, see Metzger, *op. cit.*, Part I.

Other examples of these nonsubstantive differences between successive paradigms can be retrieved from the history of any science in almost any period of its development. For the moment let us be content with just two other and far briefer illustrations. Before the chemical revolution, one of the acknowledged tasks of chemistry was to account for the qualities of chemical substances and for the changes these qualities underwent during chemical reactions. With the aid of a small number of elementary "principles"—of which phlogiston was one—the chemist was to explain why some substances are acidic, others metalline, combustible, and so forth. Some success in this direction had been achieved. We have already noted that phlogiston explained why the metals were so much alike, and we could have developed a similar argument for the acids. Lavoisier's reform, however, ultimately did away with chemical "principles," and thus ended by depriving chemistry of some actual and much potential explanatory power. To compensate for this loss, a change in standards was required. During much of the nineteenth century failure to explain the qualities of compounds was no indictment of a chemical theory.[9]

Or again, Clerk Maxwell shared with other nineteenth-century proponents of the wave theory of light the conviction that light waves must be propagated through a material ether. Designing a mechanical medium to support such waves was a standard problem for many of his ablest contemporaries. His own theory, however, the electromagnetic theory of light, gave no account at all of a medium able to support light waves, and it clearly made such an account harder to provide than it had seemed before. Initially, Maxwell's theory was widely rejected for those reasons. But, like Newton's theory, Maxwell's proved difficult to dispense with, and as it achieved the status of a paradigm, the community's attitude toward it changed. In the early decades of the twentieth century Maxwell's insistence upon the existence of a mechanical ether looked more and more like lip service, which it emphatically had not been, and the attempts to design such an ethereal medium were abandoned. Scientists no

[9] E. Meyerson, *Identity and Reality* (New York, 1930), chap. x.

longer thought it unscientific to speak of an electrical "displacement" without specifying what was being displaced. The result, again, was a new set of problems and standards, one which, in the event, had much to do with the emergence of relativity theory.[10]

These characteristic shifts in the scientific community's conception of its legitimate problems and standards would have less significance to this essay's thesis if one could suppose that they always occurred from some methodologically lower to some higher type. In that case their effects, too, would seem cumulative. No wonder that some historians have argued that the history of science records a continuing increase in the maturity and refinement of man's conception of the nature of science.[11] Yet the case for cumulative development of science's problems and standards is even harder to make than the case for cumulation of theories. The attempt to explain gravity, though fruitfully abandoned by most eighteenth-century scientists, was not directed to an intrinsically illegitimate problem; the objections to innate forces were neither inherently unscientific nor metaphysical in some pejorative sense. There are no external standards to permit a judgment of that sort. What occurred was neither a decline nor a raising of standards, but simply a change demanded by the adoption of a new paradigm. Furthermore, that change has since been reversed and could be again. In the twentieth century Einstein succeeded in explaining gravitational attractions, and that explanation has returned science to a set of canons and problems that are, in this particular respect, more like those of Newton's predecessors than of his successors. Or again, the development of quantum mechanics has reversed the methodological prohibition that originated in the chemical revolution. Chemists now attempt, and with great success, to explain the color, state of aggregation, and other qualities of the substances used and produced in their laboratories. A similar rever-

[10] E. T. Whittaker, *A History of the Theories of Aether and Electricity*, II (London, 1953), 28–30.

[11] For a brilliant and entirely up-to-date attempt to fit scientific development into this Procrustean bed, see C. C. Gillispie, *The Edge of Objectivity: An Essay in the History of Scientific Ideas* (Princeton, 1960).

sal may even be underway in electromagnetic theory. Space, in contemporary physics, is not the inert and homogenous substratum employed in both Newton's and Maxwell's theories; some of its new properties are not unlike those once attributed to the ether; we may someday come to know what an electric displacement is.

By shifting emphasis from the cognitive to the normative functions of paradigms, the preceding examples enlarge our understanding of the ways in which paradigms give form to the scientific life. Previously, we had principally examined the paradigm's role as a vehicle for scientific theory. In that role it functions by telling the scientist about the entities that nature does and does not contain and about the ways in which those entities behave. That information provides a map whose details are elucidated by mature scientific research. And since nature is too complex and varied to be explored at random, that map is as essential as observation and experiment to science's continuing development. Through the theories they embody, paradigms prove to be constitutive of the research activity. They are also, however, constitutive of science in other respects, and that is now the point. In particular, our most recent examples show that paradigms provide scientists not only with a map but also with some of the directions essential for map-making. In learning a paradigm the scientist acquires theory, methods, and standards together, usually in an inextricable mixture. Therefore, when paradigms change, there are usually significant shifts in the criteria determining the legitimacy both of problems and of proposed solutions.

That observation returns us to the point from which this section began, for it provides our first explicit indication of why the choice between competing paradigms regularly raises questions that cannot be resolved by the criteria of normal science. To the extent, as significant as it is incomplete, that two scientific schools disagree about what is a problem and what a solution, they will inevitably talk through each other when debating the relative merits of their respective paradigms. In the partially circular arguments that regularly result, each paradigm will be shown to satisfy more or less the criteria that it dictates for itself and to fall short of a few of those dictated by its opponent. There are other reasons, too, for the incompleteness of logical contact that consistently characterizes paradigm debates. For example, since no paradigm ever solves all the problems it defines and since no two paradigms leave all the same problems unsolved, paradigm debates always involve the question: Which problems is it more significant to have solved? Like the issue of competing standards, that question of values can be answered only in terms of criteria that lie outside of normal science altogether, and it is that recourse to external criteria that most obviously makes paradigm debates revolutionary. Something even more fundamental than standards and values is, however, also at stake. I have so far argued only that paradigms are constitutive of science. Now I wish to display a sense in which they are constitutive of nature as well.

158

Revolutions as Changes of World View

Examining the record of past research from the vantage of contemporary historiography, the historian of science may be tempted to exclaim that when paradigms change, the world itself changes with them. Led by a new paradigm, scientists adopt new instruments and look in new places. Even more important, during revolutions scientists see new and different things when looking with familiar instruments in places they have looked before. It is rather as if the professional community had been suddenly transported to another planet where familiar objects are seen in a different light and are joined by unfamiliar ones as well. Of course, nothing of quite that sort does occur: there is no geographical transplantation; outside the laboratory everyday affairs usually continue as before. Nevertheless, paradigm changes do cause scientists to see the world of their research-engagement differently. In so far as their only recourse to that world is through what they see and do, we may want to say that after a revolution scientists are responding to a different world.

It is as elementary prototypes for these transformations of the scientist's world that the familiar demonstrations of a switch in visual gestalt prove so suggestive. What were ducks in the scientist's world before the revolution are rabbits afterwards. The man who first saw the exterior of the box from above later sees its interior from below. Transformations like these, though usually more gradual and almost always irreversible, are common concomitants of scientific training. Looking at a contour map, the student sees lines on paper, the cartographer a picture of a terrain. Looking at a bubble-chamber photograph, the student sees confused and broken lines, the physicist a record of familiar subnuclear events. Only after a number of such transformations of vision does the student become an inhabitant of the scientist's world, seeing what the scientist sees and responding as the scientist does. The world that the student then enters is not, however, fixed once and for all by the nature of the environment, on the one hand, and of science, on the other. Rather, it is determined jointly by the environment and the particular normal-scientific tradition that the student has been trained to pursue. Therefore, at times of revolution, when the normal-scientific tradition changes, the scientist's perception of his environment must be re-educated—in some familiar situations he must learn to see a new gestalt. After he has done so the world of his research will seem, here and there, incommensurable with the one he had inhabited before. That is another reason why schools guided by different paradigms are always slightly at cross-purposes.

In their most usual form, of course, gestalt experiments illustrate only the nature of perceptual transformations. They tell us nothing about the role of paradigms or of previously assimilated experience in the process of perception. But on that point there is a rich body of psychological literature, much of it stemming from the pioneering work of the Hanover Institute. An experimental subject who puts on goggles fitted with inverting lenses initially sees the entire world upside down. At the start his perceptual apparatus functions as it had been trained to function in the absence of the goggles, and the result is extreme disorientation, an acute personal crisis. But after the subject has begun to learn to deal with his new world, his entire visual field flips over, usually after an intervening period in which vision is simply confused. Thereafter, objects are again seen as they had been before the goggles were put on. The assimilation of a previously anomalous visual field has reacted upon and changed the field itself.[1] Literally as well as metaphorically, the man accustomed to inverting lenses has undergone a revolutionary transformation of vision.

The subjects of the anomalous playing-card experiment discussed in Section VI experienced a quite similar transformation. Until taught by prolonged exposure that the universe contained

[1] The original experiments were by George M. Stratton, "Vision without Inversion of the Retinal Image," *Psychological Review*, IV (1897), 341–60, 463–81. A more up-to-date review is provided by Harvey A. Carr, *An Introduction to Space Perception* (New York, 1935), pp. 18–57.

anomalous cards, they saw only the types of cards for which previous experience had equipped them. Yet once experience had provided the requisite additional categories, they were able to see all anomalous cards on the first inspection long enough to permit any identification at all. Still other experiments demonstrate that the perceived size, color, and so on, of experimentally displayed objects also varies with the subject's previous training and experience.[2] Surveying the rich experimental literature from which these examples are drawn makes one suspect that something like a paradigm is prerequisite to perception itself. What a man sees depends both upon what he looks at and also upon what his previous visual-conceptual experience has taught him to see. In the absence of such training there can only be, in William James's phrase, "a bloomin' buzzin' confusion."

In recent years several of those concerned with the history of science have found the sorts of experiments described above immensely suggestive. N. R. Hanson, in particular, has used gestalt demonstrations to elaborate some of the same consequences of scientific belief that concern me here.[3] Other colleagues have repeatedly noted that history of science would make better and more coherent sense if one could suppose that scientists occasionally experienced shifts of perception like those described above. Yet, though psychological experiments are suggestive, they cannot, in the nature of the case, be more than that. They do display characteristics of perception that *could* be central to scientific development, but they do not demonstrate that the careful and controlled observation exercised by the research scientist at all partakes of those characteristics. Furthermore, the very nature of these experiments makes any direct demonstration of that point impossible. If historical example is to make these psychological experiments seem rele-

---

[2] For examples, see Albert H. Hastorf, "The Influence of Suggestion on the Relationship between Stimulus Size and Perceived Distance," *Journal of Psychology*, XXIX (1950), 195–217; and Jerome S. Bruner, Leo Postman, and John Rodrigues, "Expectations and the Perception of Color," *American Journal of Psychology*, LXIV (1951), 216–27.

[3] N. R. Hanson, *Patterns of Discovery* (Cambridge, 1958), chap. i.

vant, we must first notice the sorts of evidence that we may and may not expect history to provide.

The subject of a gestalt demonstration knows that his perception has shifted because he can make it shift back and forth repeatedly while he holds the same book or piece of paper in his hands. Aware that nothing in his environment has changed, he directs his attention increasingly not to the figure (duck or rabbit) but to the lines on the paper he is looking at. Ultimately he may even learn to see those lines without seeing either of the figures, and he may then say (what he could not legitimately have said earlier) that it is these lines that he really sees but that he sees them alternately *as* a duck and *as* a rabbit. By the same token, the subject of the anomalous card experiment knows (or, more accurately, can be persuaded) that his perception must have shifted because an external authority, the experimenter, assures him that regardless of what he *saw*, he was *looking at* a black five of hearts all the time. In both these cases, as in all similar psychological experiments, the effectiveness of the demonstration depends upon its being analyzable in this way. Unless there were an external standard with respect to which a switch of vision could be demonstrated, no conclusion about alternate perceptual possibilities could be drawn.

With scientific observation, however, the situation is exactly reversed. The scientist can have no recourse above or beyond what he sees with his eyes and instruments. If there were some higher authority by recourse to which his vision might be shown to have shifted, then that authority would itself become the source of his data, and the behavior of his vision would become a source of problems (as that of the experimental subject is for the psychologist). The same sorts of problems would arise if the scientist could switch back and forth like the subject of the gestalt experiments. The period during which light was "sometimes a wave and sometimes a particle" was a period of crisis— a period when something was wrong—and it ended only with the development of wave mechanics and the realization that light was a self-consistent entity different from both waves and particles. In the sciences, therefore, if perceptual switches accompany paradigm changes, we may not expect scientists to attest to these changes directly. Looking at the moon, the convert to Copernicanism does not say, "I used to see a planet, but now I see a satellite." That locution would imply a sense in which the Ptolemaic system had once been correct. Instead, a convert to the new astronomy says, "I once took the moon to be (or saw the moon as) a planet, but I was mistaken." That sort of statement does recur in the aftermath of scientific revolutions. If it ordinarily disguises a shift of scientific vision or some other mental transformation with the same effect, we may not expect direct testimony about that shift. Rather we must look for indirect and behavioral evidence that the scientist with a new paradigm sees differently from the way he had seen before.

Let us then return to the data and ask what sorts of transformations in the scientist's world the historian who believes in such changes can discover. Sir William Herschel's discovery of Uranus provides a first example and one that closely parallels the anomalous card experiment. On at least seventeen different occasions between 1690 and 1781, a number of astronomers, including several of Europe's most eminent observers, had seen a star in positions that we now suppose must have been occupied at the time by Uranus. One of the best observers in this group had actually seen the star on four successive nights in 1769 without noting the motion that could have suggested another identification. Herschel, when he first observed the same object twelve years later, did so with a much improved telescope of his own manufacture. As a result, he was able to notice an apparent disk-size that was at least unusual for stars. Something was awry, and he therefore postponed identification pending further scrutiny. That scrutiny disclosed Uranus' motion among the stars, and Herschel therefore announced that he had seen a new comet! Only several months later, after fruitless attempts to fit the observed motion to a cometary orbit, did Lexell suggest that the orbit was probably planetary.[4] When that suggestion was accepted, there were several fewer stars and one more planet in the world of the professional astronomer. A celestial body that

[4] Peter Doig, *A Concise History of Astronomy* (London, 1950), pp. 115–16.

had been observed off and on for almost a century was seen differently after 1781 because, like an anomalous playing card, it could no longer be fitted to the perceptual categories (star or comet) provided by the paradigm that had previously prevailed.

The shift of vision that enabled astronomers to see Uranus, the planet, does not, however, seem to have affected only the perception of that previously observed object. Its consequences were more far-reaching. Probably, though the evidence is equivocal, the minor paradigm change forced by Herschel helped to prepare astronomers for the rapid discovery, after 1801, of the numerous minor planets or asteroids. Because of their small size, these did not display the anomalous magnification that had alerted Herschel. Nevertheless, astronomers prepared to find additional planets were able, with standard instruments, to identify twenty of them in the first fifty years of the nineteenth century.[5] The history of astronomy provides many other examples of paradigm-induced changes in scientific perception, some of them even less equivocal. Can it conceivably be an accident, for example, that Western astronomers first saw change in the previously immutable heavens during the half-century after Copernicus' new paradigm was first proposed? The Chinese, whose cosmological beliefs did not preclude celestial change, had recorded the appearance of many new stars in the heavens at a much earlier date. Also, even without the aid of a telescope, the Chinese had systematically recorded the appearance of sunspots centuries before these were seen by Galileo and his contemporaries.[6] Nor were sunspots and a new star the only examples of celestial change to emerge in the heavens of Western astronomy immediately after Copernicus. Using traditional instruments, some as simple as a piece of thread, late sixteenth-century astronomers repeatedly discovered that comets wandered at will through the space previously reserved for the

[5] Rudolph Wolf, Geschichte der Astronomie (Munich, 1877), pp. 513–15, 683–93. Notice particularly how difficult Wolf's account makes it to explain these discoveries as a consequence of Bode's Law.

[6] Joseph Needham, Science and Civilization in China, III (Cambridge, 1959), 423–29, 434–36.

immutable planets and stars.[7] The very ease and rapidity with which astronomers saw new things when looking at old objects with old instruments may make us wish to say that, after Copernicus, astronomers lived in a different world. In any case, their research responded as though that were the case.

The preceding examples are selected from astronomy because reports of celestial observation are frequently delivered in a vocabulary consisting of relatively pure observation terms. Only in such reports can we hope to find anything like a full parallelism between the observations of scientists and those of the psychologist's experimental subjects. But we need not insist on so full a parallelism, and we have much to gain by relaxing our standard. If we can be content with the everyday use of the verb 'to see,' we may quickly recognize that we have already encountered many other examples of the shifts in scientific perception that accompany paradigm change. The extended use of 'perception' and of 'seeing' will shortly require explicit defense, but let me first illustrate its application in practice.

Look again for a moment at two of our previous examples from the history of electricity. During the seventeenth century, when their research was guided by one or another effluvium theory, electricians repeatedly saw chaff particles rebound from, or fall off, the electrified bodies that had attracted them. At least that is what seventeenth-century observers said they saw, and we have no more reason to doubt their reports of perception than our own. Placed before the same apparatus, a modern observer would see electrostatic repulsion (rather than mechanical or gravitational rebounding), but historically, with one universally ignored exception, electrostatic repulsion was not seen as such until Hauksbee's large-scale apparatus had greatly magnified its effects. Repulsion after contact electrification was, however, only one of many new repulsive effects that Hauksbee saw. Through his researches, rather as in a gestalt switch, repulsion suddenly became *the* fundamental manifestation of electrification, and it was then attraction that needed to be ex-

[7] T. S. Kuhn, *The Copernican Revolution* (Cambridge, Mass., 1957), pp. 206–9.

plained.[8] The electrical phenomena visible in the early eighteenth century were both subtler and more varied than those seen by observers in the seventeenth century. Or again, after the assimilation of Franklin's paradigm, the electrician looking at a Leyden jar saw something different from what he had seen before. The device had become a condenser, for which neither the jar shape nor glass was required. Instead, the two conducting coatings—one of which had been no part of the original device— emerged to prominence. As both written discussions and pictorial representations gradually attest, two metal plates with a non-conductor between them had become the prototype for the class.[9] Simultaneously, other inductive effects received new descriptions, and still others were noted for the first time.

Shifts of this sort are not restricted to astronomy and electricity. We have already remarked some of the similar transformations of vision that can be drawn from the history of chemistry. Lavoisier, we said, saw oxygen where Priestley had seen dephlogisticated air and where others had seen nothing at all. In learning to see oxygen, however, Lavoisier also had to change his view of many other more familiar substances. He had, for example, to see a compound ore where Priestley and his contemporaries had seen an elementary earth, and there were other such changes besides. At the very least, as a result of discovering oxygen, Lavoisier saw nature differently. And in the absence of some recourse to that hypothetical fixed nature that he "saw differently," the principle of economy will urge us to say that after discovering oxygen Lavoisier worked in a different world.

I shall inquire in a moment about the possibility of avoiding this strange locution, but first we require an additional example of its use, this one deriving from one of the best known parts of the work of Galileo. Since remote antiquity most people have seen one or another heavy body swinging back and forth on a string or chain until it finally comes to rest. To the Aristotelians,

[8] Duane Roller and Duane H. D. Roller, *The Development of the Concept of Electric Charge* (Cambridge, Mass., 1954), pp. 21–29.

[9] See the discussion in Section VII and the literature to which the reference there cited in note 9 will lead.

who believed that a heavy body is moved by its own nature from a higher position to a state of natural rest at a lower one, the swinging body was simply falling with difficulty. Constrained by the chain, it could achieve rest at its low point only after a tortuous motion and a considerable time. Galileo, on the other hand, looking at the swinging body, saw a pendulum, a body that almost succeeded in repeating the same motion over and over again ad infinitum. And having seen that much, Galileo observed other properties of the pendulum as well and constructed many of the most significant and original parts of his new dynamics around them. From the properties of the pendulum, for example, Galileo derived his only full and sound arguments for the independence of weight and rate of fall, as well as for the relationship between vertical height and terminal velocity of motions down inclined planes.[10] All these natural phenomena he saw differently from the way they had been seen before.

Why did that shift of vision occur? Through Galileo's individual genius, of course. But note that genius does not here manifest itself in more accurate or objective observation of the swinging body. Descriptively, the Aristotelian perception is just as accurate. When Galileo reported that the pendulum's period was independent of amplitude for amplitudes as great as 90°, his view of the pendulum led him to see far more regularity than we can now discover there.[11] Rather, what seems to have been involved was the exploitation by genius of perceptual possibilities made available by a medieval paradigm shift. Galileo was not raised completely as an Aristotelian. On the contrary, he was trained to analyze motions in terms of the impetus theory, a late medieval paradigm which held that the continuing motion of a heavy body is due to an internal power implanted in it by the projector that initiated its motion. Jean Buridan and Nicole Oresme, the fourteenth-century scholastics who brought the impetus theory to its most perfect formulations, are the first men

[10] Galileo Galilei, *Dialogues concerning Two New Sciences*, trans. H. Crew and A. de Salvio (Evanston, Ill., 1946), pp. 80–81, 162–66.

[1] *Ibid.*, pp. 91–94, 244.

known to have seen in oscillatory motions any part of what Galileo saw there. Buridan describes the motion of a vibrating string as one in which impetus is first implanted when the string is struck; the impetus is next consumed in displacing the string against the resistance of its tension; tension then carries the string back, implanting increasing impetus until the mid-point of motion is reached; after that the impetus displaces the string in the opposite direction, again against the string's tension, and so on in a symmetric process that may continue indefinitely. Later in the century Oresme sketched a similar analysis of the swinging stone in what now appears as the first discussion of a pendulum.[12] His view is clearly very close to the one with which Galileo first approached the pendulum. At least in Oresme's case, and almost certainly in Galileo's as well, it was a view made possible by the transition from the original Aristotelian to the scholastic impetus paradigm for motion. Until that scholastic paradigm was invented, there were no pendulums, but only swinging stones, for the scientist to see. Pendulums were brought into existence by something very like a paradigm-induced gestalt switch.

Do we, however, really need to describe what separates Galileo from Aristotle, or Lavoisier from Priestley, as a transformation of vision? Did these men really *see* different things when *looking at* the same sorts of objects? Is there any legitimate sense in which we can say that they pursued their research in different worlds? Those questions can no longer be postponed, for there is obviously another and far more usual way to describe all of the historical examples outlined above. Many readers will surely want to say that what changes with a paradigm is only the scientist's interpretation of observations that themselves are fixed once and for all by the nature of the environment and of the perceptual apparatus. On this view, Priestley and Lavoisier both saw oxygen, but they interpreted their observations differently; Aristotle and Galileo both saw pendu-

---

[12] M. Clagett, *The Science of Mechanics in the Middle Ages* (Madison, Wis., 1959), pp. 537–38, 570.

lums, but they differed in their interpretations of what they both had seen.

Let me say at once that this very usual view of what occurs when scientists change their minds about fundamental matters can be neither all wrong nor a mere mistake. Rather it is an essential part of a philosophical paradigm initiated by Descartes and developed at the same time as Newtonian dynamics. That paradigm has served both science and philosophy well. Its exploitation, like that of dynamics itself, has been fruitful of a fundamental understanding that perhaps could not have been achieved in another way. But as the example of Newtonian dynamics also indicates, even the most striking past success provides no guarantee that crisis can be indefinitely postponed. Today research in parts of philosophy, psychology, linguistics, and even art history, all converge to suggest that the traditional paradigm is somehow askew. That failure to fit is also made increasingly apparent by the historical study of science to which most of our attention is necessarily directed here.

None of these crisis-promoting subjects has yet produced a viable alternate to the traditional epistemological paradigm, but they do begin to suggest what some of that paradigm's characteristics will be. I am, for example, acutely aware of the difficulties created by saying that when Aristotle and Galileo looked at swinging stones, the first saw constrained fall, the second a pendulum. The same difficulties are presented in an even more fundamental form by the opening sentences of this section: though the world does not change with a change of paradigm, the scientist afterward works in a different world. Nevertheless, I am convinced that we must learn to make sense of statements that at least resemble these. What occurs during a scientific revolution is not fully reducible to a reinterpretation of individual and stable data. In the first place, the data are not unequivocally stable. A pendulum is not a falling stone, nor is oxygen dephlogisticated air. Consequently, the data that scientists collect from these diverse objects are, as we shall shortly see, themselves different. More important, the process by which either the individual or the community makes the transition from constrained fall to the pendulum or from dephlogisticated air to oxygen is not one that resembles interpretation. How could it do so in the absence of fixed data for the scientist to interpret? Rather than being an interpreter, the scientist who embraces a new paradigm is like the man wearing inverting lenses. Confronting the same constellation of objects as before and knowing that he does so, he nevertheless finds them transformed through and through in many of their details.

None of these remarks is intended to indicate that scientists do not characteristically interpret observations and data. On the contrary, Galileo interpreted observations on the pendulum, Aristotle observations on falling stones, Musschenbroek observations on a charge-filled bottle, and Franklin observations on a condenser. But each of these interpretations presupposed a paradigm. They were parts of normal science, an enterprise that, as we have already seen, aims to refine, extend, and articulate a paradigm that is already in existence. Section III provided many examples in which interpretation played a central role. Those examples typify the overwhelming majority of research. In each of them the scientist, by virtue of an accepted paradigm, knew what a datum was, what instruments might be used to retrieve it, and what concepts were relevant to its interpretation. Given a paradigm, interpretation of data is central to the enterprise that explores it.

But that interpretive enterprise—and this was the burden of the paragraph before last—can only articulate a paradigm, not correct it. Paradigms are not corrigible by normal science at all. Instead, as we have already seen, normal science ultimately leads only to the recognition of anomalies and to crises. And these are terminated, not by deliberation and interpretation, but by a relatively sudden and unstructured event like the gesalt switch. Scientists then often speak of the "scales falling from the eyes" or of the "lightning flash" that "inundates" a previously obscure puzzle, enabling its components to be seen in a new way that for the first time permits its solution. On other occasions the relevant illumination comes in sleep.[13] No ordinary sense of the term 'interpretation' fits these flashes of intuition through which a new paradigm is born. Though such intuitions depend upon the experience, both anomalous and congruent, gained with the old paradigm, they are not logically or piecemeal linked to particular items of that experience as an interpretation would be. Instead, they gather up large portions of that experience and transform them to the rather different bundle of experience that will thereafter be linked piecemeal to the new paradigm but not to the old.

To learn more about what these differences in experience can be, return for a moment to Aristotle, Galileo, and the pendulum. What data did the interaction of their different paradigms and their common environment make accessible to each of them? Seeing constrained fall, the Aristotelian would measure (or at least discuss—the Aristotelian seldom measured) the weight of the stone, the vertical height to which it had been raised, and the time required for it to achieve rest. Together with the resistance of the medium, these were the conceptual categories deployed by Aristotelian science when dealing with a falling body.[14] Normal research guided by them could not have produced the laws that Galileo discovered. It could only—and by another route it did—lead to the series of crises from which Galileo's view of the swinging stone emerged. As a result of those crises and of other intellectual changes besides, Galileo saw the swinging stone quite differently. Archimedes' work on floating bodies made the medium non-essential; the impetus theory rendered the motion symmetrical and enduring; and Neoplatonism directed Galileo's attention to the motion's circu-

13 [Jacques] Hadamard, *Subconscient intuition, et logique dans la recherche scientifique* (*Conférence faite au Palais de la Découverte le 8 Décembre 1945* [Alençon, n.d.]), pp. 7-8. A much fuller account, though one exclusively restricted to mathematical innovations, is the same author's *The Psychology of Invention in the Mathematical Field* (Princeton, 1949).

14 T. S. Kuhn, "A Function for Thought Experiments," in *Mélanges Alexandre Koyré*, ed. R. Taton and I. B. Cohen, to be published by Hermann (Paris) in 1963.

lar form.[15] He therefore measured only weight, radius, angular displacement, and time per swing, which were precisely the data that could be interpreted to yield Galileo's laws for the pendulum. In the event, interpretation proved almost unnecessary. Given Galileo's paradigms, pendulum-like regularities were very nearly accessible to inspection. How else are we to account for Galileo's discovery that the bob's period is entirely independent of amplitude, a discovery that the normal science stemming from Galileo had to eradicate and that we are quite unable to document today. Regularities that could not have existed for an Aristotelian (and that are, in fact, nowhere precisely exemplified by nature) were consequences of immediate experience for the man who saw the swinging stone as Galileo did.

Perhaps that example is too fanciful since the Aristotelians recorded no discussions of swinging stones. On their paradigm it was an extraordinarily complex phenomenon. But the Aristotelians did discuss the simpler case, stones falling without uncommon constraints, and the same differences of vision are apparent there. Contemplating a falling stone, Aristotle saw a change of state rather than a process. For him the relevant measures of a motion were therefore total distance covered and total time elapsed, parameters which yield what we should now call not speed but average speed.[16] Similarly, because the stone was impelled by its nature to reach its final resting point, Aristotle saw the relevant distance parameter at any instant during the motion as the distance *to* the final end point rather than as that *from* the origin of motion.[17] Those conceptual parameters underlie and give sense to most of his well-known "laws of motion." Partly through the impetus paradigm, however, and partly through a doctrine known as the latitude of forms, scholastic criticism changed this way of viewing motion. A stone moved by impetus gained more and more of it while receding from its

[15] A. Koyré, *Etudes Galiléennes* (Paris, 1939), I, 46–51; and "Galileo and Plato," *Journal of the History of Ideas*, IV (1943), 400–428.

[16] Kuhn, "A Function for Thought Experiments," in *Mélanges Alexandre Koyré* (see n. 14 for full citation).

[17] Koyré, *Etudes* . . . , II, 7–11.

starting point; distance from rather than distance to therefore became the revelant parameter. In addition, Aristotle's notion of speed was bifurcated by the scholastics into concepts that soon after Galileo became our average speed and instantaneous speed. But when seen through the paradigm of which these conceptions were a part, the falling stone, like the pendulum, exhibited its governing laws almost on inspection. Galileo was not one of the first men to suggest that stones fall with a uniformly accelerated motion.[18] Furthermore, he had developed his theorem on this subject together with many of its consequences before he experimented with an inclined plane. That theorem was another one of the network of new regularities accessible to genius in the world determined jointly by nature and by the paradigms upon which Galileo and his contemporaries had been raised. Living in that world, Galileo could still, when he chose, explain why Aristotle had seen what he did. Nevertheless, the immediate content of Galileo's experience with falling stones was not what Aristotle's had been.

It is, of course, by no means clear that we need be so concerned with "immediate experience"—that is, with the perceptual features that a paradigm so highlights that they surrender their regularities almost upon inspection. Those features must obviously change with the scientist's commitments to paradigms, but they are far from what we ordinarily have in mind when we speak of the raw data or the brute experience from which scientific research is reputed to proceed. Perhaps immediate experience should be set aside as fluid, and we should discuss instead the concrete operations and measurements that the scientist performs in his laboratory. Or perhaps the analysis should be carried further still from the immediately given. It might, for example, be conducted in terms of some neutral observation-language, perhaps one designed to conform to the retinal imprints that mediate what the scientist sees. Only in one of these ways can we hope to retrieve a realm in which experience is again stable once and for all—in which the pendulum and constrained fall are not different perceptions but rather

[18] Clagett, *op. cit.*, chaps. iv, vi, and ix.

different interpretations of the unequivocal data provided by observation of a swinging stone.

But is sensory experience fixed and neutral? Are theories simply man-made interpretations of given data? The epistemological viewpoint that has most often guided Western philosophy for three centuries dictates an immediate and unequivocal, Yes! In the absence of a developed alternative, I find it impossible to relinquish entirely that viewpoint. Yet it no longer functions effectively, and the attempts to make it do so through the introduction of a neutral language of observations now seem to me hopeless.

The operations and measurements that a scientist undertakes in the laboratory are not "the given" of experience but rather "the collected with difficulty." They are not what the scientist sees—at least not before his research is well advanced and his attention focused. Rather, they are concrete indices to the content of more elementary perceptions, and as such they are selected for the close scrutiny of normal research only because they promise opportunity for the fruitful elaboration of an accepted paradigm. Far more clearly than the immediate experience from which they in part derive, operations and measurements are paradigm-determined. Science does not deal in all possible laboratory manipulations. Instead, it selects those relevant to the juxtaposition of a paradigm with the immediate experience that that paradigm has partially determined. As a result, scientists with different paradigms engage in different concrete laboratory manipulations. The measurements to be performed on a pendulum are not the ones relevant to a case of constrained fall. Nor are the operations relevant for the elucidation of oxygen's properties uniformly the same as those required when investigating the characteristics of dephlogisticated air.

As for a pure observation-language, perhaps one will yet be devised. But three centuries after Descartes our hope for such an eventuality still depends exclusively upon a theory of perception and of the mind. And modern psychological experimentation is rapidly proliferating phenomena with which that theory can scarcely deal. The duck-rabbit shows that two men

174

with the same retinal impressions can see different things; the inverting lenses show that two men with different retinal impressions can see the same thing. Psychology supplies a great deal of other evidence to the same effect, and the doubts that derive from it are readily reinforced by the history of attempts to exhibit an actual language of observation. No current attempt to achieve that end has yet come close to a generally applicable language of pure percepts. And those attempts that come closest share one characteristic that strongly reinforces several of this essay's main theses. From the start they presuppose a paradigm, taken either from a current scientific theory or from some fraction of everyday discourse, and they then try to eliminate from it all non-logical and non-perceptual terms. In a few realms of discourse this effort has been carried very far and with fascinating results. There can be no question that efforts of this sort are worth pursuing. But their result is a language that—like those employed in the sciences—embodies a host of expectations about nature and fails to function the moment these expectations are violated. Nelson Goodman makes exactly this point in describing the aims of his *Structure of Appearance:* "It is fortunate that nothing more [than phenomena known to exist] is in question; for the notion of 'possible' cases, of cases that do not exist but might have existed, is far from clear."[19] No language thus restricted to reporting a world fully known in advance can produce mere neutral and objective reports on "the given." Philosophical investigation has not yet provided even a hint of what a language able to do that would be like.

Under these circumstances we may at least suspect that scientists are right in principle as well as in practice when they treat

---

[10] N. Goodman, *The Structure of Appearance* (Cambridge, Mass., 1951), pp. 4–5. The passage is worth quoting more extensively: "If all and only those residents of Wilmington in 1947 that weigh between 175 and 180 pounds have red hair, then 'red-haired 1947 resident of Wilmington' and '1947 resident of Wilmington weighing between 175 and 180 pounds' may be joined in a constructional definition. . . . The question whether there 'might have been' someone to whom one but not the other of these predicates would apply has no bearing . . . once we have determined that there is no such person. . . . It is fortunate that nothing more is in question; for the notion of 'possible' cases, of cases that do not exist but might have existed, is far from clear."

oxygen and pendulums (and perhaps also atoms and electrons) as the fundamental ingredients of their immediate experience. As a result of the paradigm-embodied experience of the race, the culture, and, finally, the profession, the world of the scientist has come to be populated with planets and pendulums, condensers and compound ores, and other such bodies besides. Compared with these objects of perception, both meter stick readings and retinal imprints are elaborate constructs to which experience has direct access only when the scientist, for the special purposes of his research, arranges that one or the other should do so. This is not to suggest that pendulums, for example, are the only things a scientist could possibly see when looking at a swinging stone. (We have already noted that members of another scientific community could see constrained fall.) But it is to suggest that the scientist who looks at a swinging stone can have no experience that is in principle more elementary than seeing a pendulum. The alternative is not some hypothetical "fixed" vision, but vision through another paradigm, one which makes the swinging stone something else.

All of this may seem more reasonable if we again remember that neither scientists nor laymen learn to see the world piecemeal or item by item. Except when all the conceptual and manipulative categories are prepared in advance—e.g., for the discovery of an additional transuranic element or for catching sight of a new house—both scientists and laymen sort out whole areas together from the flux of experience. The child who transfers the word 'mama' from all humans to all females and then to his mother is not just learning what 'mama' means or who his mother is. Simultaneously he is learning some of the differences between males and females as well as something about the ways in which all but one female will behave toward him. His reactions, expectations, and beliefs—indeed, much of his perceived world—change accordingly. By the same token, the Copernicans who denied its traditional title 'planet' to the sun were not only learning what 'planet' meant or what the sun was. Instead, they were changing the meaning of 'planet' so that it could continue to make useful distinctions in a world where all celestial bodies, not just the sun, were seen differently from the way they had been seen before. The same point could be made about any of our earlier examples. To see oxygen instead of dephlogisticated air, the condenser instead of the Leyden jar, or the pendulum instead of constrained fall, was only one part of an integrated shift in the scientist's vision of a great many related chemical, electrical, or dynamical phenomena. Paradigms determine large areas of experience at the same time.

It is, however, only after experience has been thus determined that the search for an operational definition or a pure observation-language can begin. The scientist or philosopher who asks what measurements or retinal imprints make the pendulum what it is must already be able to recognize a pendulum when he sees one. If he saw constrained fall instead, his question could not even be asked. And if he saw a pendulum, but saw it in the same way he saw a tuning fork or an oscillating balance, his question could not be answered. At least it could not be answered in the same way, because it would not be the same question. Therefore, though they are always legitimate and are occasionally extraordinarily fruitful, questions about retinal imprints or about the consequences of particular laboratory manipulations presuppose a world already perceptually and conceptually subdivided in a certain way. In a sense such questions are parts of normal science, for they depend upon the existence of a paradigm and they receive different answers as a result of paradigm change.

To conclude this section, let us henceforth neglect retinal impressions and again restrict attention to the laboratory operations that provide the scientist with concrete though fragmentary indices to what he has already seen. One way in which such laboratory operations change with paradigms has already been observed repeatedly. After a scientific revolution many old measurements and manipulations become irrelevant and are replaced by others instead. One does not apply all the same tests to oxygen as to dephlogisticated air. But changes of this sort are never total. Whatever he may then see, the scientist after a revolution is still looking at the same world. Further-

more, though he may previousl, have employed them differently, much of his language and most of his laboratory instruments
are still the same as they were before. As a result, postrevolu-
tionary science invariably includes many of the same manipulations, performed with the same instruments and described in
the same terms, as its prerevolutionary predecessor. If these enduring manipulations have been changed at all, the change
must lie either in their relation to the paradigm or in their concrete results. I now suggest, by the introduction of one last new
example, that both these sorts of changes occur. Examining the
work of Dalton and his contemporaries, we shall discover that
one and the same operation, when it attaches to nature through
a different paradigm, can become an index to a quite different
aspect of nature's regularity. In addition, we shall see that occasionally the old manipulation in its new role will yield different
concrete results.

Throughout much of the eighteenth century and into the
nineteenth, European chemists almost universally believed that
the elementary atoms of which all chemical species consisted
were held together by forces of mutual affinity. Thus a lump of
silver cohered because of the forces of affinity between silver
corpuscles (until after Lavoisier these corpuscles were themselves thought of as compounded from still more elementary
particles). On the same theory silver dissolved in acid (or salt
in water) because the particles of acid attracted those of silver
(or the particles of water attracted those of salt) more strongly
than particles of these solutes attracted each other. Or again,
copper would dissolve in the silver solution and precipitate
silver, because the copper-acid affinity was greater than the
affinity of acid for silver. A great many other phenomena were
explained in the same way. In the eighteenth century the theory
of elective affinity was an admirable chemical paradigm, widely
and sometimes fruitfully deployed in the design and analysis of
chemical experimentation.[20]

Affinity theory, however, drew the line separating physical

[20] H. Metzger, *Newton, Stahl, Boerhaave et la doctrine chimique* (Paris,
1930), pp. 34–68.

mixtures from chemical compounds in a way that has become unfamiliar since the assimilation of Dalton's work. Eighteenth-century chemists did recognize two sorts of processes. When mixing produced heat, light, effervescence or something else of the sort, chemical union was seen to have taken place. If, on the other hand, the particles in the mixture could be distinguished by eye or mechanically separated, there was only physical mixture. But in the very large number of intermediate cases—salt in water, alloys, glass, oxygen in the atmosphere, and so on—these crude criteria were of little use. Guided by their paradigm, most chemists viewed this entire intermediate range as chemical, because the processes of which it consisted were all governed by forces of the same sort. Salt in water or oxygen in nitrogen was just as much an example of chemical combination as was the combination produced by oxidizing copper. The arguments for viewing solutions as compounds were very strong. Affinity theory itself was well attested. Besides, the formation of a compound accounted for a solution's observed homogeneity. If, for example, oxygen and nitrogen were only mixed and not combined in the atmosphere, then the heavier gas, oxygen, should settle to the bottom. Dalton, who took the atmosphere to be a mixture, was never satisfactorily able to explain oxygen's failure to do so. The assimilation of his atomic theory ultimately created an anomaly where there had been none before.[21]

One is tempted to say that the chemists who viewed solutions as compounds differed from their successors only over a matter of definition. In one sense that may have been the case. But that sense is not the one that makes definitions mere conventional conveniences. In the eighteenth century mixtures were not fully distinguished from compounds by operational tests, and perhaps they could not have been. Even if chemists had looked for such tests, they would have sought criteria that made the solution a compound. The mixture-compound distinction was part of their paradigm—part of the way they viewed their whole

[21] *Ibid.*, pp. 124–29, 139–48. For Dalton, see Leonard K. Nash, *The Atomic-Molecular Theory* ("Harvard Case Histories in Experimental Science," Case 4; Cambridge, Mass., 1950), pp. 14–21.

field of research—and as such it was prior to any particular laboratory test, though not to the accumulated experience of chemistry as a whole.

But while chemistry was viewed in this way, chemical phenomena exemplified laws different from those that emerged with the assimilation of Dalton's new paradigm. In particular, while solutions remained compounds, no amount of chemical experimentation could by itself have produced the law of fixed proportions. At the end of the eighteenth century it was widely known that *some* compounds ordinarily contained fixed proportions by weight of their constituents. For some categories of reactions the German chemist Richter had even noted the further regularities now embraced by the law of chemical equivalents.[22] But no chemist made use of these regularities except in recipes, and no one until almost the end of the century thought of generalizing them. Given the obvious counterinstances, like glass or like salt in water, no generalization was possible without an abandonment of affinity theory and a reconceptualization of the boundaries of the chemist's domain. That consequence became explicit at the very end of the century in a famous debate between the French chemists Proust and Berthollet. The first claimed that all chemical reactions occurred in fixed proportion, the latter that they did not. Each collected impressive experimental evidence for his view. Nevertheless, the two men necessarily talked through each other, and their debate was entirely inconclusive. Where Berthollet saw a compound that could vary in proportion, Proust saw only a physical mixture.[23] To that issue neither experiment nor a change of definitional convention could be relevant. The two men were as fundamentally at cross-purposes as Galileo and Aristotle had been.

This was the situation during the years when John Dalton undertook the investigations that led finally to his famous chemical atomic theory. But until the very last stages of those investiga-

[22] J. R. Partington, *A Short History of Chemistry* (2d ed.; London, 1951), pp. 161–63.

[23] A. N. Meldrum, "The Development of the Atomic Theory: (1) Berthollet's Doctrine of Variable Proportions," *Manchester Memoirs*, LIV (1910), 1–16.

tions, Dalton was neither a chemist nor interested in chemistry. Instead, he was a meteorologist investigating the, for him, physical problems of the absorption of gases by water and of water by the atmosphere. Partly because his training was in a different specialty and partly because of his own work in that specialty, he approached these problems with a paradigm different from that of contemporary chemists. In particular, he viewed the mixture of gases or the absorption of a gas in water as a physical process, one in which forces of affinity played no part. To him, therefore, the observed homogeneity of solutions was a problem, but one which he thought he could solve if he could determine the relative sizes and weights of the various atomic particles in his experimental mixtures. It was to determine these sizes and weights that Dalton finally turned to chemistry, supposing from the start that, in the restricted range of reactions that he took to be chemical, atoms could only combine one-to-one or in some other simple whole-number ratio.[24] That natural assumption did enable him to determine the sizes and weights of elementary particles, but it also made the law of constant proportion a tautology. For Dalton, any reaction in which the ingredients did not enter in fixed proportion was *ipso facto* not a purely chemical process. A law that experiment could not have established before Dalton's work, became, once that work was accepted, a constitutive principle that no single set of chemical measurements could have upset. As a result of what is perhaps our fullest example of a scientific revolution, the same chemical manipulations assumed a relationship to chemical generalization very different from the one they had had before.

Needless to say, Dalton's conclusions were widely attacked when first announced. Berthollet, in particular, was never convinced. Considering the nature of the issue, he need not have been. But to most chemists Dalton's new paradigm proved convincing where Proust's had not been, for it had implications far wider and more important than a new criterion for distinguish-

[24] L. K. Nash, "The Origin of Dalton's Chemical Atomic Theory," *Isis*, XLVII (1956), 101–16.

ing a mixture from a compound. If, for example, atoms could combine chemically only in simple whole-number ratios, then a re-examination of existing chemical data should disclose examples of multiple as well as of fixed proportions. Chemists stopped writing that the two oxides of, say, carbon contained 56 per cent and 72 per cent of oxygen by weight; instead they wrote that one weight of carbon would combine either with 1.3 or with 2.6 weights of oxygen. When the results of old manipulations were recorded in this way, a 2:1 ratio leaped to the eye; and this occurred in the analysis of many well-known reactions and of new ones besides. In addition, Dalton's paradigm made it possible to assimilate Richter's work and to see its full generality. Also, it suggested new experiments, particularly those of Gay-Lussac on combining volumes, and these yielded still other regularities, ones that chemists had not previously dreamed of. What chemists took from Dalton was not new experimental laws but a new way of practicing chemistry (he himself called it the "new system of chemical philosophy"), and this proved so rapidly fruitful that only a few of the older chemists in France and Britain were able to resist it.[25] As a result, chemists came to live in a world where reactions behaved quite differently from the way they had before.

As all this went on, one other typical and very important change occurred. Here and there the very numerical data of chemistry began to shift. When Dalton first searched the chemical literature for data to support his physical theory, he found some records of reactions that fitted, but he can scarcely have avoided finding others that did not. Proust's own measurements on the two oxides of copper yielded, for example, an oxygen weight-ratio of 1.47:1 rather than the 2:1 demanded by the atomic theory; and Proust is just the man who might have been expected to achieve the Daltonian ratio.[26] He was, that is, a fine

[25] A. N. Meldrum, "The Development of the Atomic Theory: (6) The Reception Accorded to the Theory Advocated by Dalton," *Manchester Memoirs,* LV (1911), 1–10.

[26] For Proust, see Meldrum, "Berthollet's Doctrine of Variable Proportions," *Manchester Memoirs,* LIV (1910), 8. The detailed history of the gradual changes in measurements of chemical composition and of atomic weights has yet to be written, but Partington, *op. cit.,* provides many useful leads to it.

experimentalist, and his view of the relation between mixtures and compounds was very close to Dalton's. But it is hard to make nature fit a paradigm. That is why the puzzles of normal science are so challenging and also why measurements undertaken without a paradigm so seldom lead to any conclusions at all. Chemists could not, therefore, simply accept Dalton's theory on the evidence, for much of that was still negative. Instead, even after accepting the theory, they had still to beat nature into line, a process which, in the event, took almost another generation. When it was done, even the percentage composition of well-known compounds was different. The data themselves had changed. That is the last of the senses in which we may want to say that after a revolution scientists work in a different world.

The Road Since Structure

Thomas S. Kuhn

Massachusetts Institute of Technology

On this occasion, and in this place, I feel that I ought, and am probably expected, to look back at the things which have happened to the philosophy of science since I first began to take an interest in it over half a century ago. But I am both too much an outsider and too much a protagonist to undertake that assignment. Rather than attempt to situate the present state of philosophy of science with respect to its past — a subject on which I've little authority — I shall try to situate my present state in philosophy of science with respect to its own past — a subject on which, however imperfect, I'm probably the best authority there is.

As a number of you know, I'm at work on a book, and what I mean to attempt here is an exceedingly brief and dogmatic sketch of its main themes. I think of my project as a return, now underway for a decade, to the philosophical problems left over from the *Structure of Scientific Revolutions*. But it might better be described more generally, as a study of the problems raised by the transition to what's sometimes called the historical and sometimes (at least by Clark Glymour, speaking to me) just the "soft" philosophy of science. That's a transition for which I get far more credit, and also more blame, than I have coming to me. I was, if you will, present at the creation, and it wasn't very crowded. But others were present too: Paul Feyerabend and Russ Hanson, in particular, as well as Mary Hesse, Michael Polanyi, Stephen Toulmin, and a few more besides. Whatever a *Zeitgeist* is, we provided a striking illustration of its role in intellectual affairs.

Returning to my projected book, you will not be surprised to hear that the main targets at which it aims are such issues as rationality, relativism and, most particularly, realism and truth. But they're not primarily what the book is about, what occupies most space in it. That role is taken instead by incommensurability. No other aspect of *Structure* has concerned me so deeply in the thirty years since the book was written, and I emerge from those years feeling more strongly than ever that incommensurability has to be an essential component of any historical, developmental, or evolutionary view of scientific knowledge. Properly understood — something I've by no means always managed myself — incommensurability is far from being the threat to rational evaluation of truth claims that it has frequently seemed. Rather, it's what is needed, within a developmental perspective, to restore some badly needed bite to the whole

PSA 1990, Volume 2, pp. 3-13
Copyright © 1991 by the Philosophy of Science Association

4

notion of cognitive evaluation. It is needed, that is, to defend notions like truth and knowledge from, for example, the excesses of post-modernist movements like the strong program. Clearly, I can't hope to make all that out here: it's a project for a book. But I shall try, however sketchily, to describe the main elements of the position the book develops. I begin by saying something about what I now take incommensurability to be, and then attempt to sketch its relationship to questions of relativism, truth, and realism. In the book, the issue of rationality will figure, too, but there is no space here even to sketch its role.

Incommensurability is a notion that for me emerged from attempts to understand apparently nonsensical passages encountered in old scientific texts. Ordinarily they had been taken as evidence of the author's confused or mistaken beliefs. My experiences led me to suggest, instead, that those passages were being misread: the appearance of nonsense could be removed by recovering older meanings for some of the terms involved, meanings different from those subsequently current. During the years since, I've often spoken metaphorically of the process by which later meanings had been produced from earlier ones as a process of language change. And, more recently, I've spoken also of the historian's recovery of older meanings as a process of language learning rather like that undergone by the fictional anthropologist whom Quine misdescribes as a radical translator (Kuhn 1983a). The ability to learn a language does not, I've emphasized, guarantee the ability to translate into or out of it.

By now, however, the language metaphor seems to me far too inclusive. To the extent that I'm concerned with language and with meanings at all — an issue to which I'll shortly return — it is with the meanings of a restricted class of terms. Roughly speaking, they are taxonomic terms or kind terms, a widespread category that includes natural kinds, artifactual kinds, social kinds, and probably others. In English the class is coextensive, or nearly so, with the terms that by themselves or within appropriate phrases can take the indefinite article. These are primarily the count nouns together with the mass nouns, words which combine with count nouns in phrases that take the indefinite article. Some terms require still further tests hinging, for example, on permissible suffixes.

Terms of this sort have two essential properties. First, as already indicated, they are marked or labelled as kind terms by virtue of lexical characteristics like taking the indefinite article. Being a kind term is thus part of what the word means, part of what one must have in the head to use the word properly. Second — a limitation I sometimes refer to as the no-overlap principle — no two kind terms, no two terms with the kind label, may overlap in their referents unless they are related as species to genus. There are no dogs that are also cats, no gold rings that are also silver rings, and so on: that's what makes dogs, cats, silver, and gold each a kind. Therefore, if the members of a language community encounter a dog that's also a cat (or, more realistically, a creature like the duck-billed platypus), they cannot just enrich the set of category terms but must instead redesign a part of the taxonomy. Pace the causal theorists of reference, 'water' did not always refer to H_2O (Kuhn 1987; 1990, pp. 309-14).

Notice now that a lexical taxonomy of some sort must be in place before description of the world can begin. Shared taxonomic categories, at least in an area under discussion, are prerequisite to unproblematic communication, including the communication required for the evaluation of truth claims. If different speech communities have taxonomies that differ in some local area, then members of one of them can (and occasionally will) make statements that, though fully meaningful within that speech community, cannot in principle be articulated by members of the other. To bridge the gap between communities would require adding to one lexicon a kind-term that overlaps, shares a referent, with one that is already in place. It is that situation which the no-overlap principle precludes.

Incommensurability thus becomes a sort of untranslatability, localized to one or another area in which two lexical taxonomies differ. The differences which produce it are not any old differences, but ones that violate either the no-overlap condition, the kind-label condition, or else a restriction on hierarchical relations that I cannot spell out here. Violations of those sorts do not bar intercommunity understanding. Members of one community can acquire the taxonomy employed by members of another, as the historian does in learning to understand old texts. But the process which permits understanding produces bilinguals, not translators, and bilingualism has a cost, which will be particularly important to what follows. The bilingual must always remember within which community discourse is occurring. The use of one taxonomy to make statements to someone who uses the other places communication at risk.

Let me formulate these points in one more way, and then make a last remark about them. Given a lexical taxonomy, or what I'll mostly now call simply a lexicon, there are all sorts of different statements that can be made, and all sorts of theories that can be developed. Standard techniques will lead to some of these being accepted as true, others rejected as false. But there are also statements which could be made, theories which could be developed, within some other taxonomy but which cannot be made with this one and vice versa. The first volume of Lyons' *Semantics* (1977, pp. 237-8) contains a wonderfully simple example, which some of you will know: the impossibility of translating the English statement, "the cat sat on the mat", into French, because of the incommensurability between the French and English taxonomies for floor coverings. In each particular case for which the English statement is true, one can find a co-referential French statement, some using 'tapis', others 'paillasson,' still others 'carpette,' and so on. But there is no single French statement which refers to all and only the situations in which the English statement is true. In that sense, the English statement cannot be made in French. In a similar vein, I've elsewhere pointed out (Kuhn 1987, p. 8) that the content of the Copernican statement, "planets travel around the sun", cannot be expressed in a statement that invokes the celestial taxonomy of the Ptolemaic statement, "planets travel around the earth". The difference between the two statements is not simply one of fact. The term 'planet' appears as a kind term in both, and the two kinds overlap in membership without either's containing all the celestial bodies contained in the other. All of which is to say that there are episodes in scientific development which involve fundamental change in some taxonomic categories and which therefore confront later observers with problems like those the ethnologist encounters when trying to break into another culture.

A final remark will close this sketch of my current views on incommensurability. I have described those views as concerned with words and with *lexical* taxonomy, and I shall continue in that mode: the sorts of knowledge I deal with come in explicit verbal or related symbolic forms. But it may clarify what I have in mind to suggest that I might more appropriately speak of concepts than of words. What I have been calling a lexical taxonomy might, that is, better be called a conceptual scheme, where the "very notion" of a conceptual scheme is not that of a set of beliefs but of a particular operating mode of a mental module prerequisite to having beliefs, a mode that at once supplies and bounds the set of beliefs it is possible to conceive. Some such taxonomic module I take to be pre-linguistic and possessed by animals. Presumably it evolved originally for the sensory, most obviously for the visual, system. In the book I shall give reasons for supposing that it developed from a still more fundamental mechanism which enables individual living organisms to reidentify other substances by tracing their spatio-temporal trajectories.

6

I shall be coming back to incommensurability, but let me for now set it aside in order to sketch the developmental framework within which it functions. Since I must again move quickly and often cryptically, I begin by anticipating the direction in which I am headed. Basically, I shall be trying to sketch the form which I think any viable evolutionary epistemology has to take. I shall, that is, be returning to the evolutionary analogy introduced in the very last pages of the first edition of *Structure*, attempting both to clarify it and to push it further. During the thirty years since I first made that evolutionary move, theories of the evolution both of species and of knowledge have, of course, been transformed in ways I am only beginning to discover. I still have much to learn, but to date the fit seems extremely good.

I start from points familiar to many of you. When I first got involved, a generation ago, with the enterprise now often called historical philosophy of science, I and most of my coworkers thought history functioned as a source of empirical evidence. That evidence we found in historical case studies, which forced us to pay close attention to science as it really was. Now I think we overemphasized the empirical aspect of our enterprise (an evolutionary epistemology need not be a naturalized one). What has for me emerged as essential is not so much the details of historical cases as the perspective or the ideology that attention to historical cases brings with it. The historian, that is, always picks up a process already underway, its beginnings lost in earlier time. Beliefs are already in place; they provide the basis for the ongoing research whose results will in some cases change them; research in their absence is unimaginable though there has nevertheless been a long tradition of imagining it. For the historian, in short, no Archimedean platform is available for the pursuit of science other than the historically situated one already in place. If you approach science as an historian must, little observation of its actual practice is required to reach conclusions of this sort.

Such conclusions have by now been pretty generally accepted: I scarcely know a foundationalist any more. But for me, this way of abandoning foundationalism has a further consequence which, though widely discussed, is by no means widely or fully accepted. The discussions I have in mind usually proceed under the rubric of the rationality or relativity of truth claims, but these labels misdirect attention. Though both rationality and relativism are somehow implicated, what is fundamentally at stake is rather the correspondence theory of truth, the notion that the goal, when evaluating scientific laws or theories, is to determine whether or not they correspond to an external, mind-independent world. It is that notion, whether in an absolute or probabilistic form, that I'm persuaded must vanish together with foundationalism. What replaces it will still require a strong conception of truth, but not, except in the most trivial sense, correspondence truth.

Let me at least suggest what the argument involves. On the developmental view, scientific knowledge claims are necessarily evaluated from a moving, historically-situated, Archimedian platform. What requires evaluation cannot be an individual proposition embodying a knowledge claim in isolation: embracing a new knowledge claim typically requires adjustment of other beliefs as well. Nor is it the entire body of knowledge claims that would result if that proposition were accepted. Rather, what's to be evaluated is the desirability of a particular change-of-belief, a change which would alter the existing body of knowledge claims so as to incorporate, with minimum disruption, the new claim as well. Judgements of this sort are necessarily comparative: which of two bodies of knowledge — the original or the proposed alternative — is *better* for doing whatever it is that scientists do. And that is the case whether what scientists do is solve puzzles (my view), improve empirical adequacy (Bas van Frassen's), or increase the dominance of the ruling elite (in parody, the strong program's). I do, of course, have my own preference among these alternatives, and it makes a difference (Kuhn, 1983b). But no choice between them is relevant to what's presently at stake.

In comparative judgements of the kind just sketched, shared beliefs are left in place: they serve as the given for purposes of the current evaluation; they provide a replacement for the traditional Archimedean platform. The fact that they may — indeed probably will — later be at risk in some other evaluation is simply irrelevant. Nothing about the rationality of the outcome of the current evaluation depends upon their, in fact, being true or false. They are simply in place, part of the historical situation within which this evaluation is made. But if the actual truth value of the shared presumptions required for the evaluation is irrelevant, then the question of the truth or falsity of the changes made or rejected on the basis of that evaluation cannot arise either. A number of classic problems in philosophy of science — most obviously Duhemian holism — turn out on this view to be due not to the nature of scientific knowledge but to a misperception of what justification of belief is all about. Justification does not aim at a goal external to the historical situation but simply, in that situation, at improving the tools available for the job at hand.

To this point I have been trying to firm-up and extend the parallel between scientific and biological development suggested at the end of the first edition of *Structure*: scientific development must be seen as a process driven from behind, not pulled from ahead — as evolution from, rather than evolution towards. In making that suggestion, as elsewhere in the book, the parallel I had in mind was diachronic, involving the relation between older and more recent scientific beliefs about the same or overlapping ranges of natural phenomena. Now I want to suggest a second, less widely perceived parallel between Darwinian evolution and the evolution of knowledge, one that cuts a synchronic slice across the sciences rather than a diachronic slice containing one of them. Though I have in the past occasionally spoken of the incommensurability between the theories of contemporary scientific specialties, I've only in the last few years begun to see its significance to the parallels between biological evolution and scientific development. Those parallels have also been persuasively emphasized recently in a splendid article by Mario Biagioli of UCLA (1990). To both of us they seem extremely important, though we emphasize them for somewhat different reasons.

To indicate what is involved I must revert briefly to my old distinction between normal and revolutionary development. In *Structure* it was the distinction between those developments that simply add to knowledge, and those which require giving up part of what's been believed before. In the new book it will emerge as the distinction between developments which do and developments which do not require local taxonomic change. (The alteration permits a significantly more nuanced description of what goes on during revolutionary change than I've been able to provide before.) During this second sort of change, something else occurs that in *Structure* got mentioned only in passing. After a revolution there are usually (perhaps always) more cognitive specialties or fields of knowledge than there were before. Either a new branch has split off from the parent trunk, as scientific specialties have repeatedly split off in the past from philosophy and from medicine. Or else a new specialty has been born at an area of apparent overlap between two preexisting specialties, as occurred, for example, in the cases of physical chemistry and molecular biology. At the time of its occurrence this second sort of split is often hailed as a reunification of the sciences, as was the case in the episodes just mentioned. As time goes on, however, one notices that the new shoot seldom or never gets assimilated to either of its parents. Instead, it becomes one more separate specialty, gradually acquiring its own new specialists' journals, a new professional society, and often also new university chairs, laboratories, and even departments. Over time a diagram of the evolution of scientific fields, specialties, and sub-specialties comes to look strikingly like a layman's diagram for a biological evolutionary tree. Each of these fields has a distinct lexicon, though the differences are local, occuring only here and there. There is no *lingua franca* capable of expressing, in its entirety, the content of them all or even of any pair.

With much reluctance I have increasingly come to feel that this process of specialization, with its consequent limitation on communication and community, is inescapable, a consequence of first principles. Specialization and the narrowing of the range of expertise now look to me like the necessary price of increasingly powerful cognitive tools. What's involved is the same sort of development of special tools for special functions that's apparent also in technological practice. And, if that is the case, then a couple of additional parallels between biological evolution and the evolution of knowledge come to seem especially consequential. First, revolutions, which produce new divisions between fields in scientific development, are much like episodes of speciation in biological evolution. The biological parallel to revolutionary change is not mutation, as I thought for many years, but speciation. And the problems presented by speciation (e.g., the difficulty in identifying an episode of speciation until some time after it has occurred, and the impossibility, even then, of dating the time of its occurrence) are very similar to those presented by revolutionary change and by the emergence and individuation of new scientific specialties.

The second parallel between biological and scientific development, to which I return again in the concluding section, concerns the unit which undergoes speciation (not to be confused with a unit of selection). In the biological case, it is a reproductively isolated population, a unit whose members collectively embody the gene pool which ensures both the population's self-perpetuation and its continuing isolation. In the scientific case, the unit is a community of intercommunicating specialists, a unit whose members share a lexicon that provides the basis for both the conduct and the evaluation of their research and which simultaneously, by barring full communication with those outside the group, maintains their isolation from practitioners of other specialties.

To anyone who values the unity of knowledge, this aspect of specialization — lexical or taxonomic divergence, with consequent limitations on communication — is a condition to be deplored. But such unity may be in principle an unattainable goal, and its energetic pursuit might well place the growth of knowledge at risk. Lexical diversity and the principled limit it imposes on communication may be the isolating mechanism required for the development of knowledge. Very likely it is the specialization consequent on lexical diversity that permits the sciences, viewed collectively, to solve the puzzles posed by a wider range of natural phenomena than a lexically-homogenous science could achieve.

Though I greet the thought with mixed feelings, I am increasingly persuaded that the limited range of possible partners for fruitful intercourse is the essential precondition for what is known as progress in both biological development and the development of knowledge. When I suggested earlier that incommensurability, properly understood, could reveal the source of the cognitive bite and authority of the sciences, its role as an isolating mechanism was prerequisite to the topic I had principally in mind, the one to which I now turn.

This reference to 'intercourse', for which I shall henceforth substitute the term 'discourse', bring me back to problems concerning truth, and thus to the locus of the newly restored bite. I said earlier that we must learn to get along without anything at all like a correspondence theory of truth. But something like a redundancy theory of truth is badly needed to replace it, something that will introduce minimal laws of logic (in particular, the law of non-contradiction) and make adhering to them a precondition for the rationality of evaluations (Horwich 1990). On this view, as I wish to employ it, the essential function of the concept of truth is to require choice between acceptance and rejection of a statement or a theory in the face of evidence shared by all. Let me try briefly to sketch what I have in mind.

Ian Hacking, in an attempt (1982) to denature the apparent relativism associated with incommensurability, spoke of the way in which new "styles" introduce into science new candidates for true/false. Since that time, I've been gradually realizing (the reformulation is still in process) that some of my own central points are far better made without speaking of statements as themselves being true or as being false. Instead, the evaluation of a putatively scientific statement should be conceived as comprising two seldom-separated parts. First, determine the status of the statement: is it a candidate for true/false? To that question, as you'll shortly see, the answer is lexicon-dependent. And second, supposing a positive answer to the first, is the statement rationally assertable? To that question, given a lexicon, the answer is properly found by something like the normal rules of evidence.

In this reformulation, to declare a statement a candidate for true/false is to accept it as a counter in a language game whose rules forbid asserting both a statement and its contrary. A person who breaks that rule declares him or herself outside the game. If one nevertheless tries to continue play, then discourse breaks down; the integrity of the language community is threatened. Similar, though more problematic, rules apply, not simply to contrary statements, but more generally to logically incompatible ones. There are, of course, language games without the rule of non-contradiction and its relatives: poetry and mystical discourse, for example. And there are also, even within the declarative-statement game, recognized ways of bracketing the rule, permitting and even exploiting the use of contradiction. Metaphor and other tropes are the most obvious examples; more central for present purposes are the historian's restatements of past beliefs. (Though the originals were candidates for true/false, the historian's later restatements — made by a bilingual speaking the language of one culture to the members of another — are not.) But in the sciences and in many more ordinary community activities, such bracketing devices are parasitic on normal discourse. And these activities — the ones that presuppose normal adherence to the rules of the true/false game — are an essential ingredient of the glue that binds communities together. In one form or another, the rules of the true/false game are thus universals for all human communities. But the result of applying those rules varies from one speech community to the next. In discussion between members of communities with differently structured lexicons, assertability and evidence play the same role for both only in areas (there are always a great many) where the two lexicons are congruent.

Where the lexicons of the parties to discourse differ, a given string of words will sometimes make different statements for each. A statement may be a candidate for truth/falsity with one lexicon without having that status in the others. And even when it does, the two statements will not be the same: though identically phrased, strong evidence for one need not be evidence for the other. Communication breakdowns are then inevitable, and it is to avoid them that the bilingual is forced to remember at all times which lexicon is in play, which community the discourse is occurring within.

These breakdowns in communication do, of course, occur: they're a significant characteristic of the episodes *Structure* referred to as 'crises'. I take them to be the crucial symptoms of the speciation-like process through which new disciplines emerge, each with it own lexicon, and each with its own area of knowledge. It is by these divisions, I've been suggesting, that knowledge grows. And it's the need to maintain discourse, to keep the game of declarative statements going, that forces these divisions and the fragmentation of knowledge that results.

I close with some brief and tentative remarks about what emerges from this position as the relationship between the lexicon — the shared taxonomy of a speech community — and the world the members of that community jointly inhabit. Clearly it cannot be the one Putnam (1977, pp. 123-38) has called metaphysical realism. Insofar as the structure of the world can be experienced and the experience communicated, it is constrained by the structure of the lexicon of the community which inhabits it. Doubtless some aspects of that lexical structure are biologically determined, the products of a shared phylogeny. But, at least among advanced creatures (and not just those linguistically endowed), significant aspects are determined also by education, by the process of socialization, that is, which initiates neophytes into the community of their parents and peers. Creatures with the same biological endowment may experience the world through lexicons that are here and there very differently structured, and in those areas they will be unable to communicate all of their experiences across the lexical divide. Though individuals may belong to several interrelated communities (thus, be multilinguals), they experience aspects of the world differently as they move from one to the next.

Remarks like these suggest that the world is somehow mind-dependent, perhaps an invention or construction of the creatures which inhabit it, and in recent years such suggestions have been widely pursued. But the metaphors of invention, construction, and mind-dependence are in two respects grossly misleading. First, the world is not invented or constructed. The creatures to whom this responsibility is imputed, in fact, find the world already in place, its rudiments at their birth and its increasingly full actuality during their educational socialization, a socialization in which examples of the way the world is play an essential part. That world, furthermore, has been experientially given, in part to the new inhabitants directly, and in part indirectly, by inheritance, embodying the experience of their forebears. As such, it is entirely solid: not in the least respectful of an observer's wishes and desires; quite capable of providing decisive evidence against invented hypotheses which fail to match its behavior. Creatures born into it must take it as they find it. They can, of course, interact with it, altering both it and themselves in the process, and the populated world thus altered is the one that will be found in place by the generation which follows. The point closely parallels the one made earlier about the nature of evaluation seen from a developmental perspective: there, what required evaluation was not belief but change in some aspects of belief, the rest held fixed in the process; here, what people can effect or invent is not the world but changes in some aspects of it, the balance remaining as before. In both cases, too, the changes that can be made are not introduced at will. Most proposal for change are rejected on the evidence; the nature of those that remain can rarely be foreseen; and the consequences of accepting one or another of them often prove to be undesired.

Can a world that alters with time and from one community to the next correspond to what is generally referred to as "the real world"? I do not see how its right to that title can be denied. It provides the environment, the stage, for all individual and social life. On such life it places rigid constraints; continued existence depends on adaptation to them; and in the modern world scientific activity has become a primary tool for adaptation. What more can reasonably be asked of a real world?

In the penultimate sentence, above. the word 'adaptation' is clearly problematic. Can the members of a group properly be said to adapt to an environment which they are constantly adjusting to fit their needs? Is it the creatures who adapt to the world or does the world adapt to the creatures? Doesn't this whole way of talking imply a mutual plasticity incompatible with the rigidity of the constraints that make the world real and that made it appropriate to describe the creatures as adapted to it? These difficulties are genuine, but they necessarily inhere in any and all descriptions of undirected evolutionary processes. The identical problem is, for example, currently the subject of much discussion in evolutionary biology. On the one hand the evolutionary process gives rise to creatures more and more closely adapted to a narrower and narrower biological niche. On the other, the niche to which they are adapted is recognizable only in retrospect, with its population in place: it has no existence independent of the community which is adapted to it. (Lewontin 1978.) What actually evolves, therefore, is creatures and niches together: what creates the tensions inherent in talk of adaptation is the need, if discussion and analysis are to be possible, to draw a line between the creatures within the niche, on the one hand, and their "external" environment, on the other.

Niches may not seem to be worlds, but the difference is one of viewpoint. Niches are where *other* creatures live. We see them from outside and thus in physical interaction with their inhabitants. But the inhabitants of a niche see it from inside and their interactions with it are, to them, intentionally mediated through something like a mental representation. Biologically, that is, a niche is the world of the group which inhabits it, thus constituting it a niche. Conceptually, the world is *our* representation of *our* niche, the residence of the particular human community with whose members we are currently interacting.

The world-constitutive role assigned here to intentionality and mental representations recurs to a theme characteristic of my viewpoint throughout its long development: compare my earlier recourse to gestalt switches, seeing as understanding, and so on. This is the aspect of my work that, more than any other, has suggested that I took the world to be mind-dependent. But the metaphor of a mind-dependent world — like its cousin, the constructed or invented world — proves to be deeply misleading. It is groups and group-practices that constitute those worlds (and are constituted by them). And the practice-in-the-world of some of those groups *is* science. The primary unit through which the sciences develop is thus, as previously stressed, the group, and groups do not have minds. Under the unfortunate title, "Are species individuals?", contemporary biological theory offers a significant parallel (Hull, 1976, provides an especially useful introduction to the literature). In one sense the procreating organisms which perpetuate a species are the units whose practice permits evolution to occur. But to understand the outcome of that process one must see the evolutionary unit (not to be confused with a unit of selection) as the gene pool shared by those organisms, the organisms which carry the gene pool serving only as the parts which, through bi-sexual reproduction, exchange genes within the population. Cognitive evolution depends, similarly, upon the exchange, through discourse, of statements within a community. Though the units which exchange those statements are individual scientists, understanding the advance of knowledge, the outcome of their practice, depends upon seeing them as atoms constitutive of a larger whole, the community of practitioners of some scientific specialty.

The primacy of the community over its members is reflected also in the theory of the lexicon, the unit which embodies the shared conceptual or taxonomic structure that holds the community together and simultaneously isolates it from other groups. Conceive the lexicon as a module within the head of an individual group member. It can then be shown (though not here) that what characterizes members of the group is possession not of identical lexicons, but of mutually congruent ones, of lexicons with the same structure. The lexical structure which characterizes a group is more abstract than, different in kind from, the individual lexicons or mental modules which embody it. And it is only that structure, not its various individual embodiments, that members of the community must share. The mechanics of taxonomizing are in this respect like its function: neither can be fully understood except as grounded within the community it serves.

By now it may be clear that the position I'm developing is a sort of post-Darwinian Kantianism. Like the Kantian categories, the lexicon supplies preconditions of possible experience. But lexical categories, unlike their Kantian forebears, can and do change, both with time and with the passage from one community to another. None of those changes, of course, is ever vast. Whether the communities in question are displaced in time or in conceptual space, their lexical structures must overlap in major ways or there could be no bridgeheads permitting a member of one to acquire the lexicon of the other. Nor, in the absence of major overlap, would it be possible for the members of a single community to evaluate proposed new theories when their acceptance required lexical change. Small changes, however, can have large-scale effects. The Copernican Revolution provides especially well-known illustrations.

Underlying all these processes of differentiation and change, there must, of course, be something permanent, fixed, and stable. But, like Kant's *Ding an sich*, it is ineffable, undescribable, undiscussible. Located outside of space and time, this Kantian source of stability is the whole from which have been fabricated both creatures and their niches, both the "internal" and the "external" worlds. Experience and description are possible only with the described and describer separated, and the lexical structure which marks that separation can do so in different ways, each resulting in a different, though never wholly different, form of life. Some ways are better suited to some purposes, some to others. But none is to be accepted as true or rejected as false; none gives privileged access to a real, as against an invented, world. The ways of being-in-the-world which a lexicon provides are not candidates for true/false.

References

Biagioli, M. (1990), "The Anthropology of Incommensurability," *Studies in History and Philosophy of Science* 21: 183-209.

Hacking, I. (1982), "Language, Truth and Reason," in *Rationality and Relativism*, M. Hollis and S. Lukes (eds.). Cambridge: MIT Press, pp. 49-66.

Horwich, P. (1990), *Truth*. Oxford: Blackwell.

Hull, D.I. (1976), "Are Species Really Individual?", *Systematic Zoology* 25:174-191.

Kuhn, T.S. (1983a), "Commensurability, Comparability, Communicability," *PSA 1982, Volume Two*. East Lansing: Philosophy of Science Association, pp. 669-688.

_ _ _ _ _ _. (1983b), "Rationality and Theory Choice," *Journal of Philosophy*, 80: 563-570.

_ _ _ _ _. (1987), "What are Scientific Revolutions?" in *The Probabilistic Revolution, Volume 1: Ideas in History*, L. Krüger, L.J. Daston, and M. Heidelberger (eds.). Cambridge: MIT Press, pp. 7-22.

_ _ _ _ _. (1990), "Dubbing and Redubbing: the Vulnerabiltity of Rigid Designation," in *Scientific Theories*, Minnesota Studies in the Philosophy of Science, XIV, C.W. Savage (ed.). Minneapolis: University of Minnesota Press, pp. 298-318,

Lyons, J. (1977), *Semantics, Volume I*. Cambridge: Cambridge University Press.

Lewontin, R.C. (1978), "Adaptation," *Scientific American* 239: 212-30.

Putnam, H. (1978), *Meaning and the Moral Sciences*. London: Routledge.

EXPLANATION AND REFERENCE

I. GENERAL SIGNIFICANCE OF THE TOPIC

In this paper I try to contrast Marxist (and more broadly realist) theories of meaning with what may be called 'idealist' theories of meaning. But a word of explanation is clearly in order.

There is no Marxist 'theory of meaning' as such, but there are a series of remarks on the correspondence between concepts and things, on concepts, and on the impossibility of a priori knowledge in the writings of Engels[1] and Lenin[1] which clearly bear on problems of meaning and reference, and which constitute the starting point for such a theory. In particular, there is a passage[2] in which Engels makes the point that a concept may contain elements which are not true of the things which correspond to that concept. Engels' example is the concept *fish*. A contemporary scientific characterization of fish would include, Engels says, such properties as life under water and breathing through gills; yet lungfish and other anomalous species which lack these properties are classified as fish for scientific purposes. And Engels argues, I think correctly, that to stick to the letter of the 'definition' in applying the concept *fish* would be bad science. In short, Engels contends that:

(1) Our scientific conception (I would say 'stereotype') of a fish includes the property 'breathing through gills', but

(2) 'All fish breath through gills' is not true! (and, *a fortiori*, not analytic).

I do not wish to ascribe to Engels an anachronistic sophistication about contemporary logical issues, but without doing this it is fair to say on the basis of this argument that Engels *rejects* the model according to which such a concept as *fish* provides anything like analytically necessary and sufficient conditions for membership in a natural kind. Two further points are of importance: (1) The fact that the concept "natural kind *all* of whose members live under water, breath through gills, etc." does not strictly fit the natural kind Fish does not mean that the concept does not

Pearce and Maynard (eds.), Conceptual Change, 199–221. All rights reserved.
Copyright © 1973 by D. Reidel Publishing Company, Dordrecht-Holland.

correspond to the natural kind Fish. As Engels puts it, the concept is not exactly correct (as a description of the corresponding natural kind) but that does not make it a *fiction*. (2) The concept is continually changing as a result of the impact of scientific discoveries, but that does not mean that it ceases to correspond to the same natural kind (which is itself, of course, also changing). Again, without attributing to Engels a sophisticated theory of meaning and reference, it is fair, I think, to restate the essential gist of these two points in the following way: concepts which are not strictly true of anything may yet refer to something; and concepts in different theories may refer to the same thing. Of these two points, the second is obvious for most realists; with a few possible exceptions (e.g. Paul Feyerabend), realists have held that there are successive scientific theories about the *same* things: about heat, about electricity, about electrons, and so forth; and this involves treating such terms as 'electricity' as *transtheoretical* terms, as Dudley Shapere has called them,[3] i.e., as terms that have the same reference in different theories. The first point is more controversial; the idea that concepts provide necessary and sufficient conditions for class membership has often been attacked but, nonetheless, constantly reappears. Without it, however, the other point is moot. Bohr took it for granted that there are (at every time) numbers p and q such that the (one dimensional) position of a particle is q and the (one dimensional) momentum is p; if this was part of the meaning of 'particle' for Bohr, and in addition, "part of the meaning" means "necessary condition for membership in the extension of a term", then electrons are *not* particles in Bohr's sense, and, indeed, there are *no* particles "in Bohr's sense". (And no "electrons" in Bohr's sense of 'electron', etc.) In fact, none of the terms in Bohr's theory referred! It follows on this account that we cannot say that present electron theory is a better theory of the same particles that Bohr was referring to. I take it that this is the line of thinking that Paul Feyerabend represents. On an account like Engels', however, Bohr would have been referring to electrons when he used the word 'electron', notwithstanding the fact that some of his beliefs about electrons were mistaken, and *we* are referring to those same particles notwithstanding the fact that some of our beliefs – even beliefs included in our scientific "definition" of the term 'electron' – may very likely turn out to be equally mistaken. This seems right to me, and likewise Shapere's recent emphasis on the idea that such terms as 'electron' are *transtheoretical* seems to me right and important. The main technical contribution of this paper will be a sketch of a theory of meaning which supports Engels' and Shapere's insights. An "idealist" theory of meaning, as I am using the term, might go like this (in its simplest form): the meaning of such a sentence as 'electrons exist' is a function of certain *predictions* that can be derived from it (in a pure idealist theory, these would have to be predictions about *sensations*); these predictions are clearly a function of the *theory* in which the sentence occurs; thus 'electrons exist' has no meaning apart from this, that or the other theory, and it has a different meaning in different theories.

The question of "reference" is a harder one for an idealist: the essence of idealism is to view scientific theories and concepts as instruments for predicting sensations and not as representatives of real things and magnitudes. But a sophisticated idealist is likely to say that the question of reference is "trivial",[4] if one has a scientific language L containing the term 'electron', then one can certainly construct a metalanguage ML over it *a là* Tarski, and define "reference" in such a way that " 'electron' refers to electrons" is a trivial theorem. But if different scientific theories T_1 and T_2 are associated with different formal languages L_1 and L_2 (as they must be if the words have different meanings in T_1 and T_2), then they will be associated with different *meta*-languages ML_1 and ML_2. In ML_1 we can say " 'electron' refers to electrons", meaning that 'electron' in the sense of T_1 refers to electrons *in the sense of* T_1, and in ML_2 we can say " 'electron' refers to electrons" meaning that 'electron' in the sense of T_2 refers to electrons *in the sense of* T_2; but there is no ML in which we can even express the statement that "electron" refers to the same entities in T_1 and T_2 – or, at least, no prescription for constructing such an ML has been provided by Positivist philosophers of science. In short, just as the idealist regards 'electron' as *theory dependent*, so does he regard the semantical notions of reference and truth as theory dependent; just as the Marxist (and, more generally, the realist) regards 'electron' as *trans-theoretical*, so does he regard truth and reference as trans-theoretical.

II. THE MEANING OF PHYSICAL MAGNITUDE TERMS

A. *A Causal Account of Meaning*

My purpose here is to sketch an account of the meaning of physical magnitude terms (e.g. 'temperature', 'electrical charge'); not an account of meaning in general, although I will try to indicate similarities between what is said here about these terms and what Kripke has said about proper names and what I have said elsewhere about natural kind words. (Kripke's work has come to me second hand; even so, I owe him a large debt for suggesting the idea of causal chains as the mechanism of reference.)

On a traditional view, any term has an intension and an extension. "Knowing the meaning" is having knowledge of the intension; what it is to "know" an intension (construed, usually, as an abstract entity of some kind) is never explained. The extension of the color term 'red', for example, is the class of red things; the intension, according to Carnap, is the property Red. Carnap spoke of "grasping" the intension of terms; what it would be the "grasp" the property Red was never explained; probably Carnap would have equated it with knowing how to verify sentences of the form 'x is red', but this comes from his theory of knowledge, not his writings on semantics. In any case, understanding words is a matter of having knowledge. Full linguistic competence in connection with a word may require more knowledge than just the intension; for example, syntactical knowledge, knowledge of co-occurence regularities, etc.; but linguistic competence, like understanding, is a matter of *knowledge* – not necessarily explicit knowledge – knowledge in the wide sense, implicit as well as explicit, "knowing how" as well as 'knowing that", skills and abilities as well as facts, but all *knowledge* none the less.

According to the theory I shall present this is fundamentally wrong. Linguistic competence and understanding are not just *knowledge*. To have linguistic competence in connection with a term it is not sufficient, in general, to have the full battery of usual linguistic knowledge and skills; one must, in addition, be in the right sort of relationship to certain distinguished situations (normally, though not necessarily, situations in which the *referent* of the term is present). It is for this reason that this sort of theory is called a "causal theory" of meaning.

Coming to physical magnitude terms, what every user of the term 'electricity' knows is that electricity is a magnitude of some sort – and, in fact, not even that: electricity was thought at one time to possibly be a sort of substance, and so was heat. At any rate, speakers know that "electricity" and "heat" are putative physical *quantities* – capable of more and less, and capable of location. (I do not think that even these statements are *analytic*, but I think they have a kind of *linguistic* association with the terms in question.) In a developed semantic theory one might introduce a special semantic marker, e.g. 'physical quantity', for terms of this sort. I cannot, however, think of anything that *every* user of the term 'electricity' *has* to know except that electricity is (associated with the notion of being) a physical magnitude of some sort, and, possibly, that "electricity" (or electrical charge or charges) is capable of flow or motion. Benjamin Franklin knew that "electricity" was manifested in the form of sparks and lightning bolts; someone else might know about currents and electromagnets; someone else might know about atoms consisting of positively and negatively charged particles. They could all use the term 'electricity' without there being a discernible "intension" that they all share. I want to suggest that what they do have in common is this: that each of them is connected by a certain kind of causal chain to a situation in which a *description* of electricity is given, and generally a *causal* description – that is, one which singles out electricity as *the* physical magnitude *responsible* for certain effects in a certain way.

Thus, suppose I were standing next to Ben Franklin as he performed his famous experiment. Suppose he told me that "electricity" is a physical quantity which behaves in certain respects like a liquid (if he were a mathematician he might say "obeys an equation of continuity"); that it collects in clouds, and then, when a critical point of some kind is reached, a large quantity flows from the cloud to the earth in the form of a lightning bolt; that it runs along (or perhaps "through") his metal kite string; etc. He would have given me an *approximately correct definite description* of a physical magnitude. I could now use the term 'electricity' myself. Let us call this event – my acquiring the ability to use the term 'electricity' in this way – an *introducing event*. It is clear that each of my later uses will be causally connected to this introducing event, as long as those uses exemplify the ability I acquired in that introducing event. Even if I use the term so often that I forget when I first learned it, the intention to refer to the same magnitude that I referred to in the past by using the word links my present use to those earlier uses, and indeed the word's being in my present vocabulary at all is a causal product of earlier events – ultimately of the introducing event. If I teach the word to someone else by telling him that the word 'electricity' is the name of a physical magnitude, and by telling him certain facts about it which do not constitute a causal description – e.g., I might tell him that like charges repel and unlike charges attract, and that atoms consist of a nucleus with one kind of charge surrounded by satellite electrons with the opposite kind of charge – even if the facts I tell him do not constitute a definite description of any kind, let alone a causal description – still, the word's being in his vocabulary will be causally linked to its being in my vocabulary, and hence, ultimately, to an introducing event.

I said before that different speakers use the word 'electricity' without their being a discernible "intension" that they all share. If an "intension" is anything like a necessary and sufficient condition, then I think that this is right. But it does not follow that there are no ideas about electricity which are in some way linguistically associated with the word. Just as the idea that tigers are striped is linguistically associated with the word 'tiger', so it seems that some idea that "electricity" (i.e., electric charge or charges) is capable of flow or motion *is* linguistically associated with 'electricity'. And perhaps this is all – apart from being a physical magnitude or quantity in the sense described before – that is linguistically associated with the word.

Now then, if anyone knows that 'electricity' is the name of a physical quantity, and his use of the word is connected by the sort of causal chain I described before to an introducing event in which the causal description given was, in fact, a causal description of electricity, then we have a clear basis for saying that he uses the word to refer to electricity. Even if the causal description failed to describe electricity, if there is good reason to treat it as a mis-description *of electricity* (rather than as a description of nothing at all) – for example, if electricity was described as the physical magnitude with such-and-such properties which is responsible for such-and-such effects, where in fact electricity is not responsible for the effects in question, and the speaker intended to refer to the magnitude responsible for those effects, but mistakenly added the incorrect information "electricity has such-and-such properties" because he mistakenly thought that the magnitude responsible for those effects had those further properties – we still have a basis for saying that both the original speaker and the persons to whom he teaches the word use the word to refer to electricity.

If a number of speakers use the word 'electricity' to refer to electricity, and, in addition, they have the standard sorts of associations with the word – that it refers to a magnitude which can move or flow – then, I suggest, the question of whether it has "the same meaning" in their various idiolects simply does not arise. If a word is linguistically associated with a necessary and sufficient condition in the way that 'bachelor' is, then that sort of question *can* arise; but it does not arise, for example, in the case of proper names, and it does not arise, for a similar reason, in the case of physical magnitude terms. Thus if you know that 'Quine' is a name and I know that 'Quine' is a name and, in addition, we both refer to the same person when we use the word (even if the causal chains linking us to the referent are quite different) then the question of whether 'Quine' has the same meaning in my idiolect and in yours does not arise. More precisely: if the referent is the same, and we both associate the same minimal linguistic information with the word 'Quine', namely that it is a person's name, then the word is treated as the same word whether it occurs in your idiolect or in mine. Similarly, 'electricity' is the same word in Ben Franklin's idiolect and in mine. Of course, if you had wrong linguistic ideas about the name 'Quine' – for example, if you thought 'Quine' was a female name (not just that Quine was a woman, but that the name was restricted to females) – then there would be a difference in meaning.

This account stresses causal descriptions because physical magnitudes are invariably discovered through their effects, and so the natural way to first single out a physical magnitude is as the magnitude responsible for certain effects. Of course, the words 'responsible', 'causes', etc., do not literally have to occur in the description: *spin*, for example, was introduced by describing it as a physical magnitude having half-integral values characteristic of certain elementary particles, and giving a *law* connecting it with magnitudes previously introduced; I intend the notion of a causal description to include this case. And it is not a "necessary truth" that the description introducing a new physical magnitude should involve a notion of cause or law; but I am not trying in this paper to state "necessary truths".

Once the term 'electricity' has been introduced into someone's vocabulary (or into his "idiolect", as the dialect of a single speaker is called) whether by an introducing event, or by his learning the word from someone who learned it via an introducing event, or by his learning the word from someone linked by a chain of such transmissions to an introducing event, the referent in that person's idiolect is also fixed, even if no knowledge that that person has fixed it. And once the referent is fixed, one can use the word to formulate any number of theories about that referent (and even to formulate theoretical definitions of that referent which may be correct or incorrect scientific characterizations of that referent), without the word's being in any sense a different word in different theories. Thus the account just given fullfils the desideratum with which we started – it makes such terms as 'electricity' trans-theoretical. The "operational criteria" you can give for the presence of electricity will depend strongly on what theory you accept; but, without the illicit identification of meaning with operational criteria, it does not follow at all that *meaning* depends on the theory you accept.

The possibility of formulating definite descriptions (or even misdescriptions) of physical magnitudes depends upon the availability in our language of such "broad spectrum" notions as *physical magnitude* and *causes*; that these play a crucial role in the introduction of physical magnitude terms was argued in a previous paper.[5] In that paper, however, I did not distinguish between *defining* what I then called theoretical terms and *introducing* them. Of course, if we have available a language in which we can formulate descriptions of the referents of our various physical magnitude terms, then we can consider the various theories that we have containing those terms as so many different systems of sentences in that one language. To the extent that we can do this, we can treat the notions of reference and truth appropriate to that language as trans-theoretical notions also.

B. *Kripke's Theory of Proper Names*

I have already acknowledged a heavy indebtedness to Kripke's (unpublished) work on proper names. Since I have heard mainly secondhand reports of that work, I shall not attempt to describe it here in any great detail. But, as it has come down to me, the key idea is that a person may use a proper name to refer to a thing or person X even though he has *no* true beliefs about X. For example, suppose someone asks me who Quine is, and I falsely tell him that Quine was a Roman emperor. If he believes me, and if he goes on to use the word 'Quine' with the intention of referring to the person to whom *I* refer as Quine, then he will say such things as "Quine was a Roman emperor" – and he will be referring to a contemporary logician. Of course, he still has some true beliefs about Quine (beyond the belief that Quine is or was a person); for example, that Quine is or was named 'Quine'; but Kripke has more elaborate examples to show that even this is not always the case. On Kripke's view, the essential thing is this: that the use of a proper name to refer involves the existence of a causal chain of a certain kind connecting the user of the name (and the particular event of his using the name) to the bearer of the name.

Now then, I do not feel that one should be quite as liberal as Kripke is with respect to the causal chains one allows. I do not see much point, for example, in saying that someone is referring to Quine when he uses the name 'Quine' if he thinks that "Quine" was a Roman emperor, and that is all he "knows" about Quine; unless one has *some* beliefs about the bearer of the name which are true or approximately true, then it is at best idle to consider that the name refers to that bearer in one's idiolect. But what seems right about Kripke's account is that the knowledge an individual user of a language has need not at all fix the reference of the proper names in that individual's idiolect; the reference is fixed by the fact that that individual is causally linked to other individuals who were in a position to pick out the bearer of the name, or of some names from which the name descended. Indeed, what is important about Kripke's theory is not that the use of proper names is "causal" – what is not ? – but that the use of proper names is *collective*. Anyone who uses a proper name to refer is, in a sense, a nember of a collective which had "contact" with the bearer of the name: If it is surprising that a particular member of the collective need not have had such contact, and need not even have any good idea of the bearer of the name, it is only surprising because we think of language as private property.

The relationship of this theory of Kripke's to the above theory of physical magnitude terms should be obvious. Indeed, one might say that physical magnitude terms *are* proper names: they are proper names of *magnitudes* not *things* – however, this would be wrong, I think, since some physical magnitude terms (e.g. 'heat') are linguistically associated with rather rich information about the referent. The inportant thing about proper names is that it would be ridiculous to think that having linguistic competence can be equated in their case with knowledge of a necessary and sufficient condition – thus one is led to search for something other than the knowledge of the speaker which fixes the referent in their case.

It will be noted that I required a causal chain from the use of the phys-

ical magnitude term back to an introducing event – not back to an event
in which the physical magnitude played a significant role. The reason is
that, although no one in practice is going to be in a position to give a
definite description of a physical magnitude unless he is causally connected to such an event, the nature of *that* causal chain seems not to matter.
As long as one is in a positon to give a definite description (or even a
misdescription), one is in a position to introduce the term; and the chain
from there on is something about which much more definite statements
can be made. (In my opinion, it would be good to make a similar modification in Kripke's theory of proper names.)

C. *Natural Kind Words*

In an earlier paper[6] I presented an account of natural kind words (e.g.
'lemon') which has some relation to the present account of physical
magnitude terms. I suggested that anyone who has linguistic competence in connection with 'lemon' satisfies three conditions: (1) He
has implicit knowledge of such facts as the fact that 'lemon' is a
concrete noun, that it is the "name of a fruit", etc. – information given by
classifing the word under certain natural syntactic and semantic "markers". I criticized Jerrold Katz for the view that natural systems of semantic markers can enable us to give the exact meaning of each term (or
of *any* natural kind term); but *some* of the information associated with a
word can naturally be represented by classifying the word under such
familiar headings as 'noun', 'concrete', etc. (2) He associates the word
with a certain "stereotype" – yellow color, tart taste, thick peel, etc. (3)
He uses the word to *refer* to a certain natural kind – say, a natural kind of
fruit whose most essential feature, from a biologist's point of view, might
be a certain kind of DNA.

Two points were most important in the argument of that paper. The
first was that the properties mentioned in the stereotype (and, I would add,
the properties indicated by the semantic markers) are not being analytically predicated of each member of the extension, or, indeed, of any members
of the extension. It is not analytic that all tigers have stripes, nor that
some tigers have stripes; it is not analytic that all lemons are yellow, nor
that some lemons are yellow; it is not even analytic that tigers are animals
or that lemons are fruits. The stereotype is *associated* with the word; it is
not a necessary and sufficient condition for membership in the correspon-

ding class, nor even for being a normal member of the corresponding class. Engels' example of the word 'fish' fits right in here: what Engels was pointing out was precisely that the stereotype associated with the term 'fish' even in scientific, as opposed to lay, usage is not a necessary and sufficient condition. The second point was that speakers must be referring to a particular natural kind for us to treat them as using the same word 'lemon', or 'aluminum', or whatever. The weakness of that paper, apart from being very poorly organized and presented, is that nothing positive is said about the conditions under which a speaker who uses a word (say 'aluminum' or 'elm tree') is referring to one set of things rather than another. Clearly, the speaker who uses the word 'aluminum' need not be able to tell aluminum from molybdenum, and the speaker who use the term 'elm tree' cannot tell elm trees from beech trees if he happens to be me. But then what does determine the reference of the terms 'aluminum', and 'molybdenum' in my idiolect? In the previous paper, I suggested that the reference is fixed by a test known to experts; it now seems to me that this is just a special case of my use being causally connected to an introducing event. For natural kind words too, then, linguistic competence is a matter of knowledge plus causal connection to introducing events (and ultimately to members of the natural kind itself). And this is so for the same reason as in the case of physical magnitude terms; namely, that the use of a natural kind word involves in many cases membership in a "collective" which has contact with the natural kind, which knows of tests for membership in the natural kind, etc., only as a collective. The idea that linguistic competence in connection with a natural kind word involves more than just having the right extension or reference (where this is now explained via a causal account), but also associating the right stereotype seems to me to carry over to physical magnitude words. Natural kind words can be associated with "strong" stereotypes (stereotypes that give a strong picture of a stereotypical member – even to the point of enabling one to tell, in most cases, if something belongs to the natural kind), as in the case of 'lemon' or 'tiger', or with "weak" stereotypes (stereotypes that give no idea of what a sufficient condition for membership in the class would be) as in the case of 'molybdenum' or (unless I am a very atypical speaker) 'elm'. Similarly, it seems to me that the physical magnitude term 'temperature' is associated with a very strong stereotype, and 'electricity' with a weak one.

D. *Objections and Questions*

It is obvious that the account presented here must face certain hard questions. Without attempting to think of all of them myself, I should like to list a few that may help to launch discussion.

(1) One question that must be faced by all causal theories of meaning is how to make more precise the notion of a causal chain of the appropriate kind. How precisely can we describe the sorts of causal chains that must exist from one use of a word to a later use of the same word if we are to say that the referent or referents are the same in the two cases? And how much of a defect in these sorts of theories is it if one cannot be more precise on this point?

(2) It may seem counterintuitive that a natural kind word such as 'horse' is sharply distinguished from a term for a fictitious or non-existent natural kind such as 'unicorn', and that a physical magnitude term such as 'electricity' is sharply distinguished from a term for a fictitious or non-existent physical magnitude or substance such as "phlogiston". Indeed, I myself believe that if unicorns were found to exist and people began to discover facts about them, give non-obvious definite descriptions or approximately correct descriptions of the class of unicorns, etc., then the linguistic character of the word unicorn would change; and similarly with 'phlogiston'; but this is certain to be controversial.

(3) Some people will argue that definitions of such terms as 'electricity' (or, more precisely, 'charge') are crucial in the exact sciences, and further that such definitions should be regarded as *meaning stipulations*. I agree with the first part of this – that definitions are important in science, provided one remembers what Quine has pointed out, that "definition" is relative to a particular text or presentation, and that there is no such thing, in general, as the definition of a term "in physics" or "in biology" – only the definition in X, Y, or Z's presentation or axiomatization. I disagree with the last part – that "definitions" in science are meaning stipulations – but, again, this is certain to be controversial.

(4) Finally, there will be objections to my use of causal notions, from Humeans who expect them to be reduced away, and to my use of the term 'physical magnitude' from extensionalists and nominalists. Here I can only plead guilty to the belief that talk about what causes what, or what the laws of nature are, or what would happen if other things hap-

pened is *not* highly derived talk about mere regularities, and to the belief that the real world requires for its description not only reference to things but reference to physical magnitudes [7] – in a sense of 'physical magnitude' in which physical magnitudes exist contingently, not as a matter of logical necessity, and in which magnitudes can be synthetically identical (e.g., temperature is the same magnitude as mean molecular kinetic energy).

III. WHY POSITIVISTIC THEORY OF SCIENCE IS WRONG

My contention in this paper is not that what is wrong with positivist theory of science is positivist theory of meaning. What is wrong with positivist theory of science today, as it was wrong with the Machian theories that Lenin criticized in 1908, [8] is that it is based on an idealist or idealist-tending world view, and that that world view does not correspond to reality. However, the idealist element in contemporary positivism enters precisely through the theory of meaning; thus part of any realist critique of positivism has to include at least a sketch of rival theory. In the present section, I want to turn from the task of sketching such a rival theory, which was just completed, to the task of showing that positivistic theory of explanation broadly construed – that is, positivist theory of scientific theory – does not correspond to reality any better than the older and less sophisticated idealist theories to which it is historically the successor.

Let us for a moment review some of those older theories. The oldest theory is Bishop Berkeley's. Here one already meets what might be called the *adequacy claim*: that is, the claim that a convinced Berkelian is *entitled* to accept standard scientific theory and practice, that Berkeley can give an account of the scientific method which would justify this. Indeed, I have heard philosophers argue that acceptance of Berkeley's metaphysics would not make any difference to the scientific theories one would accept. Here one already meets an important ambiguity. One can be claiming that a Berkelian can make the move of "accepting" scientific theory in some sense other than accepting as true or approximately-true: say, accepting as a useful prediction heuristic. If this is what one means, then the claim is trivial. To be sure, Berkeley can "accept" Newtonian physics in the Pickwickian sense of "accept" as a useful scheme for making predictions. But Berkeley, to do him justice, was interested in much more: What he claimed was that an idealist could *reinterpret* (only he would not consider it *re*-interpretation, but rather *correct* interpretation) the notion of object so as to square both the layman's and the scientist's talk of objects with the idealist claim that reality consists of minds and their sensations ("spirits" and their "ideas"). The difference between the two claims is the difference between accepting the idea that social practice is the test of truth and rejecting it, between accepting the idea that the overwhelming success of scientific theory offers some reason for accepting that theory as true or approximately-true, and claiming that success in practice is *no* indication of truth. Machian positivism fails for the same reason that Berkelian idealism does: although Mach makes the claim that his construction of the world out of sensations ("Empfindungen") is compatible with lay and scientific object-talk, no demonstration at all is given that this is so. The first philosopher to both precisely state and to undertake the task of *translating* thing-language into phenomenalistic language was Carnap (in *Logische Aufbau der Welt*). And what does Carnap do? He devotes the entire book to *preliminaries*, to "reconstructions" *within* sensationalistic language (i.e. reductions of some sensation-concepts to others, not of thing-concepts to sensation-concepts), and then in the last chapter gives a sketch of the relation of thing-language to sensation-language which is *not* a translation, and which, indeed, amounts to no more than the old claim that we pick the thing-theory that is "simplest" and most useful. In short, no demonstration is given at all that the positivist is entitled to quantify over (or refer to) material things.

It is with the failure of the phenomenalist translation enterprise, that is, with the failure to find *any* interpretation of object-concepts under which the prima facie incompatibility between an idealist world-view and a materialist world-view, between a world consisting of "spirits and their ideas", or of "Empfindungen", or of total experience-slices in one "specious present", and a world consisting of fields and particles, simply *disappears* – it is with this failure that contemporary positivistic philosophy of science begins. Basically, two moves were made by the positivists after the failure of phenomenalist translation. The first was to give up construing scientific theories as systems of statements each of which had to have an intelligible interpretation (intelligible from the standpoint of what was taken as "completely understood" or "fully interpreted"), and to construe them rather as mere calculi, whose objective was to give successful predictions and otherwise to be as "simple" as possible. "Scientific theories are partially interpreted calculi". [9] The second move was to shift from phenomenalist language to "observable thing language" as one's reduction-base – i.e., to say that one was seeking an interpretation or "partial interpretation" of physical theory in "observable thing language", not in "sensationalistic language".

The second move may make it appear questionable whether positivism is still correctly characterized as an "idealist" tendency – i.e., as a tendency which regards or tends to regard the "hard facts" as just facts about actual and potential *experiences*, and all other talk as somehow just highly derived talk about actual and potential experiences. I, myself, think this characterization *is* still fundamentally correct despite the shift to "observable thing predicates" for two reasons: (1) The cut between observable things and "theoretical entities" was historically introduced as a substitute for the thing/sensation dichotomy. Indeed, the reduction of "theoretical entities" to "observable things and qualities" would hardly seem to be a natural problem to someone who did not have in the back of his head the older problem of reduction to *sensations*. The reduction of things to sensations is both a historically motivated problem and one which rests upon the sharpness of the distinction between a material thing and a sensation (of course, even this sharpness is partly an illusion, on a materialist view – substitute "material process" for material thing!), as well as the supposed "certainty" one has concerning one's own sensations. But the reduction of electrons to tables and chairs, or, more generally, of "unobservable" things to "observable" things is not historically motivated, the distinction is not sharp (Grover Maxwell asked years ago if a dust mote is something "given" when it is just big enough to see and a "construct" when it is just too small to see – can the distinction between data and constructs be a matter of size?), and one is not supposed to have certainty concerning observable things. (2) The positivists themselves frequently say that one could carry their analysis back down to the level of sensations, and that stopping with "observable thing predicates" is a matter of *convenience*.[10]

In the remainder of this section I want to show that the first move – construing scientific theories as partially interpreted calculi – does not solve the adequacy problem at all. The positivist today is no more entitled than Berkeley was to accept scientific theory and practice – that is, his own story leads to no reason to think either that scientific theory is true, or that scientific practice tends to discover truth. In a sense, this is immediate. The positivist does not claim that scientific theory is "true" in any trans-theoretic sense of "true"; the only trans-theoretic notions he has are of the order of "leads to successful prediction" and "is simple". Like the Berkelian, he has to fall back on the position that scientific theory is *useful* rather than true or approximately-true. But he does try to provide some account of the acceptability of scientific theories, even some account of their "interpretation". And he wants to maintain that in some sense the principle on which Marx says realist philosophy of science rests – that social practice is the test of truth, that the success of scientific theories is reason to think they are true or approximately-true – is right. What I want to show is that the notion of "truth" that the positivist can give us is not the one on which scientific practice is based.

A. *Truth.*

When a realistically minded scientist – that is to say, a scientist *whose practice* is realistic, not one whose official "philosophy of science" is realistic – accepts a theory, he accepts it as true (or probably true, or approximately-true, or probably approximately-true). Since he also accepts *logic*[11] he knows that certain moves *preserve truth*. For example, if he accepts a theory T_1 as true and he accepts a theory T_2 as true, then he knows that T_1 & T_2 – the *conjunction* of T_1 and T_2 – is also true, by logic, and so he accepts T_1 & T_2. If we talk about probability, we have to say that if T_1 is very highly, probably true and T_2 is very highly probably true, then the conjunction T_1 & T_2 is also highly probable (though not *as* highly as the conjuncts separately), provided that T_1 is not negatively relevant to T_2 – i.e., provided that T_2 is not only highly probable on the evidence, but also no less probable on the added assumption of T_1 (this is a judgement that must be made on the basis of what T_1 *says* and of background knowledge, of course). If we talk about approximate-truth, then we have to say that the approximations probably involved in T_1 and T_2 need to be compatible for us to pass from the approximate-truth of T_1 and T_2 to the approximate-truth of their conjunction. None of these matters is at all deep, from a realist point of view. But even if we confine ourselves to the simplest case, the case in which we can neglect the chances of error and the presence of approximations, and treat the acceptance of T_1 and T_2 as simply the acceptance of them as true, I want to suggest that the move from this acceptance to the acceptance of the conjunction is one to which one is not entitled on positivist philosophy of science. One of the simplest moves that scientists daily make, a move they make as a matter of propositional logic, a move which is central if scientific inquiry is to have any *cumulative* character at all, is totally *arbitrary* if positivist philosophy of science is right.

The difficulty is very simple. Acceptance of T_1, for a positivist, means acceptance of the calculus T_1 as leading to succesful predictions (i.e., all *observation sentences* which are theorems of T_1 are true; not all *sentences* which are theorems of T_1 are "true" in any fixed trans-theoretic sense). Similarly, the acceptance of T_2 means the acceptance of T_2 as leading to successful predictions. But from the fact that T_1 leads to successful predictions and the fact that T_2 leads to successful predictions it does not follow at all that the conjunction T_1 & T_2 leads to successful predictions. The difficulty, in a nutshell, is that the predicate which plays the *role* of truth – the predicate "leads to successful predictions" – does not have the *properties* of truth. The positivist may teach in his philosophy seminar that acceptance of a scientific theory is acceptance of it as "simple and leading to true predictions", and then go out and do science (or his students may go out and do science) by verifying theories T_1 and T_2, conjoining theories which have been previously verified, etc. – but then there is just as great a discrepancy between what he teaches in his philosophy seminar and his *practice* as there was between Berkely's teaching that the world consisted of spirits and their ideas and continuing in practice to daily rely on the material object conceptual system.

Nor does it help to bring in "simplicity". It is not obvious that the conjunction of simple theories is simple; and even if simplicity is preserved by conjunction, the conjunction of simple theories which separately lead to no false predictions may even be *inconsistent* (examples are easy to construct). More sophisticated moves have indeed been made. Thus, for Carnap truth of a theory is the same as truth of its "Ramsey sentence"[12]. But exactly the same objection applies: "truth of the Ramsey sentence" does not have the properties of truth: if T_1 has a true Ramsey sentence and T_2 has a true Ramsey sentence it does not at all follow that the conjunction does.

(For those readers familiar with Carnaps' use of the Hilbert epsilon-symbol, it may be pointed out that the difficulty comes out in very sharp form in Carnap's symbolization of his interpretation of individual theoretical terms. Thus let $T_1(P)$, $T_2(P)$ be two theories containing exactly one theoretical term P. On Carnap's own symbolization of his view[12], what P means in T_1 is $\varepsilon PT_1(P)$; what P means in T_2 is $\varepsilon PT_2(P)$; and what P means in T_1 & T_2 is $\varepsilon P[T_1(P)$ & $T_2(P)]$; this makes it explicit that P has different meanings in T_1 and T_2 and *yet a third meaning* in their conjunction.)

B. *Simplicity.*

It is easy to construct a "theory" in the positivist sense (a calculus containing some observation terms) which leads to no false predictions but which no scientist would dream of accepting. This is usually handled by saying that scientists only choose "simple" theories. Also, a simple theory may mess up science as a whole: So it is said that scientists are trying to maximize the simplicity of "total science". "Theory" means, then, "formalization of total science, or of some piece which is independent of the rest of total science". Unfortunately, no one has ever written down or ever will write down a "theory" in this sense. The fact is, that positivist philosophy of science depends on a constant slide between giving the impression that one is talking about "theories" in the customary sense – Newton's theory, Maxwell's theory, Darwin's theory, Mendel's theory – and saying, at key points of difficulty such as the one just alluded to, that one is *really* talking about a "formalization of total science", or some such thing.

The difficulty with the rule "choose the simplest theory compatible with the evidence" is that it is probably not *right*, or would probably not be right, even if one *could* formalize "total science" (at a given time). Scientists are not trying to maximize some formal property of "simplicity"; they are trying to maximize *truth* (or improve their approximation to truth, or increase the amount of approximate-truth they know without decreasing the goodness of the approximation, and so forth).

Of course, a realist might accept the rule "chose the simplest hypothesis", if it could be shown that the simplest hypothesis is always the most *probable* on the basis of the rest of his knowledge. But this is not so on any usual measure of simplicity. For example, suppose I know just three points on interstate highway 40, and those three points lie on a straight line. Suppose also that the statement 'IS 40 is straight' is logically consistent with my total knowledge. Then accepting 'IS 40 is straight' would, on the usual simplicity metrics, be accepting the simplest hypothesis. Yet I would not in fact accept 'IS 40 straight', nor would anyone with our background knowledge. Given that every other interstate highway has curves, and given the enormous length of IS 40 and the enormous impracticality of making a straight highway across the entire U.S., it is overwhelmingly probable that IS 40 is *not* straight.

Can we not say that my *total* "knowledge" is less simple if I accept 'IS 40 is straight'? Not, it seems to me, on the basis of any criterion of *simplicity* that I know of. What is obviously involved here is not *simplicity* but plausibility: What introducing the word "simplicity" does is make it look as if a calculation which is in fact the calculation of the probability of a state of affairs is in reality just a calculation of a formal property (such as number of argument places, number of primitive symbols, length and number of the axioms, perhaps shape of the curves mentioned) of an uninterpreted or semi-interpreted *calculus*. Even if the property of being the most probable hypothesis on background knowledge could be *represented* syntactically, omitting to mention that the representing property was the syntactic representation of a *probability measure*, and pretending that it was *just* a formal property (like having simple axioms), would be a way of disguising rather than revealing what was going on.

C. *Confirmation.*

Indeed, positivist philosophers of science have made attempts at formalizing the logic of confirmation. These attempts are interesting (though so far unsuccessful) researches on *any* philosophy of science. But not only do they have nothing to do with positivist theory of meaning; they are in fact *incompatible* with it. Thus when they write about meaning, positivists tell us that "theoretical terms" have different meanings in different theories; when they formalize confirmation theory, they invariably treat theories as systems of sentences in *one* language, and assume that all semantical concepts are *trans*-theoretic. Thus the positivists are engaged in formalizing *realistic* confirmation theory; not the confirmation theory (if there is one!) to which their own theory of meaning should lead.

What is going on here should be evident from Carnap's work on the foundations of mathematics. Carnap has a consistent tendency to *identify* concepts with their syntactic representations: thus, mathematical truth with theoremhood (after the discovery of Gödel's theorem, he either allowed "non-constructive rules of proof", or simply assumed set theory, and took "logical consequence" rather than derivability as the basic notion, although this trivialized the "analysis" of mathematical truth). In the same way he would have liked to identify a state of affairs having a probability of, say, .9, with the corresponding sentences having a c-value of .9 (where "c" would be a syntactically defined measure on sentences in a formalized language). Even if Carnap had found a successful "c-function", the fact is that it would have been successful because it corresponded to a reasonable probability measure over some collection of states of affairs; but this is just what Carnap's positivism did not allow him to say.

D. *Auxiliary Hypotheses*

Sometimes, as we mentioned, the positivists make it explicit that the "theories" to which their theory of science applies are "formalizations of total science", and not theories in the usual sense; but their readers do, I think, tend to come away with the impression that their model *is* a model of a scientific theory in the usual sense – especially, a physical theory. Believing this involves believing that a physical theory is a calculus, or could easily be formalized as a calculus, and that its predictions are *self-contained* – that they are deduced from the explicitly stated assumptions of the theory itself. This leads to a comparison with social sciences which is derogatory to the social sciences – for the classic social science theories are clearly *not* self-contained in this sense. For example, when Marxists write that the capitalist class controls the state – that the army and the police intervene on the side of capitalists, that the politicians who have a chance of obtaining state power through elections are tied to capitalists, etc. – what they do is make a series of generalizations about ways in which this control allegedly takes place: what happens in elections, when the army intervenes, etc. But these generalizations do not lead to predictions about specific political events (or, not to predictions in the positivist's "observation language"), without very substantive auxiliary assumptions. These auxiliary assumptions, if they could be spelled out, would be numerous and would be highly context dependent. Thus Marxist theory must either be ignored or treated as not a scientific theory at all, but only a theory sketch, and likewise for such classic theories as those of Weber or even Mill. In short, the positivist attitude tends to be that social science is science only when and to the extent that it apes *physics*. And this for the reason that the mathematical model of a scientific theory provided by the positivists is thought to clearly fit *physical* theories.

But, in fact, it fits physical theories very badly, and this for the reason that even physical theories in the usual sense – e.g., Newton's Theory of Universal Gravitation, Maxwell's theory – lead to no predictions at all without a host of auxiliary assumptions, and moreover without auxiliary assumptions that are not at all law-like, but that are, in fact, assumptions about boundary conditions and initial conditions in the case of particular systems. Thus, if the claim that the term 'gravitation', for example, had a meaning which depended on the theory were true, and the theory included such auxiliary assumptions as that "space is a hard vacuum", and "there is no tenth planet in the solar system", then it would follow that discovery that space is *not* a hard vacuum or even that there is a tenth planet would change the meaning of 'gravitation'. I think one has to be pretty idealistic in one's intuitions to find this at all plausible! It is not so implausible that knowledge of the meaning of the term 'gravitation' involves some know-ledge of the theory (although I think that this is wrong: the stereotype associated with 'gravitation' is not nearly as strong as a particular theory of gravitation), and this is probably what most readers think of when they encounter the claim that physical magnitude terms (usually called "theo-retical terms" to prejudge just the issue this paper discusses) are "theory loaded"; but the actual meaning-dependence required by positivist mean-ing theory would be a dependence not just on the *laws* of the theory, but on the particular auxiliary assumptions – for, if these are not counted as part of the theory, then the whole theory-prediction scheme collapses at the outset.

Finally, neglect of the role that auxiliary assumptions actually play in science leads to a wholly incorrect idea of how a scientific theory is con-firmed. Newton's theory of gravitation was not confirmed by checking predictions derived from it plus some set of auxiliary statements fixed in advance; rather the auxiliary assumptions had to be continually modified and expanded in the history of Celestial Mechanics. That scientific problems as often have the form of finding auxiliary hypotheses as they do of finding and checking predictions is something that has been too much neglected in philosophy of science;[13] this neglect is largely the result of the acceptance of the positivist model and its uncritical application to actual physical theories.

Harvard University

NOTES

[1] Cf. Lenin (1970) and Engels (1959).
[2] In a letter written to Conrad Schmidt in 1895; cf. Marx and Engels (1942), pp. 527–30.
[3] Cf. Shapere (1969).
[4] See, for example, the discussion by Hempel (1965), pp. 217–18. A contrasting view is sketched in Putnam (1962).
[5] Putnam (1962).
[6] Putnam (1970a).
[7] Cf. Putnam (1970b).
[8] Cf. Lenin (1970).
[9] Putnam (1962).
[10] E.g., Carnap says this on p. 63 in Carnap (1956).
[11] The role of logic in empirical science is discussed in Putnam (1971) and Putnam (1969).
[12] For details see Hempel (1965).
[13] I discuss this in 'The Corroboration of Theories', to appear in the forthcoming Popper volume in the *Library of Living Philosophers* series.

BIBLIOGRAPHY

Carnap, R., 1956, 'The Methodological Character of Theoretical Concept's, in *Minnesota Studies in the Philosophy of Science*, Vol. I (ed. by H. Feigl), Minneapolis.
Engels, F., 1959, *Anti-Dühring, Herr Eugen Dühring's Revolution in Science*, New-York.
Hempel, C., 1965, 'The Theoretician's Dilemma', in *Aspects of Scientific Explanation*, New York.
Lenin, V., 1970, *Materialism and Emperio-Criticism*, New York.
Marx, K. and Engels, F., *Selected Correspondence 1846–1895*, New York.
Putnam, H., 1962, 'What Theories Are Not', in *Logic, Methodology and Philosophy of Science* (ed. by E. Nagel, P. Suppes, and A. Tarski), Stanford.
Putnam, H., 1969, 'Is Logic Empirical?', in *Boston Studies in the Philosophy of Science*, Vol. V (ed. by R. Cohen and M. Wartofsky), New York.
Putnam, H., 1970a, 'Is Semantics Possible?', in *Language, Belief and Metaphysics*, Vol. I of *Contemporary Philosophical Thought: The International Philosophy Year Conferences at Brockport*, Albany.

Putnam, H., 1970b, 'On Properties', in *Essays in Honor of Carl G. Hempel* (ed. by N. Rescher), D. Reidel, Dordrecht, pp. 235–254.

Putnam, H., 1971, *Philosophy of Logic*, New York.

Shapere, D., 1969, 'Towards a Post-Positivistic Interpretation of Science', in *The Legacy of Logical Positivism* (ed. by P. Achenstein and S. Barker), Baltimore.

A CONFUTATION OF CONVERGENT REALISM*

LARRY LAUDAN†

University of Pittsburgh

This essay contains a partial exploration of some key concepts associated with the epistemology of realist philosophies of science. It shows that neither reference nor approximate truth will do the explanatory jobs that realists expect of them. Equally, several widely-held realist theses about the nature of inter-theoretic relations and scientific progress are scrutinized and found wanting. Finally, it is argued that the history of science, far from confirming scientific realism, decisively confutes several extant versions of avowedly 'naturalistic' forms of scientific realism.

> The positive argument for realism is that it is the only philosophy that doesn't make the success of science a miracle.
>
> -H. Putnam (1975)

1. The Problem. It is becoming increasingly common to suggest that epistemological realism is an empirical hypothesis, grounded in, and to be authenticated by its ability to explain the workings of science. A growing number of philosophers (including Boyd, Newton-Smith, Shimony, Putnam, Friedman and Niiniluoto) have argued that the theses of epistemic realism are open to empirical test. The suggestion that epistemological doctrines have much the same empirical status as the sciences is a welcome one: for, whether it stands up to detailed scrutiny or not, it marks a significant facing-up by the philosophical community to one of the most neglected (and most notorious) problems of philosophy: the status of epistemological claims.

But there are potential hazards as well as advantages associated with the 'scientizing' of epistemology. Specifically, once one concedes that epistemic doctrines are to be tested in the court of experience, it is possible that one's favorite epistemic theories may be refuted rather than confirmed. It is the thesis of this paper that precisely such a fate afflicts a form of realism advocated by those who have been in the vanguard of

*Received July 1980; Revised October 1980.

†I am indebted to all of the following for clarifying my ideas on these issues and for saving me from some serious errors: Peter Achinstein, Richard Burian, Clark Glymour, Adolf Grünbaum, Gary Gutting, Allen Janis, Lorenz Krüger, James Lennox, Andrew Lugg, Peter Machamer, Nancy Maull, Ernan McMullin, Ilkka Niiniluoto, Nicholas Rescher, Ken Schaffner, John Worrall, Steven Wykstra.

Philosophy of Science, 48 (1981) pp. 19–49.

the move to show that realism is supported by an empirical study of the development of science. Specifically, I shall show that epistemic realism, at least in certain of its extant forms, is neither supported by, nor has it made sense of, much of the available historical evidence.

2. Convergent Realism. Like other philosophical -*isms*, the term 'realism' covers a variety of sins. Many of these will not be at issue here. For instance, 'semantic realism' (in brief, the claim that all theories have truth values and that some theories—we know not which—are true) is not in dispute. Nor shall I discuss what one might call 'intentional realism' (i.e., the view that theories are generally intended by their proponents to assert the existence of entities corresponding to the terms in those theories). What I shall focus on instead are certain forms of *epistemological* realism. As Hilary Putnam has pointed out, although such realism has become increasingly fashionable, "very little is said about what realism *is*" (1978). The lack of specificity about what realism asserts makes it difficult to evaluate its claims, since many formulations are too vague and sketchy to get a grip on. At the same time, any efforts to formulate the realist position with greater precision lay the critic open to charges of attacking a straw man. In the course of this paper, I shall attribute several theses to the realists. Although there is probably no realist who subscribes to all of them, most of them have been defended by some self-avowed realist or other; taken together, they are perhaps closest to that version of realism advocated by Putnam, Boyd and Newton-Smith. Although I believe the views I shall be discussing can be legitimately attributed to certain contemporary philosophers (and will frequently cite the textual evidence for such attributions), it is not crucial to my case that such attributions can be made. Nor will I claim to do justice to the complex epistemologies of those whose work I will criticize. My aim, rather, is to explore certain epistemic claims which those who are realists might be tempted (and in some cases have been tempted) to embrace. If my arguments are sound, we will discover that some of the most intuitively tempting versions of realism prove to be chimeras.

The form of realism I shall discuss involves variants of the following claims:

R1) Scientific theories (at least in the 'mature' sciences) are typically approximately true and more recent theories are closer to the truth than older theories in the same domain;

R2) The observational and theoretical terms within the theories of a mature science genuinely refer (roughly, there are substances in the world that correspond to the ontologies presumed by our best theories);

R3) Successive theories in any mature science will be such that they 'preserve' the theoretical relations and the apparent referents of earlier theories (i.e., earlier theories will be 'limiting cases' of later theories).[1]

R4) Acceptable new theories do and should explain why their predecessors were successful insofar as they were successful.

To these semantic, methodological and epistemic theses is conjoined an important meta-philosophical claim about how realism is to be evaluated and assessed. Specifically, it is maintained that:

R5) Theses (R1)–(R4) entail that ('mature') scientific theories should be successful; indeed, these theses constitute the best, if not the only, explanation for the success of science. The empirical success of science (in the sense of giving detailed explanations and accurate predictions) accordingly provides striking empirical confirmation for realism.

I shall call the position delineated by (R1) to (R5) *convergent epistemological realism*, or CER for short. Many recent proponents of CER maintain that (R1), (R2), (R3), and (R4) are empirical hypotheses which, via the linkages postulated in (R5), can be tested by an investigation of science itself. They propose two elaborate abductive arguments. The structure of the first, which is germane to (R1) and (R2), is something like this:

1. If scientific theories are approximately true, they will typically be empirically successful;
2. If the central terms in scientific theories genuinely refer, those theories will generally be empirically successful;
3. Scientific theories are empirically successful.

4. (Probably) Theories are approximately true and their terms genuinely refer.

The argument relevant to (R3) is of slightly different form, specifically:

1. If the earlier theories in a 'mature' science are approximately true and if the central terms of those theories genuinely refer, then later more successful theories in the same science will preserve the earlier theories as limiting cases;
2. Scientists seek to preserve earlier theories as limiting cases and generally succeed.

---

[1]Putnam, evidently following Boyd, sums up (R1) to (R3) in these words:

"1) Terms in a mature science typically *refer*.

2) The laws of a theory belonging to a mature science are typically approximately true . . . I will only consider [new] theories . . . which have this property—[they] contain the [theoretical] laws of [their predecessors] as a limiting case" (1978, pp. 20–21).

3. (Probably) Earlier theories in a 'mature' science are approximately true and genuinely referential.

Taking the success of present and past theories as givens, proponents
of CER claim that *if* CER were true, it would follow that the success and
the progressive success of science would be a matter of course. Equally,
they allege that if CER were false, the success of science would be
'miraculous' and without explanation.[2] Because (on their view) CER explains the fact that science is successful, the theses of CER are thereby
confirmed by the success of science and non-realist epistemologies are
discredited by the latter's alleged inability to explain both the success of
current theories and the progress which science historically exhibits.

As Putnam and certain others (e.g., Newton-Smith) see it, the fact that
statements about reference (R2, R3) or about approximate truth (R1, R3)
function in the explanation of a contingent state of affairs, establishes
that "the notions of 'truth' and 'reference' have a causal explanatory role
in epistemology" (Putnam 1978, p. 21).[3] In one fell swoop, both epistemology and semantics are 'naturalized' and, to top it all off, we get an
explanation of the success of science into the bargain!

The central question before us is whether the realist's assertions about
the interrelations between truth, reference and success are sound. It will
be the burden of this paper to raise doubts about both I and II. Specifically, I shall argue that *four* of the five premises of those abductions are
either false or too ambiguous to be acceptable. I shall also seek to show
that, even if the premises were true, they would not warrant the conclusions which realists draw from them. Sections 3 through 5 of this essay
deal with the first abductive argument; section 6 deals with the second.

3. Reference and Success. The specifically referential side of the 'empirical' argument for realism has been developed chiefly by Putnam, who
talks explicitly of reference rather more than most realists. On the other
hand, reference is usually implicitly smuggled in, since most realists subscribe to the (ultimately referential) thesis that "the world probably contains entities very like those postulated by our most successful theories."

If R2 is to fulfill Putnam's ambition that reference can explain the success of science, and that the success of science establishes the presump-
tive truth of R2, it seems he must subscribe to claims similar to these:

---

[2]Putnam insists, for instance, that if the realist is wrong about theories being referential,
then "the success of science is a miracle". (Putnam 1975, p. 69).
[3]Boyd remarks: "scientific realism offers an *explanation* for the legitimacy of ontological commitment to theoretical entities" (Putnam 1978, Note 10, p. 2). It allegedly does
so by explaining why theories containing theoretical entities work so well: because such
entities genuinely exist.

S1) The theories in the advanced or mature sciences are successful;
S2) A theory whose central terms genuinely refer will be a successful theory;
S3) If a theory is successful, we can reasonably infer that its central terms genuinely refer;
S4) All the central terms in theories in the mature sciences do refer.

There are complex interconnections here. (S2) and (S4) explain (S1), while (S1) and (S3) provide the warrant for (S4). Reference explains success and success warrants a presumption of reference. The arguments are plausible, given the premises. But there is the rub, for with the possible exception of (S1), none of the premises is acceptable.

The first and toughest nut to crack involves getting clearer about the nature of that 'success' which realists are concerned to explain. Although Putnam, Sellars and Boyd all take the success of certain sciences as a given, they say little about what this success amounts to. So far as I can see, they are working with a largely *pragmatic* notion to be cashed out in terms of a theory's workability or applicability. On this account, we would say that a theory is successful if it makes substantially correct predictions, if it leads to efficacious interventions in the natural order, if it passes a battery of standard tests. One would like to be able to be more specific about what success amounts to, but the lack of a coherent theory of confirmation makes further specificity very difficult.

Moreover, the realist must be wary—at least for these purposes—of adopting too strict a notion of success, for a highly robust and stringent construal of 'success' would defeat the realist's purposes. What he wants to explain, after all, is why science in general has worked so well. If he were to adopt a very demanding characterization of success (such as those advocated by inductive logicians or Popperians) then it would probably turn out that science has been largely 'unsuccessful' (because it does not have high confirmation) and the realist's avowed explanandum would thus be a non-problem. Accordingly, I shall assume that a theory is 'successful' so long as it has worked well, i.e., so long as it has functioned in a variety of explanatory contexts, has led to confirmed predictions and has been of broad explanatory scope. As I understand the realist's position, his concern is to explain why certain theories have enjoyed this kind of success.

If we construe 'success' in this way, (S1) can be conceded. Whether one's criterion of success is broad explanatory scope, possession of a large number of confirming instances, or conferring manipulative or predictive control, it is clear that science is, by and large, a successful activity.

What about (S2)? I am not certain that any realist would or should endorse it, although it is a perfectly natural construal of the realist's claim that 'reference explains success'. The notion of reference that is involved here is highly complex and unsatisfactory in significant respects. Without endorsing it, I shall use it frequently in the ensuing discussion. The realist sense of reference is a rather liberal one, according to which the terms in a theory may be genuinely referring even if many of the claims the theory makes about the entities to which it refers are false. Provided that there are entities which "approximately fit" a theory's description of them, Putnam's charitable account of reference allows us to say that the terms of a theory genuinely refer.[4] On this account (and these are Putnam's examples), Bohr's 'electron', Newton's 'mass', Mendel's 'gene', and Dalton's 'atom' are all referring terms, while 'phlogiston' and 'aether' are not (Putnam 1978, pp. 20–22).

Are genuinely referential theories (i.e., theories whose central terms genuinely refer) invariably or even generally successful at the empirical level, as (S2) states? There is ample evidence that they are not. The chemical atomic theory in the 18th century was so remarkably unsuccessful that most chemists abandoned it in favor of a more phenomenological, elective affinity chemistry. The Proutian theory that the atoms of heavy elements are composed of hydrogen atoms had, through most of the 19th century, a strikingly unsuccessful career, confronted by a long string of apparent refutations. The Wegenerian theory that the continents are carried by large subterranean objects moving laterally across the earth's surface was, for some thirty years in the recent history of geology, a strikingly unsuccessful theory until, after major modifications, it became the geological orthodoxy of the 1960s and 1970s. Yet all of these theories postulated basic entities which (according to Putnam's 'principle of charity') genuinely exist.

The realist's claim that we should expect referring theories to be empirically successful is simply false. And, with a little reflection, we can see good reasons why it should be. To have a genuinely referring theory is to have a theory which "cuts the world at its joints", a theory which postulates entities of a kind that really exist. But a genuinely referring theory need not be such that all—or even most—of the specific claims it makes about the properties of those entities and their modes of interaction are true. Thus, Dalton's theory makes many claims about atoms which are false; Bohr's early theory of the electron was similarly flawed in important respects. Contra-(S2), genuinely referential theories need not be strikingly successful, since such theories may be 'massively false' (i.e., have far greater falsity content than truth content).

---

[4] Whether one utilizes Putnam's earlier or later versions of realism is irrelevant for the central arguments of this essay.

(S2) is so patently false that it is difficult to imagine that the realist need be committed to it. But what else will do? The (Putnamian) realist wants attributions of reference to a theory's terms to function in an explanation of that theory's success. The simplest and crudest way of doing that involves a claim like (S2). A less outrageous way of achieving the same end would involve the weaker,

(S2') A theory whose terms refer will usually (but not always) be successful.

Isolated instances of referring but unsuccessful theories, sufficient to refute (S2), leave (S2') unscathed. But, if we were to find a broad range of referring but unsuccessful theories, that would be evidence against (S2'). Such theories can be generated at will. For instance, take any set of terms which one believes to be genuinely referring. In any language rich enough to contain negation, it will be possible to construct indefinitely many unsuccessful theories, all of whose substantive terms are genuinely referring. Now, it is always open to the realist to claim that such 'theories' are not really theories at all, but mere conjunctions of isolated statements—lacking that sort of conceptual integration we associate with 'real' theories. Sadly a parallel argument can be made for genuine theories. Consider, for instance, how many inadequate versions of the atomic theory there were in the 2000 years of atomic 'speculating', before a genuinely successful theory emerged. Consider how many unsuccessful versions there were of the wave theory of light before the 1820s, when a successful wave theory first emerged. Kinetic theories of heat in the seventeenth and eighteenth century, developmental theories of embryology before the late nineteenth century sustain a similar story. (S2'), every bit as much as (S2), seems hard to reconcile with the historical record.

As Richard Burian has pointed out to me (in personal communication), a realist might attempt to dispense with both of those theses and simply rest content with (S3) alone. Unlike (S2) and (S2'), (S3) is not open to the objection that referring theories are often unsuccessful, for it makes no claim that referring theories are always or generally successful. But (S3) has difficulties of its own. In the first place, it seems hard to square with the fact that the central terms of many relatively successful theories (e.g., aether theories, phlogistic theories) are evidently non-referring. I shall discuss this tension in detail below. More crucial for our purposes here is that (S3) is *not strong enough* to permit the realist to utilize reference to explain success. Unless genuineness of reference entails that all or most referring theories will be successful, then the fact that a theory's terms refer scarcely provides a convincing explanation of that theory's success. If, as (S3) allows, many (or even most) referring theories can be unsuccessful, how can the fact that a successful theory's terms refer be taken to explain why it is successful? (S3) may or may not be true; but in either case it arguably gives the realist no explanatory access to scientific success.

A more plausible construal of Putnam's claim that reference plays a role in explaining the success of science involves a rather more indirect argument. It might be said (and Putnam does say this much) that we can explain why a theory is successful by assuming that the theory is true or approximately true. Since a theory can only be true or nearly true (in any sense of those terms open to the realist) if its terms genuinely refer, it might be argued that reference gets into the act willy-nilly when we explain a theory's success in terms of its truth(like) status. On this account, reference is piggy-backed on approximate truth. The viability of this indirect approach is treated at length in section 4 below so I shall not discuss it here except to observe that if the only contact point between reference and success is provided through the medium of approximate truth, then the link between reference and success is extremely tenuous.

What about (S3), the realist's claim that success creates a rational presumption of reference? We have already seen that (S3) provides no explanation of the success of science, but does it have independent merits? The question specifically is whether the success of a theory provides a warrant for concluding that its central terms refer. Insofar as this is—as certain realists suggest—an empirical question, it requires us to inquire whether past theories which have been successful are ones whose central terms genuinely referred (according to the realist's own account of reference).

A proper empirical test of this hypothesis would require extensive sifting of the historical record of a kind that is not possible to perform here. What I can do is to mention a range of once successful, but (by present lights) non-referring, theories. A fuller list will come later (see section 5), but for now we shall focus on a whole family of related theories, namely, the subtle fluids and aethers of 18th and 19th century physics and chemistry.

Consider specifically the state of aetherial theories in the 1830s and 1840s. The electrical fluid, a substance which was generally assumed to accumulate on the surface rather than permeate the interstices of bodies, had been utilized to explain *inter alia* the attraction of oppositely charged bodies, the behavior of the Leyden jar, the similarities between atmospheric and static electricity and many phenomena of current electricity. Within chemistry and heat theory, the caloric aether had been widely utilized since Boerhaave (by, among others, Lavoisier, Laplace, Black, Rumford, Hutton, and Cavendish) to explain everything from the role of heat in chemical reactions to the conduction and radiation of heat and several standard problems of thermometry. Within the theory of light, the optical aether functioned centrally in explanations of reflection, refraction, interference, double refraction, diffraction and polarization. (Of more than passing interest, optical aether theories had also made some very startling predictions, e.g., Fresnel's prediction of a bright spot at the center of the shadow of a circular disc; a surprising prediction which, when tested, proved correct. If that does not count as empirical success, nothing does!) There were also gravitational (e.g., LeSage's) and physiological (e.g., Hartley's) aethers which enjoyed some measure of empirical success. It would be difficult to find a family of theories in this period which were as successful as aether theories; compared to them, 19th century atomism (for instance), a genuinely referring theory (on realist accounts), was a dismal failure. Indeed, on any account of empirical success which I can conceive of, non-referring 19th-century aether theories were more successful than contemporary, referring atomic theories. In this connection, it is worth recalling the remark of the great theoretical physicist, J. C. Maxwell, to the effect that the aether was better confirmed than any other theoretical entity in natural philosophy!

What we are confronted by in 19th-century aether theories, then, is a wide variety of once successful theories, whose central explanatory concept Putnam singles out as a prime example of a non-referring one (Putnam 1978, p. 22). What are (referential) realists to make of this historical case? On the face of it, this case poses two rather different kinds of challenges to realism: (1) it suggests that (S3) is a dubious piece of advice in that *there can be* (and have been) *highly successful theories some central terms of which are non-referring;* and (2) it suggests that *the realist's claim that he can explain why science is successful is false at least insofar as a part of the historical success of science has been success exhibited by theories whose central terms did not refer.*

But perhaps I am being less than fair when I suggest that the realist is committed to the claim that *all* the central terms in a successful theory refer. It is possible that when Putnam, for instance, says that "terms in a mature [or successful] science typically refer" (Putnam 1978, p. 20), he only means to suggest that *some* terms in a successful theory or science genuinely refer. Such a claim is fully consistent with the fact that certain other terms (e.g., 'aether') in certain successful, mature sciences (e.g., 19th-century physics) are nonetheless non-referring. Put differently, the realist might argue that the success of a theory warrants the claim that at least some (but not necessarily all) of its central concepts refer.

Unfortunately, such a weakening of (S3) entails a theory of evidential support which can scarcely give comfort to the realist. After all, part of what separates the realist from the positivist is the former's belief that the evidence for a theory is evidence for *everything* which the theory asserts. Where the stereotypical positivist argues that the evidence selectively confirms only the more 'observable' parts of a theory, the realist generally asserts (in the language of Boyd) that:

> the sort of evidence which ordinarily counts in favor of the acceptance of a scientific law or theory is, ordinarily, evidence for the (at least approximate) truth of the law or theory as an account of the causal relations obtaining between the entities ["observation or theoretical"] quantified over in the law or theory in question. (Boyd 1973, p. 1)[5]

For realists such as Boyd, either all parts of a theory (both observational and non-observational) are confirmed by successful tests or none are. In general, realists have been able to utilize various holistic arguments to insist that it is not merely the lower-level claims of a well-tested theory which are confirmed but its deep-structural assumptions as well. This tactic has been used to good effect by realists in establishing that inductive support 'flows upward' so as to authenticate the most 'theoretical' parts of our theories. Certain latter-day realists (e.g., Glymour) want to break out of this holist web and argue that certain components of theories can be 'directly' tested. This approach runs the very grave risk of undercutting what the realist desires most: a rationale for taking our deepest-structure theories seriously, and a justification for linking reference and success. After all, if the tests to which we subject our theories only test *portions* of those theories, then even highly successful theories may well have central terms which are non-referring and central tenets which, because untested, we have no grounds for believing to be approximately true. Under those circumstances, a theory might be highly successful and yet contain important constituents which were patently false. Such a state of affairs would wreak havoc with the realist's presumption (R1) that success betokens approximate truth. In short, to be less than a holist about theory testing is to put at risk precisely that predilection for deep-structure claims which motivates much of the realist enterprise.

There is, however, a rather more serious obstacle to this weakening of referential realism. It is true that by weakening (S3) to only certain terms in a theory, one would immunize it from certain obvious counter-examples. But such a maneuver has debilitating consequences for other central realist theses. Consider the realist's thesis (R3) about the retentive character of inter-theory relations (discussed below in detail). The realist both recommends as a matter of policy and claims as a matter of fact that successful theories are (and should be) rationally replaced only by the-

---

[5]See also p. 3: "experimental evidence for a theory is evidence for the truth of even its non-observational laws". See also (Sellars 1963, p. 97).

ories which preserve reference for the central terms of their successful
predecessors. The rationale for the normative version of this retentionist
doctrine is that the terms in the earlier theory, *because it was successful*,
must have been referential and thus a constraint on any successor to that
theory is that reference should be retained for such terms. This makes
sense just in case success provides a blanket warrant for presumption of
reference. But if (S3) were weakened so as to say merely that it is reasonable to assume that *some* of the terms in a successful theory genuinely
refer, then the realist would have no rationale for his retentive theses
(variants of R3), which have been a central pillar of realism for several
decades.[6]

Something apparently has to give. A version of (S3) strong enough to
license (R3) seems incompatible with the fact that many successful theories contain non-referring central terms. But any weakening of (S3) di-
lutes the force of, and removes the rationale for, the realist's claims about
convergence, retention and correspondence in inter-theory relations.[7] If
the realist once concedes that some unspecified set of the terms of a successful theory may well not refer, then his proposals for restricting "the
class of candidate theories" to those which retain reference for the *prima
facie* referring terms in earlier theories is without foundation. (Putnam
1975, p. 22)

More generally, we seem forced to say that such linkages as there are
between reference and success are rather murkier than Putnam's and
Boyd's discussions would lead us to believe. If the realist is going to
make his case for CER, it seems that it will have to hinge on approximate
truth, (R1), rather than reference, (R2).

4. Approximate Truth and Success: the 'Downward Path'. Ignoring
the referential turn among certain recent realists, most realists continue
to argue that, at bottom, epistemic realism is committed to the view that
successful scientific theories, even if strictly false, are nonetheless 'ap-

---

[6]A caveat is in order here. *Even if* all the central terms in some theory refer, it is not
obvious that every rational successor to that theory must preserve all the referring terms
of its predecessor. One can easily imagine circumstances when the new theory is preferable
to the old one even though the range of application of the new theory is less broad than
the old. When the range is so restricted, it may well be entirely appropriate to drop reference to some of the entities which figured in the earlier theory.

[7]For Putnam and Boyd both "it will be a constraint on T_2 [i.e., any new theory in a
domain] . . . that T_2 must have this property, the property that *from its standpoint* one
can assign referents to the terms of T_1 [i.e., an earlier theory in the same domain]" (Putnam
1978, p. 22). For Boyd, see (1973, p. 8): "new theories should, *prima facie*, resemble
current theories with respect to their accounts of causal relations among theoretical entities".

231

proximately true' or 'close to the truth' or 'verisimilar'.[8] The claim generally amounts to this pair:

(T1) if a theory is approximately true, then it will be explanatorily successful; and

(T2) if a theory is explanatorily successful, then it is probably approximately true.

What the realist would *like* to be able to say, of course, is:

(T1') if a theory is true, then it will be successful.

(T1') is attractive because self-evident. But most realists balk at invoking (T1') because they are (rightly) reluctant to believe that we can reasonably presume of any given scientific theory that it is true. If all the realist could explain was the success of theories which were true *simpliciter*, his explanatory repertoire would be acutely limited. As an attractive move in the direction of broader explanatory scope, (T1) is rather more appealing. After all, presumably many theories which we believe to be false (e.g., Newtonian mechanics, thermodynamics, wave optics) were—and still are—highly successful across a broad range of applications.

Perhaps, the realist evidently conjectures, we can find an *epistemic* account of that pragmatic success by assuming such theories to be 'approximately true'. But we must be wary of this potential sleight of hand. It may be that there is a connection between success and approximate truth; *but if there is such a connection it must be independently argued for*. The acknowledgedly uncontroversial character of (T1') must not be surreptitiously invoked—as it sometimes seems to be—in order to establish (T1). When (T1')'s antecedent is appropriately weakened by speaking of approximate truth, it is by no means clear that (T1) is sound.

Virtually all the proponents of epistemic realism take it as unproblematic that if a theory were approximately true, it would deductively follow that the theory would be a relatively successful predictor and explainer of observable phenomena. Unfortunately, few of the writers of whom I am aware have defined what it means for a statement or theory to be 'approximately true'. Accordingly, it is impossible to say whether the

---

[8]For just a small sampling of this view, consider the following: "The claim of a realist ontology of science is that the only way of explaining why the models of science function so successfully . . . is that they approximate in some way the structure of the object" (McMullin 1970, pp. 63–64); "the continued success [of confirmed theories] can be explained by the hypothesis that they are in fact close to the truth . . . " (Niiniluoto forthcoming, p. 21); the claim that "the laws of a theory belonging to a mature science are typically approximately *true* . . . [provides] an *explanation* of the behavior of scientists and the success of science" (Putnam 1978, pp. 20–21). Smart, Sellars, and Newton-Smith, among others, share a similar view.

alleged entailment is genuine. This reservation is more than perfunctory. **Indeed**, on the best known account of what it means for a theory to be **approximately** true, it does *not* follow that an approximately true theory **will** be explanatorily successful.

Suppose, for instance, that we were to say in a Popperian vein that a **theory**, T_1, is approximately true if its truth content is greater than its **falsity** content, i.e.,

$$Ct_T(T_1) >> Ct_F(T_1).^9$$

(Where $Ct_T(T_1)$ is the cardinality of the set of true sentences entailed by T_1 and $Ct_F(T_1)$ is the cardinality of the set of false sentences entailed by T_1.) When approximate truth is so construed, it does *not* logically follow **that** an arbitrarily selected class of a theory's entailments (namely, some **of** its observable consequences) will be true. Indeed, it is entirely con-**ceivable** that a theory might be approximately true in the indicated sense **and** yet be such that *all* of its thus far tested consequences are *false*.[10]

Some realists concede their failure to articulate a coherent notion of **approximate** truth or verisimilitude, but insist that this failure in no way **compromises** the viability of (T1). Newton-Smith, for instance, grants **that** "no one has given a satisfactory analysis of the notion of verisimi-**litude**" (forthcoming, p. 16), but insists that the concept can be legiti-**mately** invoked "even if one cannot at the time give a philosophically **satisfactory** analysis of it." He quite rightly points out that many scien-**tific** concepts were explanatorily useful long before a philosophically co-**herent** analysis was given for them. But the analogy is unseemly, for **what** is being challenged is not whether the concept of approximate truth **is** philosophically rigorous but rather whether it is even clear enough for **us** to ascertain whether it entails what it purportedly explains. Until some-

---

[9]Although Popper is generally careful not to assert that actual historical theories exhibit **ever** increasing truth content (for an exception, see his (1963, p. 220)), other writers have **been** more bold. Thus, Newton-Smith writes that "the historically generated sequence of **theories** of a mature science" is a sequence in which succeeding theories are increasing **in** truth content without increasing in falsity content" (forthcoming, p. 2).

[10]On the more technical side, Niiniluoto has shown that a theory's degree of corrobor-**ation** co-varies with its "estimated verisimilitude" (1977, pp. 121–147 and forthcoming). **Roughly** speaking, 'estimated truthlikeness' is a measure of how closely (the content of) **a** theory corresponds to *what we take to be* the best conceptual systems that we so far have **been** able to find (1980, pp. 443ff.). If Niiniluoto's measures work it follows from the **above**-mentioned co-variance that an empirically successful theory will have a high degree **of** estimated truthlikeness. But because estimated truthlikeness and genuine verisimilitude **are** not necessarily related (the former being parasitic on existing evidence and available **conceptual** systems), it is an open question whether—as Niiniluoto asserts—the continued **success** of highly confirmed theories can be *explained* by the hypothesis that they in fact **are** close to the truth at least in the relevant respects. Unless I am mistaken, this remark **of** his betrays a confusion between 'true verisimilitude' (to which we have no epistemic **access**) and 'estimated verisimilitude' (which is accessible but non-epistemic).

one provides a clearer analysis of approximate truth than is now available, it is not even clear whether truth-likeness would explain success, let alone whether, as Newton-Smith insists, "the concept of verisimilitude is *required* in order to give a satisfactory theoretical explanation of an aspect of the scientific enterprise." If the realist would de-mystify the 'miraculousness' (Putnam) or the 'mysteriousness' (Newton-Smith[11]) of the success of science, he needs more than a promissory note that somehow, someday, someone will show that approximately true theories must be successful theories.[12]

Whether there is some definition of approximate truth which does indeed entail that approximately true theories will be predictively successful (and yet still probably false) is not clear.[13] What can be said is that, promises to the contrary notwithstanding, *none* of the proponents of realism has yet articulated a coherent account of approximate truth which entails that approximately true theories will, across the range where we can test them, be successful predictors. Further difficulties abound. Even if the realist had a semantically adequate characterization of approximate or partial truth, and even if that semantics entailed that most of the consequences of an approximately true theory would be true, he would still be without any criterion that would *epistemically* warrant the ascription of approximate truth to a theory. As it is, the realist seems to be long on intuitions and short on either a semantics or an epistemology of approximate truth.

These should be urgent items on the realists' agenda since, until we have a coherent account of what approximate truth is, central realist theses like (R1), (T1) and (T2) are just so much mumbo-jumbo.

5. Approximate Truth and Success: the 'Upward Path'. Despite the doubts voiced in section **4**, let us grant for the sake of argument that if a theory is approximately true, then it will be successful. Even granting (T1), is there any plausibility to the suggestion of (T2) that explanatory

---

[11]Newton-Smith claims that the increasing predictive success of science through time "would be totally mystifying . . . if it were not for the fact that theories are capturing more and more truth about the world" (forthcoming, p. 15).

[12]I must stress again that I am *not* denying that there *may* be a connection between approximate truth and predictive success. I am only observing that until the realists show us what that connection is, they should be more reticent than they are about claiming that realism can explain the success of science.

[13]A *non-realist* might argue that a theory is approximately true just in case all its *observable* consequences are true or within a specified interval from the true value. Theories that were "approximately true" in this sense would indeed be demonstrably successful. But, the realist's (otherwise commendable) commitment to taking seriously the theoretical claims of a theory precludes him from utilizing any such construal of approximate truth, since he wants to say that the theoretical as well as the observational consequences are approximately true.

success can be taken as a rational warrant for a judgment of approximate truth? The answer seems to be "no".

To see why, we need to explore briefly one of the connections between 'genuinely referring' and being 'approximately true'. However the latter is understood, I take it that *a realist would never want to say that a theory was approximately true if its central theoretical terms failed to refer.* If there were nothing like genes, then a genetic theory, no matter how well confirmed it was, would not be approximately true. If there were no entities similar to atoms, no atomic theory could be approximately true; if there were no sub-atomic particles, then no quantum theory of chemistry could be approximately true. In short, a necessary condition—especially for a scientific realist—for a theory being close to the truth is that its central explanatory terms genuinely refer. (An *instrumentalist*, of course, could countenance the weaker claim that a theory was approximately true so long as its directly testable consequences were close to the observable values. But as I argued above, the realist must take claims about approximate truth to refer alike to the observable and the deep-structural dimensions of a theory.)

Now, what the history of science offers us is a plethora of theories which were both successful and (so far as we can judge) non-referential with respect to many of their central explanatory concepts. I discussed earlier one specific family of theories which fits this description. Let me add a few more prominent examples to the list:

— the crystalline spheres of ancient and medieval astronomy;
— the humoral theory of medicine;
— the effluvial theory of static electricity;
— 'catastrophist' geology, with its commitment to a universal (Noachian) deluge;
— the phlogiston theory of chemistry;
— the caloric theory of heat;
— the vibratory theory of heat;
— the vital force theories of physiology;
— the electromagnetic aether;
— the optical aether;
— the theory of circular inertia;
— theories of spontaneous generation.

This list, which could be extended *ad nauseam*, involves in every case a theory which was once successful and well confirmed, but which contained central terms which (we now believe) were non-referring. Anyone who imagines that the theories which have been successful in the history of science have also been, with respect to their central concepts, genuinely referring theories has studied only the more 'whiggish' versions of the history of science (i.e., the ones which recount only those past theories which are referentially similar to currently prevailing ones).

It is true that proponents of CER sometimes hedge their bets by suggesting that their analysis applies exclusively to 'the mature sciences' (e.g., Putnam and Krajewski). This distinction between mature and immature sciences proves convenient to the realist since he can use it to dismiss any *prima facie* counter-example to the empirical claims of CER on the grounds that the example is drawn from an 'immature' science. But this insulating maneuvre is unsatisfactory in two respects. In the first place, it runs the risk of making CER vacuous since these authors generally define a mature science as one in which correspondence or limiting case relations obtain invariably between any successive theories in the science once it has passed 'the threshold of maturity'. Krajewski grants the tautological character of this view when he notes that "the thesis that there is [correspondence] among successive theories becomes, indeed, analytical" (1977, p. 91). Nonetheless, he believes that there is a version of the maturity thesis which "may be and must be tested by the history of science". That version is that "every branch of science crosses at some period the threshold of maturity". But the testability of this hypothesis is dubious at best. There is no historical observation which could conceivably *refute* it since, even if we discovered that no sciences yet possessed 'corresponding' theories, it could be maintained that eventually every science will become corresponding. It is equally difficult to *confirm* it since, even if we found a science in which corresponding relations existed between the latest theory and its predecessor, we would have no way of knowing whether that relation will continue to apply to subsequent changes of theory in that science. In other words, the much-vaunted empirical testability of realism is seriously compromised by limiting it to the mature sciences.

But there is a second unsavory dimension to the restriction of CER to the 'mature' sciences. The realists' avowed aim, after all, is to explain why science is successful: that is the 'miracle' which they allege the non-realists leave unaccounted for. The fact of the matter is that parts of science, including many 'immature' sciences, have been successful for a very long time; indeed, many of the theories I alluded to above were empirically successful by any criterion I can conceive of (including fertility, intuitively high confirmation, successful prediction, etc.). If the realist restricts himself to explaining only how the 'mature' sciences work (and recall that very few sciences indeed are yet 'mature' as the realist sees it), then he will have completely failed in his ambition to explain why science in general is successful. Moreover, several of the examples I have cited above come from the history of mathematical physics in the last century (e.g., the electromagnetic and optical aethers) and, as Putnam himself concedes, "*physics* surely counts as a 'mature' science if any science does" (1978, p. 21). Since realists would presumably insist that many of the central terms of the theories enumerated above do not genuinely refer, it follows that none of those theories could be approximately true (recalling that the former is a necessary condition for the latter). Accordingly, cases of this kind cast very grave doubts on the plausibility of (T2), i.e., the claim that nothing succeeds like approximate truth.

I daresay that for every highly successful theory in the past of science which we now believe to be a genuinely referring theory, one could find half a dozen once successful theories which we now regard as substantially non-referring. If the proponents of CER are the empiricists they profess to be about matters epistemological, cases of this kind and this frequency should give them pause about the well-foundedness of (T2).

But we need not limit our counter-examples to non-referring theories. There were many theories in the past which (so far as we can tell) were both genuinely referring and empirically successful which we are nonetheless loathe to regard as approximately true. Consider, for instance, virtually all those geological theories prior to the 1960s which denied any lateral motion to the continents. Such theories were, by any standard, highly successful (and apparently referential); but would anyone today be prepared to say that their constituent theoretical claims—committed as they were to laterally stable continents—are almost true? Is it not the fact of the matter that structural geology was a successful science between (say) 1920 and 1960, even though geologists were fundamentally mistaken about many—perhaps even most—of the basic mechanisms of tectonic construction? Or what about the chemical theories of the 1920s which assumed that the atomic nucleus was structurally homogenous? Or those chemical and physical theories of the late 19th century which explicitly assumed that matter was neither created nor destroyed? I am aware of no sense of approximate truth (available to the realist) according to which such highly successful, but evidently false, theoretical assumptions could be regarded as 'truthlike'.

More generally, the realist needs a riposte to the *prima facie* plausible claim that there is no necessary connection between increasing the accuracy of our deep-structural characterizations of nature and improvements at the level of phenomenological explanations, predictions and manipulations. It *seems* entirely conceivable intuitively that the theoretical mechanisms of a new theory, T_2, might be closer to the mark than those of a rival T_1 and yet T_1 might be more accurate at the level of testable predictions. In the absence of an argument that greater correspondence at the level of unobservable claims is more likely than not to reveal itself in greater accuracy at the experimental level, one is obliged to say that the realist's hunch that increasing deep-structural fidelity must

manifest itself pragmatically in the form of heightened experimental accuracy has yet to be made cogent. (Equally problematic, of course, is the
inverse argument to the effect that increasing experimental accuracy betokens greater truthlikeness at the level of theoretical, i.e., deep-structural, commitments.)

6. Confusions About Convergence and Retention. Thus far, I have
discussed only the static or synchronic versions of CER, versions which
make absolute rather than relative judgments about truthlikeness. Of
equal appeal have been those variants of CER which invoke a notion of
what is variously called convergence, correspondence or cumulation.
Proponents of the diachronic version of CER supplement the arguments
discussed above ((S1)-(S4) and (T1)-(T2)) with an additional set. They
tend to be of this form:

C1) If earlier theories in a scientific domain are successful and thereby,
according to realist principles (e.g.,(S3)above), approximately true, then
scientists should only accept later theories which retain appropriate portions of earlier theories;

C2) As a matter of fact, scientists do adopt the strategy of (C1) and
manage to produce new, more successful theories in the process;

C3) The 'fact' that scientists succeed at retaining appropriate parts of
earlier theories in more successful successors shows that the earlier theories did genuinely refer and that they were approximately true. And thus,
the strategy propounded in (C1) is sound.[14]

Perhaps the prevailing view here is Putnam's and (implicitly) Popper's,
according to which rationally-warranted successor theories in a 'mature'
science must (a) contain reference to the entities apparently referred
in the predecessor theory (since, by hypothesis, the terms in the earlier
theory refer), and (b) contain the 'theoretical laws' and 'mechanisms' of
the predecessor theory as limiting cases. As Putnam tells us, a 'realist'
should insist that *any* viable successor to an old theory T_1 must "contain
the laws of T_1 as a limiting case" (1978, p. 21). John Watkins, a like-
minded convergentist, puts the point this way:

> It typically happens in the history of science that when some hitherto
> dominant theory T is superceded by T^1, T^1 is in the relation of cor-
> respondence to T [i.e., T is a 'limiting case' of T^1] (1978, pp.
> 376–377).

---

[14]If this argument, which I attribute to the realists, seems a bit murky, I challenge any
reader to find a more clear-cut one in the literature! Overt formulations of this position
can be found in Putnam, Boyd and Newton-Smith.

Numerous recent philosophers of science have subscribed to a similar view, including Popper, Post, Krajewski, and Koertge.[15] This form of retention is not the only one to have been widely discussed. Indeed, realists have espoused a wide variety of claims about what is or should be retained in the transition from a once successful predecessor (T_1) to a successor (T_2) theory. Among the more important forms of realist retention are the following cases: (1) T_2 entails T_1 (Whewell); (2) T_2 retains the true consequences or truth content of T_1 (Popper); (3) T_2 retains the 'confirmed' portions of T_1 (Post, Koertge); (4) T_2 preserves the theoretical laws and mechanisms of T_1 (Boyd, McMullin, Putnam); (5) T_2 preserves T_1 as a limiting case (Watkins, Putnam, Krajewski); (6) T_2 explains why T_1 succeeded insofar as T_1 succeeded (Sellars); (7) T_2 retains reference for the central terms of T_1 (Putnam, Boyd).

The question before us is whether, when retention is understood in *any* of these senses, the realist's theses about convergence and retention are correct.

6.1 *Do Scientists Adopt the 'Retentionist' Strategy of CER?* One part of the convergent realist's argument is a claim to the effect that scientists generally adopt the strategy of seeking to preserve earlier theories in later ones. As Putnam puts it:

preserving the *mechanisms* of the earlier theory as often as possible, which is what scientists try to do . . . That scientists try to do this . . . is a fact, and that this strategy has led to important discoveries . . . is also a fact (1978, p. 20).[16]

In a similar vein, Szumilewicz (although not stressing realism) insists that many eminent scientists made it a main heuristic requirement of their

[15]Popper: "a theory which has been well corroborated can only be superseded by one . . . [which] *contains* the old well-corroborated theory—or at least a good approximation to it" (1959, p. 276).
Post: "I shall even claim that, as a matter of empirical historical fact, [successor] theories [have] always explained the *whole* of [the well-confirmed part of their predecessors]" (1971, p. 229).
Koertge: "nearly all pairs of successive theories in the history of science stand in a correspondence relation and . . . where there is no correspondence to begin with, the new theory will be developed in such a way that it comes more nearly into correspondence with the old" (1973, p. 176–177). Among other authors who have defended a similar view, one should mention (Fine 1967, p. 231 ff.), (Kordig 1971, pp. 119–125), (Margenau 1950) and (Sklar 1967, pp. 190–224).
[16]Putnam fails to point out that it is also a fact that many scientists do *not* seek to preserve earlier mechanisms and that theories which have not preserved earlier theoretical mechanisms (whether the germ theory of disease, plate tectonics, or wave optics) have led to important discoveries is also a fact.

research programs that a new theory stand in a relation of 'correspondence' with the theory it supersedes (1977, p. 348). If Putnam and the other retentionists are right about the strategy which most scientists have adopted, we should expect to find the historical literature of science abundantly provided with (a) proofs that later theories do indeed contain earlier theories as limiting cases, or (b) outright rejections of later theories which fail to contain earlier theories. Except on rare occasions (coming primarily from the history of mechanics), one finds neither of these concerns prominent in the literature of science. For instance, to the best of my knowledge, literally no one criticized the wave theory of light because it did not preserve the theoretical mechanisms of the earlier corpuscular theory; no one faulted Lyell's uniformitarian geology on the grounds that it dispensed with several causal processes prominent in catastrophist geology; Darwin's theory was not criticized by most geologists for its failure to retain many of the mechanisms of Lamarckian 'evolutionary theory'.

For all the realist's confident claims about the prevalence of a retentionist strategy in the sciences, I am aware of *no* historical studies which would sustain as a *general* claim his hypothesis about the evaluative strategies utilized in science. Moreover, insofar as Putnam and Boyd claim to be offering "an explanation of the [retentionist] behavior of scientists" (Putnam 1978, p. 21), they have the wrong explanandum, for if there is any widespread strategy in science, it is one which says, "accept an empirically successful theory, regardless of whether it contains the theoretical laws and mechanisms of its predecessors".[17] Indeed, one could take a leaf from the realist's (C2) and claim that the success of the strategy of assuming that earlier theories do not generally refer shows that it is true that earlier theories generally do not!

(One might note in passing how often, and on what evidence, realists imagine that they are speaking for the scientific majority. Putnam, for instance, claims that "realism is, so to speak, 'science's philosophy of science' " and that "science taken at 'face value' *implies* realism" (1978, p. 37).[18] Hooker insists that to be a realist is to take science "seriously" (1976, pp. 467–472), as if to suggest that conventionalists, instrumentalists and positivists such as Duhem, Poincaré, and Mach did not take science seriously. The willingness of some realists to attribute realist strategies to working scientists—on the strength of virtually no empirical research into the principles which *in fact* have governed scientific practice—raises doubts about the seriousness of their avowed commitment to the empirical character of epistemic claims.)

---

[17] I have written a book about this strategy, (Laudan 1977).

[18] After the epistemological and methodological battles about science during the last three hundred years, it should be fairly clear that science, taken at its face value, *implies* no particular epistemology.

6.2 *Do Later Theories Preserve the Mechanisms, Models, and Laws of Earlier Theories?* Regardless of the explicit strategies to which scientists have subscribed, are Putnam and several other retentionists right that later theories "typically" entail earlier theories, and that "earlier theories are, very often, limiting cases of later theories".[19] Unfortunately, answering this question is difficult, since "typically" is one of those weasel words which allows for much hedging. I shall assume that Putnam and Watkins mean that "most of the time (or perhaps in most of the important cases) successor theories contain predecessor theories as limiting cases". So construed, the claim is patently false. Copernican astronomy did not retain all the key mechanisms of Ptolemaic astronomy (e.g., motion along an equant); Newton's physics did not retain all (or even most of) the 'theoretical laws' of Cartesian mechanics, astronomy and optics; Franklin's electrical theory did not contain its predecessor (Nollet's) as a limiting case. Relativistic physics did not retain the aether, nor the mechanisms associated with it; statistical mechanics does not incorporate all the mechanisms of thermodynamics; modern genetics does not have Darwinian pangenesis as a limiting case; the wave theory of light did not appropriate the mechanisms of corpuscular optics; modern embryology incorporates few of the mechanisms prominent in classical embryological theory. As I have shown elsewhere,[20] loss occurs at virtually every level: the confirmed predictions of earlier theories are sometimes not explained by later ones; even the 'observable' laws explained by earlier theories are not always retained, not even as limiting cases; theoretical processes and mechanisms of earlier theories are, as frequently as not, treated as flotsam.

The point is that some of the most important theoretical innovations have been due to a willingness of scientists to violate the cumulationist or retentionist constraint which realists enjoin 'mature' scientists to follow.

There is a deep reason why the convergent realist is wrong about these matters. It has to do, in part, with the role of ontological frameworks in science and with the nature of limiting case relations. As scientists use the term 'limiting case', T_1 can be a limiting case of T_2 only if (a) *all* the variables (observable and theoretical) assigned a value in T_1 are assigned a value by T_2 and (b) the values assigned to every variable of T_1 are the same as, or very close to, the values T_2 assigns to the corresponding variable when certain initial and boundary conditions—consistent with T_2[21]—are specified. This seems to require that T_1 can be a limiting case

[19](Putnam 1978, pp. 20, 123).
[20](Laudan 1976, pp. 467–472).
[21]This matter of limiting conditions consistent with the 'reducing' theory is curious. Some of the best-known expositions of limiting case relations depend (as Krajewski has

of T_2 only if *all* the entities postulated by T_1 occur in the ontology of T_2. Whenever there is a change of ontology accompanying a theory transition such that T_2 (when conjoined with suitable initial and boundary conditions) fails to capture T_1's ontology, then T_1 cannot be a limiting case of T_2. Even where the ontologies of T_1 and T_2 overlap appropriately (i.e., where T_2's ontology embraces all of T_1's), T_1 is a limiting case of T_2 only if *all* the laws of T_1 can be derived from T_2, given appropriate limiting conditions. It is important to stress that *both* these conditions (among others) must be satisfied before one theory can be a limiting case of another. Where 'closet positivists' might be content with capturing only the formal mathematical relations or only the observable consequences of T_1 within a successor, T_2, any genuine realist must insist that T_1's underlying ontology is preserved in T_2's, *for it is that ontology above all which he alleges to be approximately true.*

Too often, philosophers (and physicists) infer the existence of a limiting case relation between T_1 and T_2 on substantially less than this. For instance, many writers have claimed one theory to be a limiting case of another when only some, but not all, of the laws of the former are 'derivable' from the latter. In other cases, one theory has been said to be a limiting case of a successor when the mathematical laws of the former find homologies in the latter but where the former's ontology is not fully extractable from the latter's.

Consider one prominent example which has often been misdescribed, namely, the transition from the classical aether theory to relativistic and quantum mechanics. It can, of course, be shown that *some* 'laws' of classical mechanics are limiting cases of relativistic mechanics. But there are other laws and general assertions made by the classical theory (e.g., claims about the density and fine structure of the aether, general laws about the character of the interaction between aether and matter, models and mechanisms detailing the compressibility of the aether) which could not conceivably be limiting cases of modern mechanics. The reason is a simple one: a theory cannot assign values to a variable which does not occur in that theory's language (or, more colloquially, it cannot assign properties to entities whose existence it does not countenance). Classical

---

observed) upon showing an earlier theory to be a limiting case of a later theory only by adopting limiting assumptions *explicitly denied by the later theory.* For instance, several standard textbook discussions present (a portion of) classical mechanics as a limiting case of special relativity, provided c approaches infinity. But special relativity is committed to the claim that c is a constant. Is there not something suspicious about a 'derivation' of T_1 from a T_2 which essentially involves an assumption inconsistent with T_2? If T_2 is correct, then it forbids the adoption of a premise commonly used to derive T_1 as a limiting case. (It should be noted that most such proofs can be re-formulated unobjectionably, e.g., in the relativity case, by letting $v \rightarrow o$ rather than $c \rightarrow \infty$.)

aether physics contained a number of postulated mechanisms for dealing
inter alia with the transmission of light through the aether. Such mechanisms could not possibly appear in a successor theory like the special
theory of relativity which denies the very existence of an aetherial medium and which accomplishes the explanatory tasks performed by the
aether via very different mechanisms.

Nineteenth-century mathematical physics is replete with similar examples of evidently successful mathematical theories which, because
some of their variables refer to entities whose existence we now deny,
cannot be shown to be limiting cases of our physics. As Adolf Grünbaum
has cogently argued, when we are confronted with two incompatible theories, T_1 and T_2, such that T_2 does not 'contain' all of T_1's ontology, then
not all the mechanisms and theoretical laws of T_1 which involve those
entities of T_1 not postulated by T_2 can possibly be retained—not even as
limiting cases—in T_2 (1976, pp. 1–23). This result is of some significance. What little plausibility convergent or retentive realism has enjoyed
derives from the presumption that it correctly describes the relationship
between classical and post-classical mechanics and gravitational theory.
Once we see that even in this *prima facie* most favorable case for the
realist (where *some* of the laws of the predecessor theory are genuinely
limiting cases of the successor), changing ontologies or conceptual frameworks make it impossible to capture many of the central theoretical laws
and mechanisms postulated by the earlier theory, then we can see how
misleading is Putnam's claim that "what scientists try to do" is to preserve

> the *mechanisms* of the earlier theory as often as possible—or to show
> that they are 'limiting cases' of new mechanisms . . . (1978, p. 20).

Where the mechanisms of the earlier theory involve entities whose existence the later theory denies, no scientist does (or should) feel any com-
punction about wholesale repudiation of the earlier mechanisms.

But even where there is no change in basic ontology, many theories
(even in 'mature sciences' like physics) fail to retain all the explanatory
successes of their predecessors. It is well known that statistical mechanics
has yet to capture the irreversibility of macro-thermodynamics as a genuine limiting case. Classical continuum mechanics has not yet been reduced to quantum mechanics or relativity. Contemporary field theory has
yet to replicate the classical thesis that physical laws are invariant under
reflection in space. If scientists had accepted the realist's constraint
(namely, that new theories must have old theories as limiting cases), neither relativity nor statistical mechanics would have been viewed as viable
theories. It has been said before, but it needs to be reiterated over and
again: *a proof of the existence of limiting relations between selected com-*

ponents of two theories is a far cry from a systematic proof that one theory is a limiting case of the other. Even if classical and modern physics stood to one another in the manner in which the convergent realist erroneously imagines they do, his hasty generalization that theory successions in all the advanced sciences show limiting case relations is patently false.[22] But, as this discussion shows, not even the realist's paradigm case will sustain the claims he is apt to make about it.

What this analysis underscores is just how reactionary many forms of convergent epistemological realism are. If one took seriously CER's advice to reject any new theory which did not capture existing mature theories as referential and existing laws and mechanisms as approximately authentic, then any prospect for deep-structure, ontological changes in our theories would be foreclosed. Equally outlawed would be any significant repudiation of our theoretical models. In spite of his commitment to the growth of knowledge, the realist would unwittingly freeze science in its present state by forcing all future theories to accomodate the ontology of contemporary ('mature') science and by foreclosing the possibility that some future generation may come to the conclusion that some (or even most) of the central terms in our best theories are no more referential than was 'natural place', 'phlogiston', 'aether', or 'caloric'.

6.3 *Could theories converge in ways required by the realist?* These instances of violations in genuine science of the sorts of continuity usually required by realists are by themselves sufficient to show that the form of scientific growth which the convergent realist takes as his explicandum is often absent, even in the 'mature' sciences. But we can move beyond these specific cases to show in principle that the kind of cumulation demanded by the realist is unattainable. Specifically, by drawing on some results established by David Miller and others, the following can be shown:

a) the familiar requirement that a successor theory, T_2, must both preserve as true the true consequences of its predecessor, T_1, and explain T_1's anomalies is contradictory;

b) that if a new theory, T_2, involves a change in the ontology or conceptual framework of a predecessor, T_1, then T_1 will have true and determinate consequences not possessed by T_2;

c) that if two theories, T_1 and T_2, disagree, then each will have true and determinate consequences not exhibited by the other.

[22]As Mario Bunge has cogently put it: "The popular view on inter-theory relations . . . that every new theory includes (as regards its extension) its predecessors . . . is philosophically superficial, . . . and it is false as a historical hypothesis concerning the advancement of science" (1970, pp. 309–310).

In order to establish these conclusions, one needs to utilize a 'syntactic' view of theories according to which a theory is a conjunction of statements and its consequences are defined *à la* Tarski in terms of content classes. Needless to say, this is neither the only, nor necessarily the best, way of thinking about theories; but it happens to be the way in which most philosophers who argue for convergence and retention (e.g., Popper, Watkins, Post, Krajewski, and Niiniluoto) tend to conceive of theories. What can be said is that if one utilizes the Tarskian conception of a theory's content and its consequences as they do, then the familiar convergentist theses alluded to in (a) through (c) make no sense.

The elementary but devastating consequences of Miller's analysis establish that virtually any effort to link scientific progress or growth to the wholesale retention of a predecessor theory's Tarskian content *or* logical consequences *or* true consequences *or* observed consequences *or* confirmed consequences is evidently doomed. Realists have not only got their history wrong insofar as they imagine that cumulative retention has prevailed in science, but we can see that—given their views on what should be retained through theory change—history could not possibly have been the way their models require it to be. The realists' strictures on cumulativity are as ill-advised normatively as they are false historically.

Along with many other realists, Putnam has claimed that "the mature sciences do converge . . . and that that convergence has great explanatory value for the theory of science" (1978, p. 37). As this section should show, Putnam and his fellow realists are arguably wrong on *both* counts. Popper once remarked that "no theory of knowledge should attempt to explain why we are successful in our attempts to explain things" (1973, p. 23). Such a dogma is too strong. But what the foregoing analysis shows is that an occupational hazard of recent epistemology is imagining that convincing explanations of our success come easily or cheaply.

6.4 Should New Theories Explain Why Their Predecessors Were Successful? An apparently more modest realism than that outlined above is familiar in the form of the requirement (R4) often attributed to Sellars— that every satisfactory new theory must be able to explain why its predecessor was successful insofar as it was successful. On this view, viable new theories need not preserve all the content of their predecessors, nor capture those predecessors as limiting cases. Rather, it is simply insisted that a viable new theory, T_N, must explain why, when we conceive of the world according to the old theory T_O, there is a range of cases where our T_O-guided expectations were correct or approximately correct.

What are we to make of this requirement? In the first place, it is clearly *gratuitous*. If T_N has more confirmed consequences (and greater conceptual simplicity) than T_O, then T_N is preferable to T_O even if T_N cannot explain why T_O is successful. Contrariwise, if T_N has fewer confirmed consequences than T_O, then T_N cannot be rationally preferred to T_O even if T_N explains why T_O is successful. In short, a theory's ability to explain why a rival is successful is neither a necessary nor a sufficient condition for saying that it is better than its rival.

Other difficulties likewise confront the claim that new theories should explain why their predecessors were successful. Chief among them is the ambiguity of the notion itself. One way to show that an older theory, T_O was successful is to show that it shares many confirmed consequences with a newer theory, T_N, which is highly successful. But this is not an 'explanation' that a scientific realist could accept, since it makes no reference to, and thus does not depend upon, an epistemic assessment of either T_O or T_N. (After all, an instrumentalist could quite happily grant that if T_N 'saves the phenomena' then T_O—insofar as some of its observable consequences overlap with or are experimentally indistinguishable from those of T_N—should also succeed at saving the phenomena.)

The intuition being traded on in this persuasive account is that the pragmatic success of a new theory, combined with a partial comparison of the respective consequences of the new theory and its predecessor, will sometimes put us in a position to say when the older theory worked and when it failed. But such comparisons as can be made in this manner do not involve *epistemic* appraisals of either the new or the old theory *qua* theories. Accordingly, the possibility of such comparisons provides no argument for epistemic realism.

What the realist apparently needs is an *epistemically* robust sense of 'explaining the success of a predecessor'. Such an epistemic characterization would presumably begin with the claim that T_N, the new theory, was approximately true and would proceed to show that the 'observable' claims of its predecessor, T_O, deviated only slightly from (some of) the 'observable' consequences of T_N. It would then be alleged that the (presumed) approximate truth of T_N and the partially overlapping consequences of T_O and T_N jointly explained why T_O was successful in so far as it was successful. But this is a *non-sequitur*. As I have shown above, the fact that a T_N is approximately true does not even explain why it is successful; how, under those circumstances, can the approximate truth of T_N explain why some theory different from T_N is successful? Whatever the nature of the relations between T_N and T_O (entailment, limiting case, etc.), the epistemic ascription of approximate truth to either T_O or T_N (or both) apparently leaves untouched questions of how successful T_O or T_N are.

The idea that new theories should explain why older theories were successful (insofar as they were) originally arose as a rival to the 'levels' picture of explanation according to which new theories fully explained—because they entailed—their predecessors. It is clearly an improvement over the levels picture (for it does recognize that later theories generally do not entail their predecessors). But when it is formulated as a general thesis about inter-theory relations, designed to buttress a realist episte-mology, it is difficult to see how this position avoids difficulties similar to those discussed in earlier sections.

7. The Realists' Ultimate 'Petitio Principii'. It is time to step back a moment from the details of the realists' argument to look at its general strategy. Fundamentally, the realist is utilizing, as we have seen, an ab-ductive inference which proceeds from the success of science to the con-clusion that science is approximately true, verisimilar, or referential (or any combination of these). This argument is meant to show the sceptic that theories are not ill-gotten, the positivist that theories are not reducible to their observational consequences, and the pragmatist that classical ep-istemic categories (e.g., 'truth', 'falsehood') are a relevant part of meta-scientific discourse.

It is little short of remarkable that realists would imagine that their critics would find the argument compelling. As I have shown elsewhere (1978), ever since antiquity critics of epistemic realism have based their scepticism upon a deep-rooted conviction that the fallacy of affirming the consequent is indeed fallacious. When Sextus or Bellarmine or Hume doubted that certain theories which saved the phenomena were warrant-able as true, their doubts were based on a belief that the exhibition that a theory had some true consequences left entirely open the truth-status of the theory. Indeed, many non-realists have been non-realists precisely because they believed that false theories, as well as true ones, could have true consequences.

Now enters the new breed of realist (e.g., Putnam, Boyd and Newton-Smith) who wants to argue that epistemic realism can reasonably be pre-sumed to be true by virtue of the fact that it has true consequences. But this is a monumental case of begging the question. The non-realist refuses to admit that a *scientific* theory can be warrantedly judged to be true simply because it has some true consequences. Such non-realists are not likely to be impressed by the claim that a *philosophical* theory like re-alism can be warranted as true because it arguably has some true con-sequences. If non-realists are chary about first-order abductions to avowedly true conclusions, they are not likely to be impressed by second-order abductions, particularly when, as I have tried to show above, the premises and conclusions are so indeterminate.

But, it might be argued, the realist is not out to convert the intransigent sceptic or the determined instrumentalist.[23] He is perhaps seeking, rather, to show that realism can be tested like any other scientific hypothesis, and that realism is at least as well confirmed as some of our best scientific theories. Such an analysis, however plausible initially, will not stand up to scrutiny. I am aware of no realist who is willing to say that a *scientific* theory can be reasonably presumed to be true or even regarded as well confirmed just on the strength of the fact that its thus far tested consequences are true. Realists have long been in the forefront of those opposed to *ad hoc* and *post hoc* theories. Before a realist accepts a scientific hypothesis, he generally wants to know whether it has explained or predicted more than it was devised to explain; he wants to know whether it has been subjected to a battery of controlled tests; whether it has successfully made novel predictions; whether there is independent evidence for it.

What, then, of realism itself as a 'scientific' hypothesis?[24] Even if we grant (contrary to what I argued in section 4) that realism entails and thus explains the success of science, ought that (hypothetical) success warrant, by the realist's own construal of scientific acceptability, the acceptance of realism? Since realism was devised in order to explain the success of science, it remains purely *ad hoc* with respect to that success. If realism has made some novel predictions or been subjected to carefully controlled tests, one does not learn about it from the literature of contemporary realism. At the risk of apparent inconsistency, the realist repudiates the instrumentalist's view that saving the phenomena is a significant form of evidential support while endorsing realism itself on the transparently instrumentalist grounds that it is confirmed by those very facts it was invented to explain. No proponent of realism has sought to show that realism satisfies those stringent empirical demands which the realist himself minimally insists on when appraising scientific theories. The latter-day realist often calls realism a 'scientific' or 'well-tested' hypothesis, but seems curiously reluctant to subject it to those controls which he otherwise takes to be a *sine qua non* for empirical well-foundedness.

---

[23]I owe the suggestion of this realist response to Andrew Lugg.

[24]I find Putnam's views on the 'empirical' or 'scientific' character of realism rather perplexing. At some points, he seems to suggest that realism is both empirical and scientific. Thus, he writes: "If realism is an explanation of this fact [namely, that science is successful], realism must itself be an over-arching scientific *hypothesis*" (1978, p. 19). Since Putnam clearly maintains the antecedent, he seems committed to the consequent. Elsewhere he refers to certain realist tenets as being "our highest level empirical generalizations about knowledge" (p. 37). He says moreover that realism "could be false", and that "facts are relevant to its support (or to criticize it)" (pp. 78–79). Nonetheless, for reasons he has not made clear, Putnam wants to deny that realism is either scientific or a hypothesis (p. 79). How realism can consist of doctrines which 1) explain facts about the world, 2) are empirical generalizations about knowledge, and 3) can be confirmed or falsified by evidence and yet be neither scientific nor hypothetical is left opaque.

8. Conclusion. The arguments and cases discussed above seem to warrant the following conclusions:

1. The fact that a theory's central terms refer does not entail that it will be successful; and a theory's success is no warrant for the claim that all or most of its central terms refer.

2. The notion of approximate truth is presently too vague to permit one to judge whether a theory consisting entirely of approximately true laws would be empirically successful; what is clear is that a theory may be empirically successful even if it is not approximately true.

3. Realists have no explanation whatever for the fact that many theories which are not approximately true and whose 'theoretical' terms seemingly do not refer are nonetheless often successful.

4. The convergentist's assertion that scientists in a 'mature' discipline usually preserve, or seek to preserve, the laws and mechanisms of earlier theories in later ones is probably false; his assertion that when such laws are preserved in a successful successor, we can explain the success of the latter by virtue of the truthlikeness of the preserved laws and mechanisms, suffers from all the defects noted above confronting approximate truth.

5. Even if it could be shown that referring theories and approximately true theories would be successful, the realists' argument that successful theories are approximately true and genuinely referential takes for granted precisely what the non-realist denies (namely, that explanatory success betokens truth).

6. It is not clear that acceptable theories either *do* or *should* explain why their predecessors succeeded or failed. If a theory is better supported than its rivals and predecessors, then it is not epistemically decisive whether it explains why its rivals worked.

7. If a theory has once been falsified, it is unreasonable to expect that a successor should retain either all of its content *or* its confirmed consequences *or* its theoretical mechanisms.

8. Nowhere has the realist established—except by fiat—that non-realist epistemologists lack the resources to explain the success of science.

With these specific conclusions in mind, we can proceed to a more global one: it is not yet established—Putnam, Newton-Smith and Boyd notwithstanding—that realism can explain *any* part of the success of science. What is very clear is that realism *cannot*, even by its own lights, explain the success of those many theories whose central terms have evidently not referred and whose theoretical laws and mechanisms were not approximately true. The inescapable conclusion is that insofar as many realists are concerned with explaining how science works and with assessing the adequacy of their epistemology by that standard, they have thus far failed to explain very much. Their epistemology is confronted by anomalies which seem beyond its resources to grapple with.

It is important to guard against a possible misinterpretation of this essay. *Nothing* I have said here refutes the possibility in principle of a realistic epistemology of science. To conclude as much would be to fall prey to the same inferential prematurity with which many realists have rejected in principle the possibility of explaining science in a non-realist way. My task here is, rather, that of reminding ourselves that there *is* a difference between wanting to believe something and having good reasons for believing it. All of us would like realism to be true; we would like to think that science works because it has got a grip on how things really are. But such claims have yet to be made out. Given the *present* state of the art, it can only be wish fulfilment that gives rise to the claim that realism, and realism alone, explains why science works.

REFERENCES

Boyd, R. (1973), "Realism, Underdetermination, and a Causal Theory of Evidence", *Noûs 7:* 1–12.
Bunge, M. (1970), "Problems Concerning Intertheory Relations", Weingartner, P. and Zecha, G. (eds.), *Induction, Physics and Ethics:* 285–315. Dordrecht: Reidel.
Fine, A. (1967), "Consistency, Derivability and Scientific Change", *Journal of Philosophy 64:* 231ff.
Grünbaum, Adolf (1976), "Can a Theory Answer More Questions than One of its Rivals?", *British Journal for Philosophy of Science 27:* 1–23.
Hooker, Clifford (1974), "Systematic Realism", *Synthese 26:* 409-497.
Koertge, N. (1973), "Theory Change in Science", Pearce, G. and Maynard, P. (eds.), *Conceptual Change:* 167-198. Dordrecht: Reidel.
Kordig, C. (1971), "Scientific Transitions, Meaning Invariance, and Derivability", *Southern Journal of Philosophy:* 119–125.
Krajewski, W. (1977), *Correspondence Principle and Growth of Science.* Dordrecht: Reidel.
Laudan, L. (1976), "Two Dogmas of Methodology", *Philosophy of Science 43:* 467–472.
Laudan, L. (1977), *Progress and its Problems.* California: University of California Press.
Laudan, L. (1978), "Ex-Huming Hacking", *Erkenntnis 13:* 417–435.
Margenau, H. (1950), *The Nature of Physical Reality.* New York: McGraw-Hill.
McMullin, Ernan (1970), "The History and Philosophy of Science: A Taxonomy", Stuewer, R. (ed.), *Minnesota Studies in the Philosophy of Science V:* 12–67. Minneapolis: University of Minnesota Press.
Newton-Smith, W. (1978), "The Underdetermination of Theories by Data", *Proceedings of the Aristotelian Society:* 71–91.
Newton-Smith, W. (forthcoming), "In Defense of Truth".
Niiniluoto, Ilkka (1977), "On the Truthlikeness of Generalizations", Butts, R. and Hintikka, J. (eds.), *Basic Problems in Methodology and Linguistics:* 121–147. Dordrecht: Reidel.
Niiniluoto, Ilkka (1980), "Scientific Progress", *Synthese 45:* 427–62.
Popper, K. (1959), *Logic of Scientific Discovery.* New York: Basic Books.
Popper, K. (1963), *Conjectures and Refutations.* London: Routledge & Kegan Paul.
Popper, K. (1972), *Objective Knowledge.* Oxford: Oxford University Press.
Post, H. R. (1971), "Correspondence, Invariance and Heuristics: In Praise of Conservative Induction", *Studies in the History and Philosophy of Science 2:* 213–255.
Putnam, H. (1975), *Mathematics, Matter and Method, Vol. 1.* Cambridge: Cambridge University Press.

Putnam, H. (1978), *Meaning and the Moral Sciences*. London: Routledge & Kegan Paul.
Sellars W. (1963), *Science, Perception and Reality*. New York: The Humanities Press.
Sklar, L. (1967), "Types of Inter-Theoretic Reductions", *British Journal for Philosophy of Science 18*: 190–224.
Szumilewicz, I. (1977), "Incommensurability and the Rationality of the Development of Science", *British Journal for Philosophy of Science 28*: 348.
Watkins, John (1978), "Corroboration and the Problem of Content-Comparison", Radnitzky and Andersson (eds.), *Progress and Rationality in Science*: 339–378. Dordrecht: Reidel.

Realism, Approximate Truth, and Philosophical Method

1. Introduction

1.1. Realism and Approximate Truth

Scientific realists hold that the characteristic product of successful scientific research is knowledge of largely theory-independent phenomena and that such
knowledge is possible (indeed actual) even in those cases in which the relevant
phenomena are not, in any non-question-begging sense, observable (Boyd 1982).
The characteristic philosophical arguments for scientific realism embody the
claim that certain central principles of scientific methodology require a realist explication. In its most completely developed form, this sort of abductive argument
embodies the claim that a realist conception of scientific inquiry is required in order to justify, or to explain the reliability with respect to instrumental knowledge
of, all of the basic methodological principles of mature scientific inquiry (Boyd
1973, 1979, 1982, 1983, 1985a, 1985b, 1985c; Byerly and Lazara 1973; Putnam
1972, 1975a, 1975b).

The realist who offers such arguments is not committed to the view that rationally applied scientific method will always lead to progress towards the truth, still
less to the view that such progress would have the exact truth as an asymptotic
limit (Boyd 1982, 1988). Nevertheless it would be difficult to defend scientific
realism without portraying the central developments of twentieth-century physical science, for example, as involving a dialectical and progressive interaction of
theoretical and methodological commitments (Boyd 1982, 1983).

A defense of realism along these lines requires that two things. In the first
place, the realist must be able to defend a historical thesis regarding the recent
history of relevant sciences according to which their intellectual achievements involve *approximate* theoretical knowledge and according to which theoretical progress within them has been (to a large extent) a process of (not necessarily converging) *approximation.* No realist conception that does not treat theoretical
knowledge and theoretical progress as involving approximations to the truth is

even prima facie compatible with the actual history of science. The realist must, therefore, employ a conception of approximate theoretical knowledge and of theoretical progress through approximation that makes historical sense of the recent development of scientific theories.

Secondly, the realist must be able to establish that her historical appeal to approximate theoretical knowledge and to theoretical progress by successive approximation is appropriate by philosophical as well as by historical standards. Neither the realist's historical account nor her appeal to it in the defense of scientific realism as a philosophical thesis should be undermined by any of the distinctly philosophical considerations characteristic of anti-realist positions in the philosophy of science. Important challenges to scientific realism arise from doubts that a realist conception of approximate truth and of the growth of approximate knowledge is available that satisfies both of these constraints. The appropriate realist responses to these challenges and the philosophical implications of those responses are the subject of the present essay.

1.2. Challenges to a Realist Treatment of Approximation

A number of philosophers (realists included) have had serious concerns about the realist's ability to provide an adequate account of the development of scientific theories as involving the growth of approximate theoretical knowledge. The *locus classicus* of objections to realism reflecting these concerns is surely Laudan 1981 (see also Fine 1984). That there should be such concerns is, in significant measure, a reflection of the striking difference between the depth of our understanding of the notion of (exact) truth and that of our understanding of approximate truth.

Since the work of Tarski in the 1930s we have had a systematic, general, and topic-and-context-independent mathematical and philosophical theory of (exact) truth. By contrast there is no generally accepted general and systematic theory of approximate truth. We have available from the various special sciences a very large number of well-worked-out examples of particular instances of approximation but the details in theses cases depend not only on contingent and often esoteric facts about the relevant natural phenomena, but also upon the particular context of application within which the approximate theories and models are to be applied. In part because of the complexities created by such topic and context dependence, we do not have as clear a general understanding of what the epistemological relevance of appeals to approximate truth should be. Moreover, as we shall see, the dependence of the relevant details upon a posteriori theoretical claims raises special problems of philosophical method when an appeal to conception of approximate truth is to be made in the course of a defense of scientific realism.

I have argued elsewhere (Boyd 1982, 1983, 1985a, 1985b, 1985c, 1988) that the scientific realist must adopt distinctly naturalistic conceptions of philosophical methodology and of central issues in epistemology and metaphysics. My aim in the present paper will be to show how the distinctly naturalistic arguments for realism that I have developed in the papers cited can be extended to provide an adequate realist treatment of approximate truth.

Instead of replying to particular anti-realist arguments in the literature, I shall respond to four objections that capture, I believe, the deep philosophical concerns that the realist's conception of approximate theoretical knowledge properly occasion. My expectation is that the responses to those objections will provide an adequate basis for a realist's response to other objections regarding her conception of approximate truth and approximate knowledge. The objections I shall consider are these:

1. (*The historical objection*) Realists are simply mistaken as a matter of historical fact: many important scientific advances seem to have been grounded in what (by realist standards) were deep errors in background theories. Approximately true background theoretical knowledge is thus not required to explain reliability of scientific practices.

2. (*The triviality objection*) The the realist might reply (following Hardin and Rosenberg 1982, for example) about many of the advances in question that the relevant background theories were *to some extent* or *in some respects* approximately true.

Here the realist's philosophical project is in danger of being reduced to *triviality*. The problem is that we lack altogether a general theory of approximation: we have no general characterization of what it is for a sentence to be approximately true, to be approximately true to a specified degree or in a specified respect, or to be more nearly true (in specific respects or in general) than some other sentence. If we had such a general theory then the realist could appeal to it in refining the thesis that the relevant historical episodes reflect some respects of approximation to the truth. As it is we are faced with the fact that *any* consistent theory is approximately true in some respects or other, and *any* sequence of such theories will reflect progress towards the truth in some respects or other.

3. (*The contrivance objection*) The realist might next reply by distinguishing between relevant and irrelevant respects of approximation to the truth regarding matters theoretical, and by claiming that the growth of scientific knowledge characteristically involves the former. Here the realist avoids triviality at the expense of a contrived or ad hoc conception of approximate truth, indeed at the expense of both contrivance (objection 3) and circularity (see objection 4).

The contrivance in question arises from the important difference just mentioned between extant theories of truth and of approximate truth respectively. In the case of truth *simpliciter* Tarski's strategy for defining truth (Tarski 1951) provides a uniform treatment that is largely independent of the particular subject matter or of the particular historical episodes or context of application under consideration. By contrast, our conception of relevant approximation reflects considerations specific to the particular theory or theories, historical settings, and contexts of application under consideration.

Thus, for example, if the realist sees relativistic mechanics as growing out of previously acquired approximate theoretical knowledge her conception of the relevant respects of approximation reflected in Newtonian mechanics will emphasize numerical accuracy for systems of particles with relative velocities low with respect to that of light, the identification of, and the development of reliable measurement procedures for, various physical magnitudes, and the central role assigned to certain fundamental laws. It will de-emphasize, for example, numerical accuracy for high relative velocities, or of soundness of the Newtonian theoretical conception of space and time.

Here the distinctions between relevant and irrelevant respects of approximation reflect judgements, based on current theoretical conceptions, about the respects in which Newtonian mechanics happened to be approximately true, and similarly theory-dependent judgments about the role that such approximations played in the successful development of relativistic mechanics. Since we lack a general theory of approximation, the realist's appeal to relevant respects of approximation in response to the triviality objection will always have to be grounded in just this sort of topic-and-episode-sensitive conception.

We can now see why the realist's treatment of respects of approximation will involve an ad hoc or contrived element. For each of the episodes of scientific inquiry typically considered by philosophers of science there is a standard realist picture (or, at any rate, a narrow range of such pictures) of how the relevant approximations to the truth have gone and what contributions, if any, they have made to the subsequent growth of scientific knowledge. The realist, in defining the relevant sense(s) of approximation, will rely on such a picture. But such a picture merely reflects the realist research tradition in the history and philosophy of science. Since there is no topic-and-episode-neutral conception of relevant approximation with respect to which her proposed definitions may be assessed, the realist will simply be presupposing the soundness of the "findings" of her own tradition when she defines the difference(s) between relevant and irrelevant respects of approximation. It is no surprise – and certainly no basis for an abductive argument for realism – that the realist can construct a realist account of approximate truth when she is permitted to beg questions in so thoroughgoing a way.

4. (*The circularity objection*). There is some precedent in scientific inquiry, especially historical inquiry, for explanatory concepts that lack topic-and-episode-neutral general specifications of the sort alluded to above: sometimes theoretical considerations that resist incorporation into a fully general definition can justify the (topic-and-episode-nonneutral) ways in which such concepts are applied in particular cases. Let us suppose for the sake of argument that this is the case with respect to the employment of the concept of approximate truth in the various historical explanations of scientific progress (or its absence) that are

offered in the realist tradition. Even if the realist's accounts of the relevant episodes are thus methodologically acceptable *as explanations in the history of science*, they will involve an unacceptable circularity if they are understood to address the *philosophical* issue between scientific realists and anti-realists.

Here's why: Any realist explanation of the growth of knowledge and of reliable methodology in a particular field must involve an account of the kinds of epistemically relevant causal interactions that exist(ed) between members of the relevant scientific community and the features of the world that were (or are) the alleged objects of their study. Thus for example, a realist account of such developments in atomic theory will incorporate a causal account of how scientists gain(ed) epistemic access to various subatomic particles and the realist's claim that atomic theory is about such unobservable theory-independent particles will depend on that account (see sections 2.1.3 and 2.1.4). The realist's account of epistemic access to subatomic particles will be grounded in the best available theory of such particles together with related contemporary physical theories.

Suppose now that the realist's explanation of the development of some field, including the relevant account of epistemic access, is advanced in defense of realism as a philosophical thesis. Plainly the resulting defense of realism is cogent only if the realist's explanation, and her account of epistemic access in particular, are understood *realistically*. For example, only if the account of epistemic access to subatomic particles is understood realistically is the realist's case that atomic theory has an unobservable and theory-independent subject matter advanced. But, on the realist's own account, her explanation and the account of epistemic access it incorporates are ordinary scientific theories themselves grounded in the very research tradition regarding which a defense of realism is sought. To insist on a realistic interpretation of the realist's explanation would thus *presuppose* realism regarding the tradition in question. Thus the realist's appeal to her explanation of the development of instrumentally reliable methodology in an abductive argument for realism as a philosophical thesis is question-beggingly circular.

1.3. An Argumentative Strategy

The challenges we are considering seem to fall into two classes: The first three represent an essentially prephilosophical critique of the realist's historical explanations: they deny that the realist's conception of the role of approximate truth regarding theoretical matters in the growth of scientific knowledge represents the best explanation for the relevant episodes in the history of science. The fourth offers a distinctly philosophical challenge: it argues that even if the *realist's* account of the growth of scientific knowledge does provide the best explanation, inductive inference to *realism* begs the philosophical question at issue.

After some philosophical preliminaries, I propose to respond to the challenges in two distinct stages corresponding to these two classes . In the first stage of my response, I treat the characteristic realist explanatory appeal to approximate truth as an ordinary piece of historical explanation. I identify a general methodological problem of *parametric specification* in explanatory contexts of which the deeper problems raised by the first three challenges are special cases, and I identify the generally appropriate solution to that problem. I then indicate why it is plausible that the realist's explanatory appeal to approximate truth satisfies the methodological demands dictated by the solution in question.

With respect to the fourth challenge, I assume for the sake of argument that the realist's historical explanations have been confirmed and I inquire whether they are to be understood realistically or whether instead such an understanding – which is essential to the realist's case – begs the question against the anti-realist. Here too I argue that the methodological question regarding the realist's appeal to approximate truth – in this case a question about *philosophical method* – is a special case of a more general methodological question about the appropriate interaction between philosophical considerations and empirical findings in the philosophy of science. I define the notion of a large-scale *philosophical package* and I indicate why the incorporation of realistically understood scientific theories into the realist philosophical package is compatible with (and indeed required by) an adequate *and noncircular* defense of the realist package against rival philosophical conceptions.

On now to the philosophical preliminaries.

2. Philosophical Preliminaries

2.1. The Abductive Argument for Scientific Realism

The challenges we are considering arise in the context of a class of abductive arguments for realism according to which we must recognize approximate knowledge of unobservable (and appropriately mind-independent) "theoretical entities" in order to adequately explain the growth of even instrumental knowledge in recent science. To assess the realist's arguments and the appeals to the notion of approximate truth embodied in them, we need an understanding of just what those arguments are. In what follows of this section I'll indicate, in broad outline, how the abductive arguments for realism go.

2.1.1. Objective Knowledge from Theory Dependent Method

By the "instrumental reliability" of a scientific theory I mean the extent of its capacity to make approximately true observational predictions about observable phenomena – the extent of its approximate empirical adequacy. By the "instrumental reliability" of some body of methods I mean the extent to which their practice is conducive to the acceptance of instrumentally reliable theories. The abductive arguments for scientific realism take place in a dialectical situation in which scientific realists and their philosophical opponents largely agree that the methods of actual recent scientific practice are significantly instrumentally reliable.

The abductive arguments for realism are in the first instance directed against the empiricist who denies the possibility of "theoretical" knowledge – knowledge of "unobservables." Against the empiricist the realist argues that only by accepting the reality of approximate theoretical knowledge can we adequately explain the (uncontested) instrumental reliability of apparently theory-dependent scientific methods. In the present paper I shall focus my attention primarily on the dispute between realists and empiricists, reserving attention to the corresponding dispute between realists and constructivists largely to a later paper. I discuss the realism-constructivism dispute briefly in section 2.4 and briefly discuss the distinctly constructivist version of the circularity objection in section 4.3.

The case for realism lies largely in the recognition of the extraordinary role that theoretical considerations play in actual (and patently successful) scientific practice. To take the most striking example, scientists routinely modify or extend operational "measurement" or "detection" procedures for "theoretical" magnitudes or entities on the basis of new theoretical developments. The reliability and justifiability of this sort of methodology is perfectly explicable on the realist's conception of measurement and of theoretical progress. Accounts of the revisability of operational procedures that are compatible with an empiricist position appear inadequate to explain the way in which theory-dependent revisions of "measurement" and "detection" procedures make a positive methodological contribution to the progress of science.

There are two important consequences of the realist explanation for the reliability of the methodology in question. First, scientific research, when it is successful, is *cumulative by successive (but not necessarily convergent) approximations to the truth*. Second, this cumulative development is possible because *there is a dialectical relationship between current theory and the methodology for its improvement*. The approximate truth of current theories explains why our existing measurement procedures are (approximately) reliable. That reliability, in turn, helps to explain why our experimental or observational investigations are successful in uncovering new theoretical knowledge, which, in turn, may produce improvements in measurement techniques, etc.

Theory dependence of methods and the consequent dialectical interaction of theory and method are entirely general features of all aspects of scientific methodology – principles of experimental design, choices of research problems, standards for the assessment of experimental evidence and for assessing the quality and methodological import of explanations, principles governing theory choice, and rules for the use of theoretical language. In all cases there is a pattern of dialectical interaction between accepted theories and associated methods of just the sort exemplified in the case of the theory dependence of measurement and detection procedures. Moreover, this pattern of theory dependence contributes to the reliability of scientific methodology rather than detracting from it (Boyd 1972, 1973, 1979, 1980, 1982, 1983, 1985a, 1985b, 1985c; Kuhn 1970, Putnam 1972, 1975a, 1975b; Van Fraassen 1980).

According to the realist, the only scientifically plausible explanation for the reliability of a scientific methodology that is so theory dependent is a thoroughgoingly realistic explanation: Scientific methodology, dictated by currently accepted theories, is reliable at producing further knowledge *precisely because, and to the extent that, currently accepted theories are relevantly approximately true.* Scientific method provides a paradigm-dependent paradigm-modification strategy: a strategy for modifying or amending our existing theories and methods in the light of further research that is such that its methodological principles at any given time will themselves depend upon the theoretical picture provided by the currently accepted theories. If the body of accepted theories is itself relevantly sufficiently approximately true, then this methodology operates to produce a subsequent dialectical improvement both in our knowledge of the world and in our methodology itself. It is not possible, according to the realist, to explain even the instrumental reliability of actual recent scientific practice without invoking this explanation and without adopting the realistic conception of scientific knowledge that it entails (Boyd 1972, 1973, 1979, 1982, 1983, 1985a, 1985b, 1985c).

2.1.2. Projectability, Evidence, Theoretical Plausibility and the Evidential Indistinguishability Thesis

If the realist's abductive argument is correct, a dramatic rethinking of our notion of scientific evidence is required. Consider the question of the "degree of confirmation" of a theory given a body of observational evidence. To a very good first approximation, a theory receives significant evidential support from a body of successful predictions (or other evidentially favorable observations) just in case (a) the theory is itself "projectable" (see Goodman 1973), (b) the observations in question pit the theory's predictions (or, in other contexts, its explanations) against those of its projectable rivals; and (c) in the relevant experiments or observational settings, there have been suitable controls for those possible artifactual influences that are themselves suggested by projectable theories of those settings (Boyd 1982, 1983, and especially 1985a).

Central to the realist's argument is the observation that projectability judgments are, in fact, judgments of theoretical plausibility: we treat as projectable those proposals that relevantly resemble our existing theories (where the determination of the relevant respects of resemblance is itself a theoretical issue). The reliability of this conservative preference is explained by the approximate truth of existing theories, and one consequence of this explanation is that *judgments of theoretical plausibility are evidential.* The fact that a proposed theory is plausible in the light of previously confirmed theories is some evidence for its (approximate) truth. Judgments of theoretical plausibility are matters of inductive inference from (partly) theoretical premises to theoretical conclusions; precisely these inferences justify, and explain the reliability of, "inductive inference to the best explanation" (Boyd 1972, 1973, 1979, 1982, 1983, 1985a, 1985b, 1985c) .

The claim that judgments of theoretical plausibility are evidential affords the realist a reply to the deepest empiricist argument against realism. The empiricist appeals (tacitly or explicitly) to a principle that I have called the *evidential indistinguishability thesis.* In its most plausible form it holds that for any two empirically equivalent total sciences, the empirical support or disconfirmation that one receives, given a given body of observational data, will be just the same as that received by the other. The empiricist's conclusion that knowledge of unobservables is impossible is a straightforward application of this thesis, which can be thought of as an empiricist analysis of the claim that all scientific knowledge is empirical knowledge. The realist accepts the latter claim but rejects the empiricist analysis. Instead, the realist holds, evidential considerations regarding theoretical plausibility are indirectly experimental and can serve to distinguish total sciences that embody or naturally extend the current total science (that are favored by those considerations) from empirically equivalent total sciences which significantly depart from the prevailing total science (which such considerations reject as unprojectable). [See Boyd 1982, 1983, and section 2.2.]

2.1.3 . Natural Definitions

Locke speculates at several places in Book IV of the *Essay* (see, e.g., IV, iii, 25) that when kinds of substances are defined, as empiricism requires, by purely conventional "nominal essences," it will be impossible to have a general science of, say, chemistry. There is no reason to believe that kinds defined by nominal essences will reflect actual causal structure and thus be apt for the formulation or confirmation of general knowledge of substances. Only if we are able to sort substances according to their hidden real essences will systematic general knowledge of substances be possible.

Locke was right (at any rate so the realist thinks). Only when kinds (properties, relations, magnitudes, etc.) are defined by natural rather than conventional definitions is it possible to obtain the theory-dependent solutions to the problem of projectability just described (Putnam 1975a; Quine 1969a; Boyd 1979, 1982, 1983). It is thus central to the realist's abductive argument that most scientific terms be seen as possessing natural rather than conventional definitions. Such terms are defined in terms of properties, relations, etc., that render the kinds (etc.) to which they refer appropriate to particular sorts of scientific or practical reasoning. In the case of such terms, proposed definitions are always in principle revisable in the light of new evidence or new theoretical developments, and it is possible for people to refer to the same kind (property, magnitude, etc.) by a term while disagreeing about what its correct a posteriori natural definition is. This last consequence of the naturalistic conception of definitions is essential to the realist's dialectical conception of the development of scientific knowledge and methods. The realist will (at least typically) need to portray developments in which mature scientific communities change their conception of the definitions of kinds, relations, magnitudes, etc., as dialectical advances (or, if things go badly, setbacks) rather than as changes of subject matter (Putnam 1972, 1975a, 1975b; Boyd 1979, 1980, 1988). (For more on naturalistic definitions see section 2.5.)

2.1.4. Reference and Epistemic Access

If the traditional empiricist account of definition is to be abandoned for scientific terms in favor of a naturalistic account, then a naturalistic conception of reference is required for such terms. An account of the appropriate sort is provided by recent causal theories of reference (see, e.g., Feigl 1956, Kripke 1972, Putnam 1975a). The reference of a term is established by causal connections of the right sort between the use of the term and (instances of) its referent.

The connection between naturalistic theories of reference and of knowledge (see section 2.2) is quite intimate: reference is itself an epistemic notion and the sorts of causal connections that are relevant to reference are just those that are involved in the reliable regulation of belief (Boyd 1979, 1982). *Roughly,* and for nondegenerate cases, a term t refers to a kind (property, relation, etc.) k, just in case there exist causal mechanisms whose tendency is to bring it about, over time, that what is predicated of the term t will be approximately true of k. In such a case, we may think of the properties of k as regulating the use of t, and we may think of what is said using t as providing us with socially coordinated *epistemic access* to k. t refers to k (in nondegenerate cases), just in case the socially coordinated use of t provides significant epistemic access to k, and not to other kinds (properties, etc.) (Boyd 1979, 1982). The mechanisms of reference *just are* the mechanisms of reliable belief regulation .

Thus, just as the realist conception requires, two different terms, or the same term in two historically different settings, may afford epistemic access to, and thus may refer to, the same kind (property, etc.) even though the definitions associated with them by the relevant linguistic communities are quite different or even inconsistent.

One further feature of the naturalistic conception of reference is important to an understanding to the realist's conception of the growth of approximate knowledge. In many scientifically important cases the use of a term may afford epistemic access to more than one kind (property, relation, . . .), but our knowledge may be insufficient for us to recognize that this is so, and we may consequently have a conception of, as it seems to us, one kind (etc.) that conflates information regarding several distinct kinds.

Field (1973, 1974) calls the relation thus established between a term and several kinds (etc.) *partial denotation*, and he calls the revision of language usage to eliminate such cases of ambiguity *denotational refinement*. On the realist's conception of the growth of approximate knowledge one sort of approximate knowledge is that represented by a body of sentences involving a partially denoting term when what is predicated of that term in these sentences represents methodologically important approximations to the truth regarding one or more of the relevant *partial denotata* considered individually. In such cases, one characteristic form of subsequent improvement in approximation is the discovery of the ambiguity and the consequent denotational refinement (see Boyd 1979).

2.2. Naturalism and Radical Contingency in Epistemology

Modern epistemology has been largely dominated by "foundationalist" conceptions: all knowledge is seen as grounded in certain foundational beliefs that have an epistemically privileged position. Other true beliefs are instances of knowledge only if they can be justified by appeals to foundational knowledge. It is an a priori question which beliefs fall in the privileged class. Similarly, the basic inferential principles that are legitimate for justifying nonfoundational knowledge claims can be justified a priori; it is moreover an a priori question about a given inference whether it meets the standards set by those principles or not. We may fruitfully think of foundationalism as consisting of two parts, *premise foundationalism* which holds that all knowledge is justifiable from an a priori specifiable core of foundational beliefs, and *inference foundationalism*, which holds that principles of justifiable inference are reducible to inferential principles that are *a priori justifiable* and whose application is *a priori checkable*.

Recent work in "naturalistic epistemology" (see, e.g., Armstrong 1973; Goldman 1967, 1976; Quine 1969b) strongly suggests that foundationalism is fundamentally mistaken. For the typical case of perceptual knowledge, there seem to be neither premises nor inferences; instead perceptual knowledge obtains when perceptual beliefs are produced by epistemically reliable mechanisms. Even where premises and inferences occur, it seems to be the reliable production of belief that distinguishes cases of knowledge from other cases of true belief. A variety of naturalistic considerations suggest that there are no beliefs that are epistemically privileged in the way foundationalism seems to require.

I have argued (see Boyd 1982, 1983, 1985a, 1985b,1985c) that the abductive defense of scientific realism requires an even more thoroughgoing naturalism in epistemology and, consequently, an even more thoroughgoing rejection of foundationalism. In particular *all* of the significant methodological principles of inductive inference in science are profoundly theory dependent. They are a reliable guide to the truth only because, and to the extent that, the relevant background theories are relevantly approximately true. They are not reducible to some more basic rules whose reliability as a guide to the truth is independent of the truth of background theories. Since it is a contingent empirical matter which background theories are approximately true, the justifiability of scientific principles of inference rests ultimately on a contingent matter of empirical fact, just as the epistemic role of the senses rests upon the contingent empirical fact that the senses are reliable detéctors of external phenomena. Thus inference foundationalism is radically false; there are no a priori justifiable rules of nondeductive inference, and it is an a posteriori question about any such inference whether or not it is justifiable. The epistemology of empirical science is an empirical science (Boyd 1982, 1983, 1985a, 1985b).

One consequence of this radical contingency of scientific methods is important to the realist's conception of the growth of approximate knowledge. The emergence of successful modern scientific methodology as we know it depended upon the logically, epistemically, and historically contingent emergence of a relevantly approximately true theoretical tradition. It is not possible to understand the initial emergence of such a tradition as the consequence of some more abstractly conceived scientific or rational methodology that itself is theory independent. There is no such methodology. The theoretical innovations that established the first successful paradigm within a particular scientific discipline must be thought of as the beginnings of successful methodology within the field, not as consequences of it (for a further discussion see Boyd 1982).

Note that radical contingency in epistemology is central to the realist's case against empiricism. Against the evidential indistinguishability thesis the realist argues that plausibility judgments grounded in the current total science afford evidential distinctions between empirically equivalent total sciences. But, according to the realist's account, it is not the *currency* of the current total science that makes plausibility judgments with respect to it epistemically reliable but its approximate truth. That a time should have arisen in which totaı sciences embodied relevant approximations to the truth is of course radically contingent. Thus central to the realist's rebuttal to empiricism are the epistemological principles that reflect that contingency.

2.3. Metaphysics and 'Metaphysics'

Logical positivists employed the term 'metaphysics' for the sort of inquiry about "unobservables" that verificationism led them to reject. Most of what has traditionally fallen under that term was 'metaphysics' in the positivists' sense, but so was inquiry about, e.g., the atomic structure of matter. If scientific realism is right, then it follows that scientists routinely do successful 'metaphysics'. With respect to metaphysics (as philosophers and others ordinarily use the term) the situation is more complex.

If scientific realism is true for any of the standard reasons then scientists have discovered the real essences of chemical kinds (Kripke 1971, 1972) and have thus done some real metaphysics. Moreover, the fact that scientific knowledge of unobservables is possible makes it a serious question whether or not scientific findings have (or will have) resolved some traditional metaphysical questions. Certainly the recent near consensus in favor of a materialist conception of mind reflects a realist understanding of the possibility of experimental metaphysics. Nevertheless it does not follow from scientific realism that scientists routinely tend to get the right answers to the distinctly metaphysical questions that are the special concern of philosophers even when their methods lead them to adopt theories that reflect answers to such question.

In particular, when a scientific realist proposes to explain the reliability of the scientific methods employed at a particular historical moment by appealing to the approximate truth of the background theories accepted at that time, she need not hold that the metaphysical conceptions embodied in those theories represent a good approximation by philosophical standards. Two examples will illustrate the point.

Consider the way in which the reliability of the methods by which Darwin's account in the *Origin* was assessed is to be explained by reference to the approximate truth of much of the prevailing background biological theory. A great deal was known, for example, about species – not just facts about particular species, but about anatomical, behavioral, genetic and biogeographical generalizations that can only be formulated in terms of the notion of a species. The realist will hold that the approximations to the truth embodied in this lore of species is part of what explains the reliability of the research methods in biology employed by Darwin and his contemporaries.

Prior to Darwin's work the prevailing conception made species membership in the first instance a property of individuals; after Darwin we have correctly seen a species as in the first instance a family of populations. The background biological theories of Darwin's era got it profoundly wrong about the metaphysics of species. Nevertheless, the classificatory practices of pre-Darwinian biologists were reliable enough to serve to establish the rich and significantly accurate lore about species upon which the reliability of methodology in early evolutionary theory crucially depended – or, at any rate, so the realist may reasonably maintain.

Similarly, the realist will want to explain the reliability of the methods by which physicists assessed early developments in quantum theory by appealing to respects in which the prequantum theory of, say, atoms and subatomic particles was approximately true. She will appeal to the correct identification of various subatomic particles and of (many of) the fundamental physical magnitudes, to the availability of reliable procedures for the detection of those particles and for the measurement of various of their physical properties, and to the classical insights reflected both in the formulation of the equation for the time-evolution of quantum mechanical systems and in the techniques employed in practice in picking the appropriate Hamiltonian for quantum mechanical systems.

Indeed she will want to portray much of the early development of quantum theory as the gradual extension of the range of phenomena for which an adequate quantum mechanical treatment had been provided. On such an account, at any given stage, in the early development of quantum theory, the proposed models for physical systems were always a mixture of distinctly quantum mechanical components together with essentially classical (or relativistic) components awaiting later quantum mechanical reformulation. The realist will want to explain the reliability and justifiability of this sort of development by appealing to the respects of approximation to the truth of classical mechanics itself and of the successive stages in the development of the quantum theory.

Consider now the classical conception of atomic phenomena understood as a contribution to philosophical metaphysics. Arguably the metaphysical component of that conception is some sort of mechanistic atomism: a picture of discrete and unproblematically individuated particles and their associated fields interacting in a deterministic fashion without action at a distance. Our current quantum mechanical conception of matter rejects each component of this picture: for the atomist's discrete particles we substitute entities with wavelike features for which particlelike individuation is sometimes impossible; we reject determinism; and we acknowledge that there are nonlocal effects that would surely be precluded by the classical philosophical rejection of action at a distance. Classical conceptions of the atomic world were, let us agree, poor approximations to the truth in metaphysics. Does this preclude their having been good enough approximations in other respects to sustain the realist's account of the development of quantum theory?

Plainly not. Whatever other objections there may be to the realist's account, it is not a cogent objection that the classical conception that her account treats as relevantly approximately true is not good metaphysics. All she need do is to explain how the metaphysical errors in the classical conception failed to vitiate the methodological contribution of its genuine insights. To this end she might, e.g., appeal to the respects in which subatomic particles are (classical) particlelike, to the determinism of the time-evolution of quantum mechanical systems prior to measurement, and to the wide variety of phenomena that do not significantly exhibit the effects of nonlocal "action at a distance." Perhaps in the case of the development of evolutionary theory and certainly in the case of quantum mechanics, the realist's account will have scientists doing 'metaphysics' with some significant success; in neither case must she portray them as doing good metaphysics.

The cases just discussed illustrate an additional point. In each case, if the metaphysical criticism of the earlier theoretical tradition is sound, then it embodied, in addition to metaphysical errors, errors about the logical form of certain key propositions. Conspecificity is a relation between populations, not individuals; so pre-Darwinian biology embodied a mistake about the logical type of propositions regarding species membership. Similarly, quantum mechanics requires that we think of the classically acknowledged physical magnitudes as corresponding to Hermitian operators rather than to vector- or scalar-valued functions; in consequence classical mechanics is mistaken about the logical form of, e.g., attributions of position or momentum to particles. Neither error undermines the contribution that the approximate truth of the earlier theory is said to have made to the methodology by which the latter theory was developed and confirmed. The realist need attribute to successful background theories neither metaphysical success nor logical exactitude. Approximation need not be philosophically clean. (Note that the distinctly realist naturalistic semantic conceptions are operative in this discussion. What evolutionary theory and quantum mechanics have taught us is that, as we might say, "there are no classical species" and "there are no classical particles." Only naturalistic alternatives to the empiricist conceptions of definitions and reference permit the realist to say – as the account just given requires – that nevertheless Darwinian species and the particle-like phenomena acknowledged by quantum mechanics were the subjects of the relevant classical investigations.)

2.4. Realism Causation and Mind Independence

The realist conception of science contrasts with various neo-Kantian constructivist conceptions according to which when scientific theories address fundamental questions there is a deep element of social construction of reality reflected in what they say. It is sometimes said that realists and constructivists differ over the extent to which the reality studied by scientists is "mind independent" or is "theory independent." In order to understand the demands placed on the realist by what we have called the "circularity objection," we require some understanding of what is distinctly realist about the realist's explanatory appeal to approximate truth of theoretical presuppositions, given tnat the constructivist shares with the realist the conviction that scientific progress involves theoretical as well as instrumental knowledge and that scientific methods are deeply theory dependent. In the present essay I'll touch on this issue only briefly.

2.4.1. Defining Mind Independence

The realist and the constructivist each reject the Humean and verificationist claim that reference to hidden mechanisms, essences, and causal powers is, on "rational reconstruction," eliminable from the findings of science. They agree scientists' methods and conceptions are determined by ineliminably metaphysical conceptions about the basic sorts of mechanisms, processes, and forces that operate to produce the phenomena under study and that this dependence is not merely a psychological quirk of the "context of invention" to be rationally reconstructed away in the "context of confirmation." They agree too in rejecting the eliminative Humean or regularity account of the causal powers and relations discovered by scientific inquiry. So where does the difference lie, what is the import of the question of the mind or theory independence of reality given that both parties reject empiricism?

The answer, subject to an important qualification, is that the realist denies, while the constructivist affirms, that the adoption of theories, paradigms, conceptual frameworks, perspectives – or the having of associated interests, intentions, purposes, etc. – in some way constitutes, or contributes to the constitution of, the causal powers of, and the causal relations between, the objects scientists study in the context of those theories, interests, etc. Of course (here is the qualification) the realist does not deny that the adoption of theories, etc., and the having of projects or interests, are themselves causal phenomena and thus contribute *causally* to the establishment of, for example, those causal factors that are explanatory in, for example, the history, philosophy and sociology of science and that in consequence the adoption of a theory in such a discipline could contribute causally to the causal powers and relations that are the subject matter of the theory itself. What the realist denies is that there is some further sort of contribution (logical, conceptual, socially constructive, or the like) which the adoption of theories or the having of interests makes to the establishment of causal powers and relations.

Thus the realist denies the *noncausal* contribution of minds and (the adoption of) theories to the establishment of causal powers and relations, whereas the constructivist insists that such a contribution is fundamental. While the present paper focuses primarily on the realist's abductive argument against empiricism, it is important to note two constraints that a suitably developed realist explanation of the reliability of scientific methods must meet if there is to be any prospect of its serving as the basis for a rebuttal to constructivism. In the first place, the definitions of natural kinds, categories, etc., to which the realist's explanation makes essential reference are, in a certain sense, interest dependent. The properties and causal powers that are relevant to explanation or prediction depend on the practical or theoretical projects being undertaken. Thus appropriateness of definitions and conceptual frameworks depends upon the interests with respect to which they are to be employed. The realist must acknowledge this fact in a way which is compatible with denying that the interest dependence in question involves any noncausal contribution of the adoption of interests or projects to the causal powers of the objects of scientific study.

Similarly, as Quine and others have reminded us, even when an agenda of interests and projects is fixed, there may be several ways of defining kinds and categories – of "cutting the world at its joints" – that are equally adequate to the task of reflecting explanatorily significant causal relations (even as the realist understands those relations). It may sometimes happen that the theoretical commitments of two such frameworks will appear to involve conflicting metaphysical conceptions. The choice between such frameworks will be, for the realist, arbitrary. Thus the realist's account of approximation must not treat one such framework as more nearly approximately true than the others, despite apparent metaphysical conflicts; certainly it must not treat the adoption of one rather than another as contributing noncausally to the establishment of causal relations or to similar settling of matters metaphysical.

It is by no means uncontroversial that arbitrariness and the interest dependence of kinds can be treated in the way the realist requires. For the purposes of the present essay I'll assume that an appropriate realist treatment is possible, while acknowledging that, in an essay in which constructivism rather then empiricism was the primary target, the question would require more extensive treatment. Two other issues regarding mind independence deserve our brief attention.

2.4.2. Mind Independence and the Causal Role of Minds

We have seen that the realist acknowledges the causal role of mental phenomena (since, e.g., she explains the reliability of scientific method by reference to the causal powers of approximately true beliefs) and differs from constructivists only in that she denies such phenomena a noncausal role in constituting causal structure. Nevertheless there are cases in which the attribution of a plainly causal role to mental phenomena has been seen as supporting constructivism. Two such cases deserve attention. First, scholars who are impressed by the social role of ideology often claim that "human nature" and the "natures" of various socially defined groups are "social constructions," and often they appear to mean by this, at least in the first instance, that the actual psychological capacities and tendencies exhibited by people generally or by members of socially defined groups are significantly determined by the ideologically established beliefs about psychological tendencies and capacities that are accepted in their own culture – determined in such a way as to tend to make their psychologies conform to the ideology.

Interestingly, many who make such claims seem to take this mode of social construction to be appropriate to a constructivist conception of reality and of knowledge. Plainly this is not so. Whatever the independent evidence for constructivism, the fact that culturally transmitted stereotypes causally influence the actual psychological makeup of those stereotyped provides no evidence of the sort of *non*causal determination of causal structure by minds or theories that the constructivist requires.

The second case concerns solutions to the problem of defining the notion of measurement in quantum mechanics. According to one important conception, part of what characterizes measurements is that they are epistemically relevant interactions so that measurement is defined in terms of knowledge – that is in terms of something (one component of which is) mental – and it is a special sort of interaction with a knowing system that produces discontinuous changes in physical state and results in sharp values for measured quantities. It is sometimes added that the explanation for the fact that measurements are not governed by

Schrödinger's equation is that they involve interactions between a physical system (whose isolated time evolution is governed by that equation) and a nonphysical mind. Whether or not the second suggestion is adopted, it is sometimes suggested that the special role of knowing systems thus identified refutes realism because it shows that the phenomena studied by scientists – in particular the results of their experimental measurements – are mind dependent. Reflection shows that this interpretation (even in its dualist version) simply assigns a distinctive causal role to certain mental phenomena. No noncausal social construction of causal structure is suggested. Indeed, the development of quantum mechanics might well be cited as the most dramatic recent demonstration of our *inability* to define causal reality in accordance with our conceptual schemes (for an excellent discussion see McMullin 1984).

2.5. Homeostatic Property-Cluster Definitions, Realism and Bivalence

There is an established practice of identifying realism regarding a body of inquiry with the view that all of the sentences in the vocabulary employed within it have determinate mind-independent truth values and such a conception of realism places a significant constraint on any realist account of the growth of approximate knowledge. We have just seen that the requirement of mind independence must be carefully qualified. Moreover, the role in approximation that the realist assigns to partial denotation and to denotational refinement (see 2.1.4) precludes any understanding according to which scientific statements must have determinate truth value: statements involving partially denoting expressions might be true on one denotational refinement and false on another.

There is a quite different way in which a realist conception of scientific language predicts failures of bivalence, and it is important to our understanding of the realist's explanatory project both because it reflects another dimension of dialectical complexity in the realist's account of approximation and because it provides the philosophical machinery for a deeper analysis of the underlying notion of scientific rationality.

The sorts of essential definition of substances anticipated by Locke and reflected in the currently accepted natural definitions of chemical kinds by molecular formulas (e.g., "water = H_2O") appear to specify necessary and sufficient conditions for membership in the kind in question. Recent *non*naturalistic property-cluster or criterial attribute theories in the "ordinary language" tradition suggest the possibility of definitions that do not provide necessary and sufficient conditions. Instead, some terms are said to be defined by a collection of properties such that the possession of an adequate number of those properties is sufficient for falling within the extension of the term. It is supposed to be a conceptual (and thus an a priori) matter what properties belong in the cluster and which combinations of them are sufficient for falling under the term. However,

it is usually insisted that the kinds corresponding to such terms are "open tex-
tured," so that there is some indeterminacy in extension legitimately associated
with property-cluster or criterial attribute definitions. The "imprecision" or
"vagueness" of such definitions is seen as a perfectly appropriate feature of ordinary linguistic usage, in contrast to the artificial precision suggested by rigidly
formalistic positivist conceptions of proper language use.

I doubt that there are any terms whose definitions actually fit the ordinary-
language model, because I doubt that there are any significant "conceptual truths"
at all. I believe however that terms with somewhat similar definitions are commonplace in the special sciences that study complex phenomena. Here's what I
think often happens (I formulate the account for monadic property terms; the account is intended to apply in the obvious way to the cases of terms for polyadic
relations, magnitudes, etc):

(i) There is a family F of properties that are contingently clustered in nature
in the sense that they co-occur in an important number of cases.

(ii) Their co-occurrence is, at least typically, the result of what may be
metaphorically (sometimes literally) described as a sort of *homeostasis*. Either
the presence of some of the properties in F tends (under appropriate conditions)
to favor the presence of the others, or there are underlying mechanisms or
processes that tend to maintain the presence of the properties in F, or both.

(iii) The homeostatic clustering of the properties in F is causally important:
there are (theoretically or practically) important effects that are produced by a
conjoint occurrence of (many of) the properties in F together with (some or all
of) the underlying mechanisms in question.

(iv) There is a kind term *t* that is applied to things in which the homeostatic
clustering of most of the properties in F occurs.

(v) *t* has no analytic definition; rather all or part of the homeostatic cluster F,
together with some or all of the mechanisms that underlie it, provide the natural
definition of *t*. The question of just which properties and mechanisms belong in
the definition of *t* is an a posteriori question–often a difficult theoretical one.

(vi) Imperfect homeostasis is nomologically possible or actual: some thing
may display some but not all of the properties in F; some but not all of the relevant
underlying homeostatic mechanisms may be present.

(vii) In such cases, the relative importance of the various properties in F and
of the various mechanisms in determining whether the thing falls under *t*–if it
can be determined at all–is a theoretical rather than a conceptual issue.

(viii) Moreover, there will be many cases of extensional vagueness that are
such that they are not resolvable even given all the relevant facts and all the true
theories. There will be things that display some but not all of the properties in
F (and/or in which some but not all of the relevant homeostatic mechanisms operate) such that no rational considerations dictate whether or not they are to be
classed under *t*, assuming that a dichotomous choice is to be made.

(ix) The causal importance of the homeostatic property cluster F together with the relevant underlying homeostatic mechanisms is such that the kind or property denoted by t is a natural kind (see section 2.1.3).

(x) No refinement of usage that replaces t by a significantly less extensionally vague term will preserve the naturalness of the kind referred to. Any such refinement would either require that we treat as important distinctions that are irrelevant to causal explanation or to induction, or that we ignore similarities that are important in just these ways.

(xi) The homeostatic property cluster that serves to define *t* is not individuated extensionally. Instead, the property cluster is individuated like a (type or token) historical object or process: certain changes over time (or in space) in the property cluster or in the underlying homeostatic mechanisms preserve the identity of the defining cluster. In consequence, the properties that determine the conditions for falling under *t* may vary over time (or space), *while* t *continues to have the same definition.* The historicity of the individuation criterion for the definitional property cluster reflects the explanatory or inductive significance (for the relevant branches of theoretical or practical inquiry) of the historical development of the property cluster and of the causal factors that produce it, and considerations of explanatory and inductive significance determine the appropriate standards of individuation for the property cluster itself. The historicity of the individuation conditions for the property cluster is thus essential for the naturalness of the kind to which t refers.

The paradigm cases of natural kinds – biological species – are examples of homeostatic-cluster kinds. The appropriateness of any particular biological species for induction and explanation in biology depends upon the imperfectly shared and homeostatically related morphological, physiological and behavioral features that characterize its members. The definitional role of mechanisms of homeostasis is reflected in the role of interbreeding in the modern species concept; for sexually reproducing species, the exchange of genetic material between populations is thought by some evolutionary biologists to be essential to the homeostatic unity of the other properties characteristic of the species, and it is thus reflected in the species definition that they propose (see Mayr 1970). The *necessary* indeterminacy in extension of species terms is a consequence of evolutionary theory, as Darwin observed: speciation depends on the existence of populations that are intermediate between the parent species and the emerging one. Any "refinement" of classification that artificially eliminated the resulting indeterminacy in classification would obscure the central fact about speciation upon which the cogency of evolutionary theory depends.

Similarly, the property-cluster and homeostatic mechanisms that define a species must be individuated nonextensionally as a processlike historical entity. It is universally recognized that selection for characters that enhance reproductive isolation from related species is a significant factor in phyletic evolution, and it is one which necessarily alters over time the species's defining property cluster and homeostatic mechanisms (Mayr 1970).

It follows that a consistently developed scientific realism *predicts* indeterminacy for those natural kind or property terms that refer to complex homeostatic phenomena; such indeterminacy is a necessary consequence of "cutting the world at its (largely mind-independent) joints" (contrast, e.g., Putnam 1983 on "metaphysical realism" and vagueness). Realists' accounts of approximation need not honor bivalence even when partial denotation is not at issue. Similarly, scientific realism predicts the existence of nonextensionally individuated definitional clusters for at least some natural kinds, and thus it treats as legitimate vehicles for the growth of approximate knowledge linguistic practices that would, from a more traditional empiricist perspective, look like diachronic inconsistencies in the standards for the application of such natural kind terms.

Moreover, the homeostatic-cluster conception of definitions may permit a more perspicuous formulation of the central explanatory thesis of scientific realism. I have argued elsewhere (Boyd 1979, 1982, 1983) for an understanding of knowledge and of reference according to which (although I did not use this terminology) the relations 'x knows that y' and 'x refers to y' possess homeostatic property-cluster definitions. I will suggest in section 3.7 that scientific rationality has a homeostatic property-cluster definition and that the realist's explanation for the reliability of scientific methods is best understood as the crucial component in an explanation of the homeostatic unity of scientific rationality.

Not all challenges to realism that arise from considerations about bivalence require in rebuttal an appeal to the possibility of actual bivalence failure. For example, the measurement problem in quantum mechanics is sometimes put by saying that quantum systems lack determinate values of classical magnitudes prior to measurement, and the problem is to characterize the interactions that relieve the indeterminacy with respect to a particular magnitude. Sometimes the alleged indeterminacy prior to measurement is seen as an indication of the failure of realism. Realism is seen as predicting determinacy for (premeasurement) values of classical magnitudes.

In response the realist need not appeal to the possibility of a realist explanation for failures of bivalence. There are two ways of understanding the claim about a physical system that it possesses a determinate value of a classical magnitude, a determinate component of orbital angular momentum, for example. On the first understanding, that claim is understood to incorporate the classical *mis*conception of the logical status of statements about angular momentum, in which case the statement is always false, in however many respects special cases of such statements may also have been usefully approximately true. Alternatively, the statement may be interpreted as attributing to the system an eigenstate of the relevant operator, in which case it need not be false, but it has, depending on the system in question, some determinate truth value. On careful analysis there is no bivalence failure here to explain.

3. Approximate Truth and Parametric Specification: The Realist's Explanation as Ordinary Science

3.1. The Status of the Realist's Explanation

Recall that the argumentative strategy proposed in section 1.3 calls for us to first assess the evidence for the realist's explanation for the instrumental reliability of scientific methods considered as an ordinary scientific hypothesis. If the realist's explanation appears well confirmed, then there will remain the further and more distinctly philosophical task of determining whether or not, with respect to the realist's explanation itself, it is legitimate to adopt the realist interpretation without which no defense of a realist position in the philosophy of science is forthcoming.

This approach presupposes that the realist's explanation has the form of an ordinary causal explanation in science subject to confirmation or disconfirmation by ordinary scientific standards. Two considerations might suggest that it does not. First, some philosophical explanations of epistemic matters seem noncausal; this is true, for example, of some transcendental explanations and of some "ordinary language" analyses of notions like "evidence," "reliable," "justification," and the like. Secondly, there are ways of thinking of the notions of truth and approximate truth (disquotational analyses, for example) that make them noncausal.

The realist's conception of the epistemology and semantics of scientific theories does not raise any of these problems. Truth is definable from "primitive denotation" (Field 1974), and denotation, on the realist's account, is an epistemic and thus a causal matter; truth is correspondence truth and correspondence is a matter of complex causal interactions. Similarly, to talk of respects of approximation to the truth is to talk of respects of similarity and difference between actual causal situations and certain possible ones. It is philosophically challenging to give a general account of the nature of such comparisons with counterfactual possibilities, but such comparisons are so routine a feature of ordinary causal reasoning in science (including reasoning about the reliability of particular methods) that there is no reason to suppose that they raise difficulties in the present context.

Likewise the explanatory claims of the realist are perfectly ordinary causal claims. Under certain sorts of historical and social circumstances individually and socially held beliefs are said to exhibit a particular causal power – a tendency to generate methods that are (causally) conducive to the establishment of approximate knowledge – when they are in causally relevant ways approximately true. However controversial, this is an ordinary causal thesis about the interactions of scientific communities and the rest of the world. We may reasonably inquire about how it fares by ordinary scientific standards of evidence. It is to this issue that we now turn our attention.

3.2. Does the History of Science Immediately Refute the Realist's Explanation?

According to the historical objection, the realist's explanation for the reliability of scientific method is refuted by the fact that there have been episodes in the history of science during which methodological practices were successful, but during which the relevant background theories were not, by contemporary standards, approximately true as the realist's explanation requires. The realist's response comes in two parts.

First, the realist's explanation does not require that scientists, even during periods of mature inquiry, be especially good at doing metaphysics. The realist need not necessarily show about any episode in the history of science that the relevant background theories are close to the truth on metaphysical matters. The realist's position is not compromised by any respects of error in earlier background theories that do not undermine her appeal to the specific respects of approximation regarding unobservable phenomena that are crucial to her explanation of the reliability of methods during that episode.

Second, the realist's account of the methods of science predicts that there will be early stages in the history of any currently mature science in which the relevant background theories will have been too far from the truth to ensure the sort of reliability of methods that is characteristic of mature sciences. This conclusion is a consequence of the radical contingency in epistemology dictated by the realist explanation for the reliability of scientific methods and, in particular, of the claim that it is, in an epistemically important sense, accidental that the earliest relevantly approximately true theories arise within any scientific discipline. Of course I do not mean that no historical explanations are possible for particular early successes, but only that, according to the realist, the explanation cannot involve appeal to the operation of rational methods with anything like the reliability of the methods of (what from the contemporary point of view are) theoretically more mature stages in the same sciences.

In sum, the realist's explanation is vulnerable to straightforward refutation by the phenomenon of successful science guided by deeply false background theories only if (a) the relevant historical episodes involve the operation of methods that exhibit the profound and routine reliability of judgments of projectability and related matters characteristic of the most mature sciences in the twentieth century, and (b) the respects of falsity in the relevant background theories are not merely deep but such as to preclude an explanation of that reliability by appeal to the respects in which those theories are approximately true. The tendency in recent empiricist philosophy of science towards realism reflects precisely the opposite conception: philosophers were tempted by realism precisely because they thought they could see how to offer a realist explanation of the reliability of methodological practices in highly successful science, and they lost their confidence in alternative empiricist "rational reconstructions" of those methods. In any event what I envision as the realist's reply to the historical objection is simply that there aren't actual cases satisfying (a) and (b). Realism is, after all, supposed to be an empirical thesis, and here is one of the empirical claims upon which it rests.

3.3. Triviality, Contrivance, and the Methodology of Parametric Specification

Against the charge of immediate historical falsification, the realist replies by insisting, as the logic of her explanations dictates anyway, that her thesis is that background theories in mature sciences must be seen as approximately true in relevant respects. As we saw in section 1.2, the realist now faces the challenge that her explanations are trivial: that any consistent theory is true in some respects, and that she has offered no general theory of the relative importance of respects of approximate truth. Here the reply is the obvious one that the respects of approximation that are important are those that are required to sustain the realist's distinctive explanation of the reliability of scientific methods and that it is with respect to these that approximations to the truth are claimed. The reply is successful just in case the charge of contrivance can be met: just in case, that is, the realist can argue that, even in the absence of a general context- and episode-neutral account of degrees of approximation, her appeal to respects of approximation appropriate to her own theoretical project does not constitute an ad hoc and thus methodologically inappropriate contrivance.

In order to assess the prospects for the realist's explanations we need to know what distinguishes such contrivances from methodologically appropriate appeals to context-specific specifications of causal variables. Fortunately the question is not esoteric; frequently, especially in the context of historical explanations, we confirm theories that appeal to context-dependent specifications of causal parameters and the methodology for avoiding ad hoc theorizing is well understood. Consider for example explanations in evolutionary theory. There are a variety of possible evolutionary mechanisms – individual selection, kin selection, genetic drift, selection for pleiotropically linked traits, etc. – for no one of which does evolutionary theory provide a context-independent prediction of its influence in any particular evolutionary episode. Moreover, in particular evolutionary episodes several of these factors may operate, and there is no context-independent way of predicting their relative influence. Still, the modern evolutionary explanation for the diversity of life is well confirmed. What methodological principles permit us to treat the explanations provided by evolutionary theory as appropriate rather than ad hoc, and as appropriate for "inductive inference to the best explanation"?

The answer is pretty clear. What we require of the various individual evolutionary explanations for particular features of living organisms is that they cohere not only with each other but with the independent results of inquiry in the related scientific disciplines: geology, genetics, developmental biology, animal behavior, atmospheric sciences, oceanography, anthropology, etc. This requirement of integration of the various particular explanations into the broader framework of scientific knowledge constitutes our methodological safeguard against the possibility that the apparent explanatory successes of evolutionary theory are reflections of mere contrivance. This pattern is quite general: particular explanations provide evidence for a broader theory whose explanatory resources they exploit just in case theory-dependent evidential standards, including requirements of theoretical integration, dictate the acceptance of the particular explanations, and just in case the success of those individual explanations lends inductive support for the causal claims of the broader theory (Boyd 1985b).

Exactly the same standards apply, of course, to the realist's broad explanation for the reliability of scientific methods. The charge of contrivance is met just in case the realist's explanations for the reliability of methods in particular episodes, including the context-dependent specifications of respects of approximation they contain, are independently supported by scientific evidence, and in particular that they pass the test of coherence with the rest of established scientific theory, and (this is the easier part) just in case these particular realist explanations lend inductive support to the broader realist explanatory picture of scientific epistemology.

3.4. The Local Coherence of Realism

Are the individual realist explanations for the reliability of specific scientific methods well confirmed and do they in particular cohere appropriately with the rest of science? Do they inductively support the general realist conception of the growth of approximate knowledge? At an important level of analysis the answer to both questions must be "obviously yes."

The particular realist explanations of the reliability of methods fall roughly into two categories. In the first category are the theoretical explanations for the reliability of particular measurement and computational procedures and for the reliability of various sorts of controls and other features of the design of experimental and observational studies. In the second category are the theoretical explanations for the reliability of the judgments of projectability which determine the broader outlines of rational experimental and observational inquiry. Explanations in either category may be either static or dialectical. By a static explanation I understand an explanation that explains the reliability of some piece of methodology by appeal to the approximate truth of some theories that have been long established at the time of the relevant methodological judgments; dialectical explanations explain the reliability of some novel feature of methodology or of some revision of a previously established methodological practice by appealing to changes in theoretical outlook that bring about a closer approximation to the truth along relevant lines.

At any given time in the history of recent science, individual realist explanations in the first category both static and dialectical look just like well-established pieces of boringly normal science: they are the sorts of claims that are routinely made explicit in the methods sections of papers in the empirical sciences, in which scientists explain the appropriateness of research design. Most explanations of the static sort and almost all of the dialectical ones will embody reference to context-specific degrees and respects of approximation in the current theoretical conception or its immediate predecessors. Those explicit pieces of scientific theorizing are not produced in service of any philosophical or historical project, realist or otherwise. In the better established sciences they are apparently as well confirmed as anything gets; certainly there is no evidence that they fail to cohere with the rest of established science. That, after all, is what made such pieces of ordinary science so disturbing to empiricists. The prospect that they are vulnerable to the contrivance objection is vanishingly remote.

Scientists seem rarely to investigate explicitly the causal question of the reliability of particular projectability judgments *under that description*. They do however offer justifications for their own methodological judgments, critiques of such judgments by others, and proposals for changes in such judgments. Such justifications are made explicit in published papers, in referees' reports, in grant proposals, in the introductory parts of experimental papers, and in theoretical papers and books and the judgments they justify are *in fact* judgments of projectability of the sort to which the realist explanation refers. It is all but the consensus position among students of the logic of scientific inference (e.g., Hanson 1958, Kuhn 1970, Quine 1969a; Van Fraassen 1980) that ordinary scientific standards of reasoning treat these projectability judgments as inductive inferences from background theories, just as realism requires. Here again the justifications in question routinely appeal to context-specific respects of approximation, especially in cases in which they mirror realist explanations of the dialectical sort. There is again no prospect that scientists' reasoning in such cases is contrived to serve a philosophical purpose nor is there any reason to hold that the requirement of coherence with the rest of science is not honored in such reasoning – indeed it is in reasoning of this sort that the requirement of coherence finds its expression in ordinary science!

Here then is the phenomenon of *local coherence*: the explicit and near-explicit findings of ordinary science examined synchronically seem to strongly confirm, if only tacitly, the particular explanations for the reliability of projectability judgments on which the realist's explanatory enterprise rests and they appear to do so in a way that subjects the context-dependent judgments of relevant respects of approximation which they contain to the appropriate requirement of coherence.

Do the particular realist explanations we are considering, taken together, inductively support the realist conception of scientific epistemology developed in part 2? Here we cannot defer to any particular science except philosophy, but we can observe that the whole tendency to take realism seriously as an alternative to logical empiricism from the mid-1950s on reflects the extremely widespread judgment among philosophers of science that the actual practices of science *appear* to require a realist explanation. I conclude that, if we examine the question *pre*philosophically, there appears to be very good reason to hold that the realist's explanation for the reliability of scientific methods is well confirmed as a scientific hypothesis and in particular, that there is no reason to think that the realist's approach to the problem of parametric specification is any more in doubt than, say, that of the evolutionary biologists who must also rely on specifications not given antecedently by a context-independent formula.

We turn now to the question of what the distinctly philosophical dimension is to the confirmation of the realist's explanatory hypothesis. The elaborate machinery rehearsed in part 2 indicates that a lot is going on philosophically. Some of it is relevant only to the question of circularity, but much is relevant also to a defense of realism as a scientific thesis in the methodological climate created by the philosophical disputes over realism.

3.5. What's Distinctly Philosophical? I: Diachronic Patterns of Inference and of Language Use

Central to the realist's abductive argument for realism is the claim that no alternative exists that adequately explains the reliability of scientific methods or justifies their use. It is possible to *imagine* that a case along these lines for realism – or at any rate against the verificationist insistence that knowledge of unobservables is impossible – could be made by the synchronic examination of only a few episodes in the history of science for which only realist explanations and justifications seem available. Nevertheless the deep plausibility of empiricist epistemological principles, especially the evidential indistinguishability thesis, is so great that it is doubtful that realism about a few isolated cases would, even as a scientific hypothesis, be rationally acceptable. Instead the plausibility of any individual realist explanation seems to rest upon diachronic considerations that provide additional and crucial support for the general realist explanation of the reliability of scientific methods. In effect, the role of these diachronic considerations is to establish that the individual synchronic-realist explanations can be coherently integrated into a scientifically acceptable historical conception of the reliability of scientific methodology.

In particular there are two patterns in the history of science whose recognition

279

is a distinctive contribution of philosophers and historians in making the case for the realist's explanations. In the first place, there is the utterly commonplace phenomenon of *mutual ratification* between consecutive stages in the development of scientific disciplines. It is routine in the case of theoretical innovations that (a) the new and innovative theoretical proposal is such that the only justification scientists have for accepting it, given the relevant evidence, is that it resolves some scientific problem or question *while preserving certain key features of the earlier theoretical conceptions;* and (b) the new proposal ratifies the earlier conception as approximately true in just those respects that justify their role in its own acceptance. Moreover the patterns of mutual ratification are characteristically seen to be *retrospectively sustained*: although later theoretical innovations typically require a revision in our estimates of the degrees and respects of approximation of both the earlier innovative proposals and their predecessors, the initially discernable relation of mutual ratification is typically sustained as a very good first approximation to the evidentially and methodologically important relations between the innovation and its predecessor theories. It is the ubiquity of this sort of *retrospectively sustained mutual ratification* and the difficulty in "rationally reconstructing" it away with respect to the justification of theoretical innovations that has made the case for realism so plausible.

A second pattern concerns the use of scientific language. The realist conception of projectability requires that the categories that scientists employ in formulating general laws and causal claims typically reflect underlying causal structures rather than conventionally specified nominal essences, and many of the changes in classificatory practice for which individual realist explanations are forthcoming seem to indicate an attempt to obtain a fit between categories and causal structure. It is essential to the case for realism that this pattern in scientific language use be sustained: that the diachronic linguistic behavior of scientists involves an apparent disposition to take the definitions of scientific kinds, relations, magnitudes, etc., to be revisable in the light of new data and new theoretical developments. Thus the identification of just such a pattern of *apparent essentialism* in the actual linguistic practices in scientific communities is an important distinctly philosophical contribution to the case for the realist's explanation of the reliability of scientific method.

3.6. What's Distinctly Philosophical? II: Epistemological, Metaphysical, and Semantic Underpinnings

The ubiquitous patterns of retrospectively sustained mutual ratification and apparent essentialism constitute philosophical reasons to accept the realist's explanation, and the recognition of those patterns was a central factor in the emergence of contemporary scientific realism. Still, their effect would not have been so significant were it not for more theoretical attempts to understand their philosophical import. The obvious examples here are causal theories of reference and associated naturalistic conceptions of definition. Had it not proven possible to articulate these distinctly philosophical theories, then it might have been rational to hold that the apparently rational theory-and-evidence-driven revision of definitions in science was only apparent, or only apparently rational. The initial case for the realist explanation would have been crucially undermined.

Analogous considerations hold for the epistemological dimension. Both the realist explanations for the reliability of scientific methods in particular cases and the view that the ubiquity of the pattern of mutual ratification supports the broader realist explanation entail that evidential considerations in science are deeply theory dependent. Were it not possible to provide a realist epistemological framework that incorporates this conclusion – and in particular were it not possible to articulate that framework so as to refute the evidential indistinguishability thesis and make palatable the consequent abandonment of foundationalism – then it would have not been rational to take either the particular explanations or the pattern of mutual ratification as significant support for the realist explanation. Thus the development of a nonfoundationalist realist treatment of projectability judgments and the incorporation of that treatment into an independently developing tradition of nonfoundationalist naturalism in epistemology proves to have been essential for the rational acceptance of the realist explanation.

On to metaphysics. The causal theory of reference and the naturalistic conceptions within epistemology with which realist anti-foundationalism can be profitably assimilated all appear to reflect a distinctly non-Humean conception of causal relations. The cogency of these fundamental elements in the defense of the realist's explanation depend therefore (at least prima facie) on the successful articulation of a non-Humean conception of causation (e.g., Boyd 1985b; Mackie 1974; Shoemaker 1980).

Acceptance of the realist's explanation as a scientific theory does not entail the acceptance of scientific realism, since the realist's explanation might itself be interpreted nonrealistically. What I have been suggesting is that nevertheless the realist's explanation is sufficiently novel in its apparent epistemological, semantic, and metaphysical implications that the articulation of just the sort of broader realistic and naturalistic conceptions of (scientific and other) knowledge, of language, and of metaphysics indicated in part 2 is essential for the defense of that explanation.

I think that the picture just presented captures the current case that the realist's explanation for the reliability of scientific methods is a well-confirmed scientific theory, context-dependent specifications of respects of approximation notwithstanding. An even broader philosophical setting for that case is available if we exploit the distinctly naturalistic conception of homeostatic property-cluster definitions outlined in section 2.5

3.7. Realism and the Homeostatic Character of Scientific Rationality

I argued in section 2.5 that lots of natural kinds, properties, etc. possess homeostatic property-cluster definitions, and I suggested that knowledge and reference are among them. I want now to suggest a similar homeostatic cluster treatment of scientific rationality itself. Ordinarily we think of scientific rationality as being exhibited in two different features of the practice of science: the high level of deliberative rationality in the reasoning of researchers, and the spectacular successes of scientific research in understanding and predicting natural phenomena. If foundationalism is mistaken, as it surely seems to be, then the first of these features does not logically entail the second, and the realist explanation may be thought of as explaining why (and when) they reliably co-occur. Here is a kind of homeostasis of the two distinct components of scientific rationality.

Once it is recognized that this co-occurrence is a causal matter, then it is easy to see that at a finer level of analysis there is a family of similar sorts of co-occurrences requiring explanation. The methodological norms in a particular subdiscipline are set not only by the background theoretical findings in that subdiscipline but as well by findings from other subdisciplines and from quite different disciplines altogether. That the methodological norms determined by such a wide range of theories should be unified enough to be a practical guide to successful scientific research requires explanation. Why aren't the resulting methodological norms characteristically irreconcilably conflicting, for instance?

Similarly, scientists working largely independently within different disciplines frequently converge on the same solution to a problem they may not have recognized that they have in common. Why should this happen? Likewise, it often happens that largely independently developing disciplines become ripe for interdisciplinary work, and their largely independently developed theories and methodologies prove (with some difficult but not impossible negotiation) to be integrable. Why is this so frequently possible?

What I propose is that we think of scientific rationality as being defined by the homeostasis of all of these various components of scientific practice and that we should think of the realist explanation of the coincidence of deliberative rationality and theoretical and empirical success as the first step toward a more general realist explanation of the relevant homeostasis. It is even possible that this project could be extended fruitfully to incorporate a naturalistic conception of moral rationality (Boyd 1988; Brink 1984, 1989; Miller 1984b; Railton 1986; Sturgeon 1984a, 1984b).

If the proposal of the present section were to prove successful it would prima facie provide further support for the realist explanation and for the philosophical naturalism that underwrites it. However, we still need to know whether the realist's explanation should itself be understood realistically or whether instead, as the circularity objection suggests, that would simply beg the question against the anti-realist.

4. Meeting the Circularity Objection

4.1. Circularity and Philosophical Packages

According to the circularity objection, the realist's explanation for the success of scientific methods, even if well confirmed, cannot without begging the question be interpreted realistically and thus cannot without circularity be treated as confirming scientific realism. The problem posed by this objection is faced not only by the particular defense of realism under consideration but by almost any plausible defense of scientific realism.

The reason is simple: in all but the most trivial cases the defense of realism regarding one or more theories or traditions will require the defense of a theory of epistemic contact that spells out the sort of epistemically relevant causal relations that are supposed to obtain between the subject matter of the theories or traditions and the behavior of the relevant inquirers. Because the realist thesis and the theory of epistemic contact that supports it are causal theses, their confirmation will always depend upon the confirmation of theories (or, for very simple cases, commonplaces) about the causal powers of the entities that are the putative subject matter of the theory or tradition in question. The confirmation of specific theories of epistemic contact will, in turn, depend in part upon theoretical considerations grounded in the best available theories of the relevant subject matter. Such theories will be a vital background assumption against which the evidence for the realist thesis is judged. As we have seen, the theory of epistemic contact, and (thus) the theories upon which its confirmation in turn depends, will themselves have to be understood realistically if they are to help to validate the realist thesis itself. But of course these theories will, in any plausible case, be subject to the same anti-realist assessments as the theory or tradition about which realism is initially in question. Indeed if that theory is a well-established contemporary theory it may *itself* provide the foundations for the relevant theory of epistemic contact! Is this not a point at which the defense of realism begs the question against anti-realists?

Here the answer is "no." If theories of epistemic contact by themselves constituted the sole argument of the realist against anti-realism, if for example the *sole* argument in favor of realism in atomic theory consisted of the articulation of an apparently well-confirmed theory of epistemic contact between scientists and atoms, their properties and their constituent parts, then the question would indeed be begged by the assumption that that theory itself should be understood realistically. The actual role of theories of epistemic contact is quite different. The issue of realism arises in the form we have been discussing only in the case of a theory or tradition of inquiry about which there is a prima facie case that it possess a theory-independent (even if unobservable) subject matter. The prima facie case for realism will thus rest upon the apparent confirmation of a (realistically understood) theory of epistemic contact. In the special case of realism defended along the lines proposed here, that theory of contact is the one embodied in the realist's explanation for the reliability of scientific methods. The defense of realism, however, depends not upon the theory of epistemic contact *alone* but upon the ability of realists to incorporate suitably elaborated versions of it into an epistemological, semantic and metaphysical conception of the theory or tradition in question (a *philosophical package*) that is superior to that those available to defender of the various anti-realist conceptions.

Thus, for example, the defense of realism regarding the tradition of atomic theory depends upon the best-confirmed atomic theories providing the basis for an apparently realistic theory of epistemic contact, but it depends as well upon additional, more explicitly philosophical considerations, which legitimize the realist treatment of such a theory. On the version of scientific realism presented here, these additional considerations are of two sorts. First, it is argued that only on a realist construal of atomic theory generally, and of the relevant theory of epistemic contact in particular, is it possible to avoid skepticism about the possibility of purely instrumental knowledge in physics and chemistry: knowledge of a sort acknowledged by empiricists and constructivists as well as by realists. Secondly, it is argued that the picture that emerges from a realist treatment of atomic theory is consonant in its departures from foundationalism and in its treatment of scientific language with other quite independently defensible developments in epistemology and semantic theory.

In such a dialectical setting, the dependence of the realist theories of epistemic contact upon a realist understanding of the theory or tradition in question (or of some closely related theory or tradition) need not constitute begging the question against the anti-realist. Fairness to the case for realism requires that realism be understood in a context provided by a realist interpretation of the apparently best-confirmed realist theories of epistemic contact and of the apparently best-confirmed substantive theories of the alleged (theory-independent) subject matter in question.

Importantly, just the same understanding of the issue is required by fairness to the case *against* realism. Both the empiricists' and the constructivists' anti-realist arguments depend upon the assumption that the realist accepts the prevailing theoretical conception and its associated methodology. The realist is understood to take the properties of the putative socially unconstructed referents of the terms of a theory or tradition to be, at least approximately, those required by (a realist understanding of) the apparently best-confirmed theories of the presumed subject matter and to accept the methodology dictated by them as approximately reliable. On those assumptions (but not without them) the empiricist can reason that the realist's position commits her to the possibility of investigating the properties of unobservable phenomena, and thus to an epistemological position against which the empiricist has very powerful arguments.

The constructivist anti-realist similarly assumes that the realist accepts a realist interpretation of the prevailing theoretical and methodological conceptions. Only on such an understanding is it clear that the realist is committed to the possibility of investigating a theory-independent reality using theory-dependent methods – just the possibility that the constructivist critique of realism rejects. Thus an adequate treatment of the controversy between realists and either of their standard opponents requires that we accept that the philosophical package offered in defense of realism contains the apparently best-confirmed theories of the alleged subject matter, realistically understood, and in particular that it be understood as incorporating an associated realistically understood conception of epistemic contact.

Once it is seen that no question is begged against the anti-realist by adopting a realist interpretation of the realist's explanation for the reliability of scientific methodology, we are left with the question: Suppose that the realist's explanation is well confirmed, then why would a realist philosophical package incorporating a realist version of that explanation be superior to an empiricist package incorporating the explanation instrumentally interpreted or to a constructivist package incorporating the realist's explanation understood as a piece of social construction? My main aim in the present essay is to show that the realist's appeal to a distinctively realist explanation for the growth of approximate knowledge, incorporating an appropriate context-and-episode-dependent account of relevant respects of approximation, does not involve any triviality, contrivance, or begging of the question – not to finish once and for all the task of defending realism. I will therefore indicate only briefly the outlines of the considerations that seem to me to justify a preference for the realist package over the two alternatives in question.

4.2. Against the Empiricist Package

The key argument for scientific realism according to the program presented here is that realism as a scientific hypothesis presents the only scientifically acceptable explanation for the realiability of scientific methods. The empiricist might be unimpressed by the demand for explanation in this case (Fine 1984, Van Fraassen 1980). Still the realist can also argue that accepting the realist explanation provides as well the only justification we have for accepting the instrumental findings of science (Boyd 1983, 1985a). One possible empiricist response is that we can justify accepting the inductive deliverances of an apparently realistic scientific method as a result of the second-order induction about induction whose conclusion is that reasoning like a realist in science is instrumentally reliable.

285

Since this conclusion is only about observables, the empiricist can accept it and employ it to justify accepting currently accepted theories as empirically adequate.

Against this rebuttal I have argued (Boyd 1983, 1985a) that the induction in question is demonstrably just as theory dependent as any other in science and is thus unavailable to the empiricist who is adopting the proposed strategy. Here is a possible reply: We justify the second-order induction by a third-order induction about inductions about induction, the third-order induction by appeal to a fourth-order induction, etc. For the nth case the justification for the relevant projectability judgments is provided not by apparently realistic theoretical considerations but by the n + first-order induction.

If I am right this last response is what the incorporation of the realist's explanation into an empiricist philosophical package would require if that package were to provide any even remotely plausible account of the justification of (instrumental) scientific knowledge. I claim that the resulting philosophical package would prove to be only remotely plausible in consequence. Here we have not just infinite regress but infinite ascent: each level of inductive inference is justified by appeal to a more abstract and problematical level of inductive inference. Given that the realist's package already incorporates an alternative, less speculative, and independently justified naturalistic epistemology I predict that it will prove superior.

4.3. Against the Constructivist Package

Response to the sort of constructivist philosophical package that might be constructed so as to include the realist's explanation for the reliability of scientific methods is substantially more difficult. Constructivism is a richer philosophical program than empiricism, and at the same time it incorporates features (often just the ones that add to its richness) whose consistency is disputable. Rather than even beginning to sort out all of the issues that a thoroughgoing realist response to constructivism would have to address, I will just indicate briefly how two quite standard objections to constructivism might be brought to bear on the proposed package.

In the first place, any adequate philosophical package will have to incorporate versions of most of the apparently best-established scientific and methodological findings. The suggestion outlined in section 2.4, that the establishment of social institutions and linguistic conventions does not contribute noncausally to the causal powers of the objects studied by participants in those institutions and conventions, has very deep roots in quite diverse features of our understanding both of causation and of social phenomena. Thus any constructivist philosophical package will be prima facie vulnerable at any point at which it incorporates a distinctly constructivist conception of the social construction of causal relations. The proposed constructivist package would incorporate this doubtful feature into its version of the naturalistic account of the reliability of scientific methods and thus in to the very center of its basic epistemology. It is doubtful therefore that the proposed package will afford as satisfactory a treatment of absolutely central epistemological issues as its realist rivals.

A second standard objection to constructivism is that the historical fact of anomalies indicates that the world scientists study does not have a structure logically, socially, or conceptually determined by the paradigms or theories they accept. It is beyond the scope of this essay to examine the variants on this objection and the range of possible replies. It cannot be doubted however that it does pose a serious challenge to the acceptability of any constructivist package. Since there are anomalies in methodological matters that exactly parallel those in theoretical matters, the incorporation of a doctrine of social construction of the reliability of scientific method seems hardly to strengthen the constructivist philosophical package.

I conclude that the resources exist for a spirited defense of a realist philosophical package against empiricist and constructivist alternatives, and in particular that the incorporation of a realist interpretation of the realist's explanation of the reliability of scientific methodology strengthens rather than (as the circularity challenge suggests) weakens the realist package.

References

Armstrong, D.M. 1973. *Belief, Truth and Knowledge.* Cambridge: Cambridge University Press.

Boyd, R. 1972. Determinism, Laws and Predictability in Principle. *Philosophy of Science* 39: 431-50.

———. 1973. Realism, Underdetermination and a Causal Theory of Evidence. *Noûs* 7: 1-12.

———. 1979. "Metaphor and Theory Change." In *Metaphor and Thought,* ed. A. Ortony. Cambridge: Cambridge University Press.

———. 1980. "Materialism Without Reductionism: What Physicalism Does Not Entail." In *Readings In Philosophy of Psychology,* vol.1, ed. N. Block. Cambridge: Harvard University Press.

———. 1982. Scientific Realism and Naturalistic Epistemology. In *PSA 1980. Volume Two,* eds. P.D. Asquith and R.N. Giere. East Lansing: Philosophy of Science Association.

———. 1983. On the Current Status of the Issue of Scientific Realism. *Erkenntnis* 19: 45-90.

———. 1985a. "Lex Orendi est Lex Credendi." In *Images of Science: Scientific Realism Versus Constructive Empiricism,* eds. Churchland and Hooker. Chicago: University of Chicago Press.

———. 1985b. "Observations, Explanatory Power, and Simplicity." In *Observation, Experiment, and Hypothesis In Modern Physical Science,* eds. P Achinstein and O. Hannaway. Cambridge: MIT Press.

———. 1985c. The Logician's Dilemma. *Erkenntnis* 22: 197-252.

———. 1988. "How to be a Moral Realist." In *Moral Realism,* ed. G. Sayre McCord. Ithaca: Cornell University Press.

Brink, D. 1984. Moral Realism and the Skeptical Arguments from Disagreement and Queerness. *Australasian Journal of Philosophy* 62.2: 111-25.

———. 1989. *Moral Realism and the Foundations of Ethics.* Cambridge: Cambridge University Press.

Byerly and Lazara. 1973. Realist Foundations of Measurement. *Philosophy of Science* 40: 10-28.

Carnap, R. 1934. *The Unity of Science* Trans. M. Black. London: Kegan Paul.

Feigl, H. 1956. "Some Major Issues and Developments in the Philosophy of Science of Logical Empiricism." in *Minnesota Studies in the Philosophy of Science*, vol. 1, *The Foundations of Science and the Concepts of Psychology and Psychoanalysis*, eds. H. Feigl and M. Scriven. Minneapolis: University of Minnesota Press.

Field, H. 1973. Theory Change and the Indeterminacy of Reference. *Journal of Philosophy* 70: 462-81.

——. 1974. Tarski's Theory of Truth. *Journal of Philosophy* 69: 347-75.

Fine, A. 1984. "The Natural Ontological Attitude." In *Scientific Realism*, ed. J. Leplin. Berkeley: University of California Press.

Goldman, A. 1967. A Causal Theory of Knowing. *Journal of Philosophy* 64: 357-72.

——. 1976. Discrimination and Perceptual Knowledge. *Journal of Philosophy* 73: 771-91.

Goodman, N. 1973. *Fact Fiction and Forecast*, 3d ed. Indianapolis and New York: Bobbs-Merrill.

Hanson, N. R. 1958. *Patterns of Discovery*. Cambridge: Cambridge University Press

Hardin, C., and Rosenberg, A. 1982. In Defense of Convergent Realism. *Philosophy of Science* 49: 604-15.

Kripke, S.A. 1971. "Identity and Necessity." In Identity and Individuation, ed. M.K. Munitz. New York: New York University Press.

——. 1972. Naming and Necessity. In *The Semantics of Natural Language*, eds. D. Davidson and G. Harman. Dordrecht: D. Reidel.

Kuhn, T. 1970. *The Structure of Scientific Revolutions*, 2d ed. Chicago: University of Chicago Press.

Laudan, L. 1981. A Confutation of Convergent Realism. *Philosophy of Science* 48:218-49.

Mackie, J. L. 1974. *The Cement of the Universe*. Oxford: Oxford University Press.

Mayr, E. 1970. *Populations, Species and Evolution*. Cambridge: Harvard University Press.

Mc Mullin, E. 1984. "A Case for Scientific Realism." In *Scientific Realism*, ed. J. Leplin. Berkeley: University of California Press.

Miller, R. 1984a. *Analyzing Marx*. Princeton: Princeton University Press.

——. 1984b. Ways of Moral Learning. Philosophical Review 94:507-56.

Nagel, E. 1961. *The Structure of Science*. New York: Harcourt Brace.

Putnam, H. 1962. "The Analytic and the Synthetic." In *Minnesota Studies in the Philosophy of Science, vol. 3, Realism and Reason*, eds. H. Feigl and G. Maxwell. Minneapolis: University of Minnesota Press.

——. 1972. "Explanation and Reference." In *Conceptual Change*, eds. G. Pearce and P. Maynard. Dordrecht: Reidel.

——. 1975a. "The Meaning of Meaning." In *Mind, Language and Reality*, ed. H. Putnam. Cambridge: Cambridge University Press.

——. 1975b. "Language and Reality." In *Mind, Language and Reality*. Cambridge: Cambridge University Press.

——. 1983. "Vagueness and Alternative Logic." In *Realism and Reason*. Cambridge: Cambridge University Press.

Quine, W.V.0. 1969a. "Natural Kinds." *Ontological Relativity and Other Essays*. New York: Columbia University Press.

——. 1969b. "Epistemology Naturalized." In *Ontological Relativity and Other Essays*. New York: Columbia University Press.

Railton, P. 1986. Moral Realism. *Philosophical Review* 95:163-207.

Rawls, J. 1971. *A Theory of Justice*. Cambridge: Harvard University Press.

Shoemaker, S. 1980. "Causality and Properties." *Time and Cause*, In ed. P. van Inwagen. Dordrecht: D. Reidel.

Sturgeon, N. 1984a. "Moral Explanations." In *Morality, Reason and Truth*, eds. D. Copp and D. Zimmerman. Totowa, N.J.: Rowman and Allanheld.

——. 1984b. Review of *Moral Relativism* and *Virtues and Vices*, by P. Foot. Journal of Philosophy 81: 326–33.

Tarski, A. 1951. "The Concept of Truth in Formalized Languages." In *Logic, Semantics and Metamathematics.* New York: Oxford University Press.

Van Fraassen, B. 1980. *The Scientific Image.* Oxford: Oxford University Press.

Varieties of Progress

1. A Problem Reposed

In conceiving of science as progressive we envisage it as a sequence of consensus practices that get better and better with time. Improvement need not be constant. Like the fortunes of a firm, the qualities of consensus practices may fluctuate. But, if a scientific field is making progress, there should be a general upward trend, comparable to the generally rising graph of profits that the firm's president proudly displays at board meetings.

The challenge is, of course, to say what "better and better" means in the scientific case, what corresponds to the accumulating dollars, yen, marks, etc. It is important to be clear from the start that the *first* task is to pinpoint those relations that should obtain among the members of a sequence of consensus practices if that sequence is to be progressive. To ask *how we know* that a field of science is making progress is to pose a separate question. There is an important connection between the questions: a definition of progress would be of little value if there were good reasons for thinking that it is impossible for us ever to know that we are making progress in the specified sense. Despite the connection, we should not confuse the issues, for example by lambasting a proposal for defining the notion of progress with the complaint that it has not yet been explained how we know that we are making progress. In my treatment, the definition will come first (this chapter). A response to various kinds of worries about concepts that I use will be presented later (Chapter 5).

Because one species of concern is so prevalent, it is worth forestalling related criticisms before we begin. I shall sometimes make free use of the concept of truth, explicating certain types of scientific progress as involving the attainment of truth. Moreover, I shall make judgments about the record of various scientists, past and present, in attaining truth, judgments based on the understanding of nature furnished by our current science. If the uses I make of the concept of truth are incoherent, then my account of progress will be wrong. Similarly, if the judgments about past and present attainment of truth that I make are incorrect or unjustified, then my errors will infect either my analysis of scientific progress or my claims about the ways in which science has made progress. Thus there are important objections that I shall

290

ultimately have to overcome. However, before those objections can be properly posed and properly addressed, the account of progress must first be presented, so that it can become clear just where and to what extent I rely on conceptions that others find dubious. In Chapter 5 I shall try to give critical voices the hearing they deserve. Meanwhile, I hope that my (wild? naive? optimistic? flagrant?) claims about truth in this chapter can be read with patience by those who are moved by the criticisms, as I formulate a position whose coherence and tenability can later be debated.

Let me turn now to a second qualification. Sometimes in the history of science, fields split, merge, or give birth to hybrid progeny. In such cases the community relations within science (as a whole) are altered, and there are new consensus practices that have some kinship with more than one prior consensus practice, or old consensus practices that bear important relations to more than one later consensus practice. For most of this chapter, I shall ignore the special complications of such cases. I shall consider progress within a field, supposing that there is a single sequence of communities, uniquely determined by the fact that some members of an earlier community are veterans in the immediate successor community, and that we examine the relations among the consensus practices of the communities in this sequence.

We judge the progressiveness of a sequence of practices by thinking about the pairwise relations between practices. Just as we evaluate the fortunes of a firm over a period of time by looking at the returns in successive time segments, so too with fields of science. Things are looking up if for *most* pairs of adjacent time segments the later returns are greater than the earlier returns and if any subsequence in which the returns successively diminish is followed by a subsequence in whose final segment the returns surpass any level previously attained. Similarly, we can count a sequence of practices as progressive if *most* members of the sequence are progressive with respect to their predecessor and if any nonprogressive subsequences are followed by a subsequence whose final member is progressive with respect to any earlier practice in the sequence. Obviously, more determinate conceptions can be introduced by eliminating the vague reference to "most" practices in the sequence. For example, we can define *strict* progress in a sequence of practices by demanding that each successor practice in the sequence be progressive with respect to its predecessor. A sequence may be said to be weakly progressive just in case its final member is progressive with respect to its first member.[1] The fundamental point, however, is that any of these ways of characterizing a notion of progressiveness for sequences is dependent on an antecedent notion of progressiveness as a binary relation among practices.

Because practices are multidimensional it is possible that a change from P_1 to P_2 should be progressive along some dimensions but not along others. I shall not suppose that there is some way in which the dimensions are weighted so as to yield an overall measure of progress. Instead, I shall distinguish varieties of progress. Combining this idea with the discussion of the last

1. I am grateful to Kim Sterelny for suggesting this formulation.

paragraph, let us say that the sequence of practices $P_1, \ldots, P_n$ is *broadly progressive* just in case for every pair of adjacent members there is a component of practice with respect to which the change from the earlier to the later is progressive *and* the change from P_1 to P_n is progressive with respect to every component of practice.

The fundamental project, then, is to define a family of relations that hold between two practices P_i and P_j when, for some component of practice, the change from P_i to P_j would be progressive. Our definitions should be subject to criteria of adequacy in that they need to make it clear how achieving a progressive sequence of practices would realize (or realize more closely) our goals. The account that follows will presuppose that there are goals for the project of inquiry that all people share—or ought to share. The varieties of progress I describe will be understood in terms of the greater achievement of these goals.

In the terms of the last chapter, the goals in question are impersonal. Some are epistemic, others nonepistemic. We need a specification of impersonal goals for science, goals that can ultimately be defended as worthy of universal endorsement. On the basis of this, we must identify binary relations of progressiveness among practices, which hold with respect to the various components of practice. Third, we should be able to show that when the later practices of fields of science stand in the relations of progressiveness to earlier practices, then the fields in question are realizing the shared impersonal goals (or realizing them more fully).

2. The Practical and the Cognitive

One theme recurs in the history of thinking about the goals of science: science ought to contribute to "the relief of man's estate," it should enable us to control nature—or perhaps, where we cannot control, to predict, and so adjust our behavior to an uncooperative world—it should supply the means for improving the quality and duration of human lives, and so forth. At first, this conception of science as promoting impersonal nonepistemic goals looks relatively easy to understand. Suppose that, at some particular time, we have a practical project of doing something, and that this project is stalled because we cannot answer a particular question or make a particular device. If the development of science enables us, at some later time, to answer the question or to make the device, then it appears that we have made *practical progress*.

Unfortunately, precisely because our ideas about what we want to achieve may be modified in the wake of a scientific accomplishment, the notion of practical progress proves far more difficult than we might have thought. I believe that we can only make sense of it in the light of a very general account of human flourishing, one that lies beyond the scope of this book. By contrast, the notion of cognitive progress, which initially appears much more ethereal and elusive, can be approached far more easily because of the relative narrowness of the set of impersonal epistemic goals.[2] Although the task of providing an account of practical progressiveness is an important one, it should follow the more tractable project of understanding cognitive progress.

3. Cognitive Progress: An Overview

The most obvious pure epistemic goal is truth. Indeed, talk of truth—or approximation to truth[3]—has dominated philosophical discussions of scientific

2. Thus we can conceive of ourselves as recognizing the value of a large number of fundamental impersonal goals, mostly nonepistemic. Different people may give very different weights to these goals. Moreover, there may be substantial disagreement with respect to the weights assigned to the set of epistemic goals vis-à-vis the nonepistemic complement. All this is compatible with a very small degree of variation concerning the *relative* weights of the epistemic goals. My picture of the situation is illustrated by the following (obviously artificial) suggestion. Assume that there are two fundamental epistemic goals E_1 and E_2, and a much larger set of fundamental (nonepistemic) practical goals $P_1, \ldots, P_n$. Each person's conception of value is given by an assignment of weights, $u_1, u_2, w_1, \ldots, w_n$. I imagine that

(i) The *absolute* values of all the weights are variable from person to person.
(ii) The *relative* values of the w_i vary from person to person.
(iii) The *relative* values of the u_i are constant: i.e., u_1/u_2 takes some value k which is the same for (almost) all people.

To sum up this idea in a simple way, we may suppose that people differ in their practical values, differ in their understanding of the value of the epistemic, but, insofar as they commit themselves to any epistemic projects, share the same fundamental value system. Here I generalize the Kuhnian idea (1962/70 42, 1977 322) that scientists are bound together by shared values, by proposing that, in our dedication to inquiry, we endorse a shared system of fundamental goals. (To repeat, defense of this apparently optimistic view will come in Chapter 5.) Note that my supposition would automatically be correct if, instead of two fundamental epistemic goals, E_1 and E_2, there were just one.

3. The core of the problem originally posed by David Miller (1974) is this. Consider two theories (axiomatic deductive systems) T_1 and T_2. Let C_{T1} and C_{T2} be the sets of true members of T_1 and T_2, F_{T1} and F_{T2} the sets of false members of T_1 and T_2. According to the approach taken by Popper, which has an obvious motivation, we can count T_1 as having greater verisimilitude than T_2 provided that C_{T1} includes C_{T2} and F_{T2} includes F_{T1} with at least one of the inclusions being proper (and similarly, with substitution of indices for T_2's having greater verisimilitude than T_1). Assume now that T_1 neither includes nor is included in T_2. Then there are a statement s that is in T_1 but not T_2 and a statement r that is in T_2 but not T_1. Consider cases:

(a) Both s and r are true: then s is in C_{T1} but not C_{T2} and r is in C_{T2} but not C_{T1}; so neither does C_{T1} include C_{T2} nor does C_{T2} include C_{T1}; hence, by Popper's criterion, the theories are incomparable.
(b) s is true and r is false: then s is in C_{T1} but not C_{T2}, so C_{T2} cannot include C_{T1}; let f be any statement in F_{T1} (there must be such statements unless we have the trivial case in which all of T_1 is true); then $(s \& f)$ is in F_{T1} but not in F_{T2}; hence F_{T2} cannot include F_{T1}; it follows that neither theory can have greater verisimilitude.
(c) s is false and r is true: r is in C_{T2} but not in C_{T1}; let f be any statement in F_{T2}; then $(r \& f)$ is in F_{T2} but not in F_{T1}; so it cannot be that the required inclusions hold in

progress. But, in my judgment, truth is not the important part of the story. Truth is very easy to get.[4] Careful observation and judicious reporting will enable you to expand the number of truths you believe. Once you have some truths, simple logical, mathematical, and statistical exercises will enable you to acquire lots more. Tacking truths together is something any hack can do (see Goldman 1986 123, Levi 1982 47, and, for the locus classicus of the point, Popper 1959 chapter 5). The trouble is that most of the truths that can be acquired in these ways are boring. Nobody is interested in the minutiae of the shapes and colors of the objects in your vicinity, the temperature fluctuations in your microenvironment, the infinite number of disjunctions you can generate with your favorite true statement as one disjunct, or the probabilities of the events in the many chance setups you can contrive with objects in your vicinity. What we want is *significant* truth. Perhaps, as I shall suggest later (Section 8), what we want is significance and *not* truth.

Scientists in the tradition that extends beyond the seventeenth century to the ancient Greeks have been moved by the impersonal epistemic aim of fathoming the structure of the world. In less aggressively realist language, what they have wanted to do (as a community) is to organize our experience of nature. Seventeenth century thinkers such as Boyle and Bacon attributed special significance to this project because, seeing nature as God's Creation, to expose the structure of nature was to recognize the divine intentions. That further link, useful, possibly even essential, though it may have been in showing how science contributes to our overall conception of what is valuable, can be discarded if we merely pursue the internal project of understanding our impersonal epistemic goals. Our task is to make more precise the murky notion of uncovering the structure of nature (or of organizing experience), of discovering, insofar as we are able, how the world works.[5]

Historians have been struck by the apparent variability of the goals professed by different scientists, working in different communities at different times. But I want to start from the opposite end. It is hardly surprising that

---

either direction, and, again, T_1 and T_2 are incomparable.

(d) Both s and r are false: then s is in F_{T_1} but not F_{T_2}, and r is in F_{T_2} but not F_{T_1}; neither theory can have greater verisimilitude.

As Miller saw, these considerations show that Popper's approach to verisimilitude cannot be employed to compare theories (axiomatic deductive systems) except under very special circumstances. Inclusion relations among sets of true and false consequences are not sufficiently fine-grained to do the work required of them. But, since the sets of consequences in question are typically infinite, it is not clear what other kinds of comparison are possible. Hence the technical difficulties.

4. Again, I ignore those skeptical complaints that insist that truth is impossible to get, postponing discussion of them until Chapter 5. However, even the skeptic should agree that *if* truth is obtainable at all, *boring* truth is relatively easily generated.

5. As I shall suggest in response to claims that the aims of science vary from context to context, the reference to our limitations is important. Different conceptions of what we can hope to achieve yield different *derivative* goals. See the discussion of (Laudan 1984) in Chapter 5.

we can find considerable community of goals among contemporary evolutionary theorists, or that we can take Dobzhansky and Darwin (to cite two major figures) to share a conception of the aims of biological science. When we expand to a triumvirate, including Owen with Dobzhansky and Darwin, some of the kinship is lost: there are formulations of the goals of biology endorsed by Darwin and Dobzhansky, to which Owen would not subscribe. Nevertheless, at a more fundamental level, Owen shares their common vision. Like them, he contends that biology should explain the diversity of living things and trace the patterns in that diversity. He differs in his understanding of how this goal is to be realized, dissenting from the Darwinian conception of history as the key to life's diversity.

We can thus expand the group of those who share goals for science, including not only those in the Darwinian tradition but some of their predecessors. How far can we go? Can we include Goethe and Linnaeus, Buffon and Ray, Theophrastus and Aristotle? Given our recognition of what Darwin and Owen share, I see no reason why not. Moreover, even if the attempt does break down at some point—say at the division between modern and premodern science—there is surely still value in articulating the goals of modern science, and understanding its conception of progress.

I shall start with two varieties of progress that have received little attention in the literature, but that will be fundamental to my account of significance. The next two sections will be devoted to articulating the notions of *conceptual progress* and *explanatory progress*. Before we take up the details, it is useful to have a map of the overall view.

As science proceeds we become better able to conceptualize and categorize our experience. We replace primitive conceptions of what belongs with what, what things are akin, by improved ideas about natural groupings. Our language develops so that we are able to refer to natural kinds and to specify our references descriptively. In addition, we are able to construct a hierarchy of nature, a picture of what depends on what. Against the background of our categories and our hierarchy, we are able to pose significant questions, questions that demand more detail about our picture of the world. Significant statements answer significant questions. Significant experiments help us to resolve significant questions. Significant instruments enable us to perform significant experiments. Methodological improvements help us to learn better how to learn, to improve our evaluation of significant statements. Thus, to sum up the general approach, significance is derivative from the background project of ordering nature, a project that is articulated in our attempts to conceptualize and to explain.

4. Conceptual Progress

Conceptual progress is made when we adjust the boundaries of our categories to conform to kinds and when we are able to provide more adequate specifications of our referents.[6] Striking examples come from the history of all sciences: 'planet,' 'electrical attraction,' 'molecule,' 'acid,' 'gene,' 'homology,' 'Down's syndrome' are all terms for which faulty modes of reference have been improved. Consider, for example, pre-Copernican usages of 'planet.' What set of objects was singled out? One answer is that, for the pre-Copernican, as for us, planets are Venus, Mars, and things like that—in which case the planets would include the Earth. Another is to declare that the planets are specified as those heavenly bodies that are sometimes observed to give rise to erratic motions against the sphere of fixed stars (the "wanderers")— in which case the planets would be Mercury, Venus, Mars, Jupiter, and Saturn. Or, allowing the specification to take account of possible perceptual limitations, we might declare that the planets comprise not only those observed to "wander" but those we would observe "wandering" if we had keener senses, that is, all the things we would count as planets of the solar system except the Earth. If we adopt the first approach to the reference of 'planet' then pre-Copernicans succeed in marking out a natural division, but they fail disastrously when they come to say what they are talking about. On either of the second approaches, they are fully aware of what they are referring to but the set of objects they single out is not a natural category. Either way, post-Copernicans do better. And so it goes, not only for this example but for the other terms on my brief list and for many more besides.

As I shall explain later, the answers of the last paragraph are not entirely adequate, but they enable us to appreciate the possibility of understanding conceptual change and conceptual progress in terms of shifts in mode of reference. The thesis that I shall defend in this section is that the conceptual shifts in science that have caused most attention (and that are supposed to be troublesome) can be understood, and understood as progressive, by recognizing them as involving improvements in the reference potentials of key terms. Recall from Section 7 of the last chapter that the reference potential of a term for a community is a compendium of the ways in which the references

6. Here I shall take over, uncritically for the time being, Plato's famous metaphor of "carving nature at the joints." So part of the task of fathoming the structure of nature will be seen as an exercise in anatomy, exposing the lineaments of the objective divisions of nature and showing which things belong together. A strong realism about the structuring of nature provides a simple way of developing the fundamental epistemic aim selected in the last section, but it brings well-known epistemological difficulties in its train. I shall suggest later (Chapter 5 Section 8) that there are other metaphysical options, which seek to explicate the notion of natural kind by focusing on the groups to which reference is made in laws or in explanatory schemata, which can be linked to the project of fathoming the structure of nature, and which can avoid some of the epistemological difficulties. As I have suggested elsewhere (Kitcher 1986, 1989) I believe that this latter approach to the metaphysical issues has important advantages. However, the treatment of many questions about progress can profitably be undertaken by leaving the notions of natural kind and objective dependency (see Section 5) temporarily unanalyzed, using them to explicate conceptions of progress, and only returning to metaphysics later. So this chapter will remain neutral between the options of strong realism and the type of weakened (Kantian) realism that I have previously espoused.

of tokens of that term can be fixed for members of the community. I shall illustrate how reference potentials may shift by analyzing an apparently problematic case, and then offer a general view about improvement of reference potentials.

Kuhn and Feyerabend achieved the important insight that different communities of scientists, working in the same field, may organize the aspects of nature that concern them in different ways (Kuhn 1962/70, 1983; Feyerabend 1963a, 1965, 1970). Different conceptualizations make differences to observation, inference, and explanation (see Chapter 7).[7] But, most famously, Kuhn and Feyerabend have suggested that the phenomenon of "conceptual incommensurability" that they identify engenders problems of communication and of reporting of evidence. I shall begin by reviewing one of Kuhn's favorite problematic cases, with a view both to showing that there is indeed an interesting conceptual shift and to revealing that it does not have the dire consequences for communication with which it is sometimes credited.

There is a very simple story about the chemical revolution of the eighteenth century. Priestley, so the story goes, employed a language containing terms— 'phlogiston,' 'principle'—that fail to refer. Lavoisier used a language containing expressions—'oxygen,' 'element'—that refer to kinds that Priestley could not have identified. So there is a conceptual advance between Priestley and Lavoisier, one that involves the replacement of expressions that fail to refer by genuinely referring expressions and the introduction of terms that single out kinds for the first time.

Kuhn saw that this story is extremely implausible. If Priestley's language was so inadequate, then it is hard to understand how his chemistry could ever have appeared successful and how he could have contributed to the development of Lavoisier's own ideas. Yet Priestley was a successful chemist, one who appeared to his contemporaries to be extending the range and power of the phlogiston theory, and he was responsible for the experiment in which he (and, following his lead, Lavoisier) isolated a new gas, the gas Lavoisier called "oxygen." At times Kuhn seems to want to solve the problem of Priestley's success by declaring that the terms we take to be nonreferring ('phlogiston' and so forth) really did refer, picking out constituents of the "different world" inhabited by Priestley and his fellow phlogistonians. Divided by the chemical revolution, Priestley and Lavoisier inhabit different worlds (or, to put it another way, their theories have different ontologies). Unfortunately, this strategy of accounting for Priestley's apparent success will only accommodate part of the historical evidence. As we insist on the "many worlds" approach, the revolutionary divide between Priestley and Lavoisier looks ever more difficult to bridge, and communication between the two men

---

7. Thus, in the writings of Kuhn (1962/1970), Feyerabend (1963a, 1965, 1970), and Hanson (1958) there are suggestions about how differences in conceptualization issue in differences in perception. For the moment I am solely concerned with the relations among scientific languages, and with the thesis of *conceptual* rather than *perceptual* incommensurability.

appears partial, at best.[8] How then do we explain the dialogue between them, the role that Priestley played in the genesis of Lavoisier's ideas, the assessments and evaluations of Lavoisier's proposals that occur in the writings of Priestley and his followers? Sensitive to the historical phenomena, Kuhn does not fully commit himself to the "many worlds" interpretation, offering instead a position whose ambiguities have excited much subsequent discussion.[9]

We have a puzzle. If the central terms of Priestley's theoretical language do not refer, then it seems impossible to ascribe to him the achievements warranted by the historical record.[10] If those terms do refer, then his "world" must contain entities that we no longer recognize. An adequate solution to the puzzle must (i) recognize Priestley's contributions to the development of chemistry, (ii) avoid populating nature with strange entities, (iii) specify the exact way in which Lavoisier made a conceptual advance.

Begin with a brief overview of features of the language of the phlogiston theory. This theory attempted to give an account of a number of chemical reactions and, in particular, it offered an explanation of processes of combustion. Substances which burn are rich in a "principle," *phlogiston*, which is imparted to the air in combustion. When we burn wood, for example, phlogiston is given to the air, leaving ash as a residue. Similarly, when a metal is heated, phlogiston is emitted, and we obtain the *calx* of the metal.

Champions of the phlogiston theory knew that, after a while, combustion in an enclosed space will cease. They explained this phenomenon by supposing that air has a limited capacity for absorbing phlogiston. By heating the red calx of mercury on its own, Priestley found that he could obtain the metal mercury, and a new kind of "air," which he called *dephlogisticated air*. (According to the phlogiston theory, the calx of mercury has been turned into the metal mercury by taking up phlogiston; since the phlogiston must have been taken from the air, the resultant air is dephlogisticated.) Dephlogisticated air supports combustion (and respiration) better than ordinary air—but this is only to be expected, since the removal of phlogiston from the air leaves the air with a greater capacity for absorbing phlogiston.

In the last decades of the eighteenth century, the phlogiston theorists became interested in the properties of a gas, which they obtained by pouring a strong acid (concentrated sulfuric acid, for example) over a metal or by passing steam over heated iron. They called the gas *inflammable air*.

8. See (Kuhn 1962/1970 149, Kuhn 1970 200, 201, 204). Perhaps the most radical statement of the "many worlds" position is at (Kuhn 1962/1970 102–103).

9. For Kuhn's own retrospective on his views, see his (1983). Discussions and critiques of his claims about conceptual incommensurability can be found in (Shapere 1964, 1965), (Scheffler 1967), (Kordig 1971), (Davidson 1973), (Field 1973), (Kitcher 1978), and (Devitt 1981). The present discussion closely follows the line of my earlier attempts in my (1978, 1982, 1983c).

10. Even if one takes Priestley's principal accomplishments to reside in his actions rather than in his words, there is still a problem in coordinating his dicta with his behavior. If we assume that his central terms do not refer, then he has to appear as serendipitous—producing a stream of babble while simultaneously doing things (isolating oxygen, synthesizing water) that somehow enable others to repeat his doings and discuss them.

If we compare the descriptions of these familiar reactions given by the phlogiston theory, and by modern elementary chemistry (which descends from the work of Lavoisier), some identifications are immediately suggested. We might naturally suppose that dephlogisticated air is oxygen and that inflammable air is hydrogen. I shall consider the merits of these identifications in a moment.

In presenting the puzzle of Priestley's language, I spoke vaguely of Priestley's "success" and his "contributions" to chemistry. Can we be more precise? One simple suggestion is that Priestley enunciated various *true statements* which had not previously been accepted.[11]

Priestley was not occupied with some work of unwitting fiction. Inside his misbegotten and inadequate language are some important new truths about chemical reactions, trying to get out. A false presupposition, the idea that something is *emitted* in combustion, infects most of the terminology. The natural approach to Priestley's language is to take the faulty central idea of the phlogiston theory, the idea that phlogiston is the substance emitted in combustion, as fixing the reference of 'phlogiston.' Assuming that some expressions used by proponents of the phlogiston theory (e.g., '*x* is emitted from *y*') are coreferential with their modern homonyms, and that compound expressions that embed nonreferential constituents are themselves nonreferring (e.g., that 'the substance obtained by removing *x* completely from *y*' refers only if *x* and *y* both refer), it follows that virtually none of Priestley's statements is true. For since the reference of 'phlogiston' is fixed through the description "the substance emitted in combustion," since 'emitted' means what we standardly take it to mean, and since nothing is *emitted* in combustion, 'phlogiston' fails to refer. Because the reference of other technical terms is fixed by descriptions that contain 'phlogiston'—dephlogisticated air is to be the substance obtained when phlogiston is removed from the air—those terms also fail to refer.[12] Hence the statements that talk about the properties of phlogiston, phlogisticated air, dephlogisticated air, and so forth, cannot be true. So it appears that we cannot honor the idea that Priestley and his followers advanced some new, true statements.

---

11. In the wake of the work of Kuhn and Feyerabend, such simple references to truth are likely to provoke mirth, scorn, or shock. It is well to remember that *part* of the reason for viewing these suggestions as simplistic depends on accepting Kuhnian or Feyerabendian theses about what occurs in episodes like the transition between Priestley and Lavoisier, already adopting one of their solutions to our puzzle. Thus a partial defense of our natural way of talking about past theorists as articulating truths is to show that we can accommodate the historical material while continuing to engage in that talk. Other parts of the defense will be supplied in Chapter 5.

12. This consequence strikes me as intuitive, but it does merit some discussion. We could defend it by supposing (following Frege) that an expression formed by extensionally embedding a nonreferring expression fails to refer or by adopting a principle of substitutability for nonreferring names: a complex expression formed by extensionally embedding a nonreferring expression preserves its reference under substitution of nonreferring expressions that belong to the same syntactic category. The general idea of substitutability principles for nonreferring expressions is due to Hartry Field.

Even a brief reading of the writings of the phlogistonians reinforces the idea of true doctrines trying to escape from flawed language. Consider Priestley's account of his first experience of breathing oxygen.

> My reader will not wonder, that, after having ascertained the superior goodness of dephlogisticated air by mice living in it, and the other tests above mentioned, I should have had the curiosity to taste it myself. I have gratified that curiosity, by breathing it, . . . The feeling of it to my lungs was not sensibly different from that of common air; but I fancied that my breast felt peculiarly light and easy for some time afterwards. (Priestley 1775/1970 II 161–162).

Surely Priestley's token of 'dephlogisticated air' refers to the substance which he and the mice breathed—namely, oxygen.

Similarly, it seems that Cavendish's uses of the terms refer in the following passage:

> From the foregoing experiments it appears, that when a mixture of inflammable and dephlogisticated air is exploded in such proportion that the burnt air is not much phlogisticated, the condensed liquor contains a little acid which is always of the nitrous kind, whatever substance the dephlogisticated air is procured from; but if the proportion be such that the burnt air is almost entirely phlogisticated, the condensed liquor is not at all acid, but seems pure water, without any addition whatever. . . . (Cavendish 1783/1961 19)

We readily understand what prompted this description. Cavendish had performed a series of experiments in which samples of hydrogen and oxygen were "exploded" together; in some cases, the oxygen obtained was not entirely pure, and the small amount of nitrogen, mixed in it, participated in a reaction to form a small amount of nitric acid; when the sample of oxygen was pure there was no formation of nitric acid, and the only reaction was the combination of hydrogen and oxygen to form water. In reporting the experiments he had done, Cavendish used the terms of the phlogiston theory to refer to the substances on which he had performed the experiments: that is, he used 'inflammable air' to refer to hydrogen, 'dephlogisticated air' to refer to oxygen, and so forth.[13]

13. These claims and my kindred remarks about Priestley will surely jar some historical sensibilities. I am using the ideas of modern science to provide a picture of the relationships between Priestley and Cavendish and the world. Isn't this Whiggish, illegitimate, question begging? I reply that without some picture of the relationship between past figures and the world, history is impossible: we cannot frame any hypotheses about what their discourse means. Thus the strategy of agnosticism leaves the historian in the presence of a parade of uninterpretable symbols, a detached text without significance. So where should we turn for a view of nature that will enable us to understand the actions and reactions of the protagonists? Is there some *other* account that would serve us better than modern science? Why not the best—the account we take for granted in planning our own activities?

One modish answer is to insist that we adopt the "actor's categories." To this it might be pertinent to inquire why we should favor a discredited account of nature rather than the one which we typically adopt. But there is a deeper point. Without an account of the relationship between the actor and nature already in hand we cannot interpret the language of past science, cannot even formulate the actor's categories. I claim that the account I

Our problems in specifying what Priestley and Cavendish were talking about arise from the presupposition that there should be a uniform mode of reference for all tokens of a single type. Once we liberate ourselves from this presupposition, adopting the notion, introduced in the last chapter, that a scientific term (type) may have a heterogeneous reference potential, we can begin to make sense of the language of the phlogiston theory, recognizing the achievements of Priestley and Cavendish and pinpointing the inadequacies of their language.

Successful ascriptions of reference should accord with a principle that Richard Grandy dubs the "principle of humanity." The principle enjoins us to attribute to the speaker whom we are trying to understand a "pattern of relations among beliefs, desires and the world [which is] as similar to ours as possible" (Grandy 1973 443).[14] In deciding on the referent of a token, we must construct an explanation of its production. That explanation, and the hypothesis about reference we choose, should enable us to trace familiar connections among Priestley's beliefs and between his beliefs and entities in the world, ascribing dominant intentions that we would expect someone in his situation to have.

Let us expose the way in which this strategy (a strategy that I take to be tacitly employed by sensitive historians) works in the case at hand. How did Priestley come to use 'dephlogisticated air' to refer to oxygen? Priestley began his discussion by talking about various attempts he had made to remove phlogiston from the air. He then recorded the details of his experiments on the red calx of mercury from which he had liberated a "new air." After a number of mistaken attempts to identify the gas, he finally managed to describe it in the terminology of the phlogiston theory.

Being now fully satisfied with respect to the *nature* of this new species of air,

---

provide in this section, far from riding roughshod over the actor's categories, actually displays those categories and enables us to see what Priestley and Cavendish meant.

Obviously, the historical sensitivities are raised by genuine worries. We should not assume that homonymous expressions have the same reference (or reference potential). We should not assume—that the account of nature that we use in interpreting the languages of the past is infallible—it is not, and the history that we develop on the basis of the picture offered by present science inherits the troubles of modern science (and possibly more besides). I hope that thoughtful historians will appreciate the ways in which my use of present science responds to the points of real concern, and will not be misled into over-reaction, based ultimately on zeal for an impossible project—that of some kind of empathic telekinesis into the minds of past scientists.

14. I should emphasize that Grandy's principle diverges from the principle of charity in *not* requiring that we maximize agreement with those whom we interpret. The underlying idea is that we attribute beliefs by supposing that our interlocutors have cognitive equipment that is similar to our own, and using what we know about the experiences they have had. Those who have received very different stimuli can be expected to have formed very different beliefs; (Grandy 1973) provides a lucid elaboration of these points. I should note that for some interpretative projects—but probably not for those which arise in doing the history of science—the principle of humanity will need refining in much the way that it corrected the principle of charity: on occasion, background evidence may suggest to us that we are not dealing with interlocutors who share cognitive equipment similar to our own.

viz. that, being capable of taking more phlogiston from nitrous air, it therefore originally contains less of this principle [i.e., phlogiston]; my next inquiry was, by what means it comes to be so pure, or philosophically speaking, to be so much dephlogisticated;... (Priestley 1775/1970 II 120)

From our perspective Priestley has misdecribed the new gas. His remarks on this occasion identify the gas obtained by heating the red calx of mercury as dephlogisticated air, *and set up a mode of reference for tokens of 'dephlo-gisticated air' that fixes the reference by the description "the substance obtained when the substance emitted in combustion is removed from the air."* When tokens are produced, with this mode of reference, they fail to refer. On other occasions, Priestley and his friends (Cavendish, for example) produce tokens of 'dephlogisticated air' with a different mode of reference. Their dominant intention is to refer to the kind of stuff that was isolated in the experiments they are reporting—to wit, oxygen. Thus, in the passages quoted earlier, Priestley describes the sensation of breathing oxygen and Cavendish proposes an account of the composition of water. *Because the referents of the tokens of 'dephlogisticated air' are fixed differently on these occasions, Priestley and Cavendish enunciate important new truths.*[15]

Of course, they do so in a misguided language. But it is not hard to pinpoint the troubles of their idiom. Not only is the reference potential of terms such as 'dephlogisticated air' (and even of 'phlogiston'; see Kitcher 1978 534) heterogeneous, but the modes of reference are connected by a faulty theoretical hypothesis. Thinking of combustion as a process of emission, Priestley and his friends are led to take it for granted that the gas liberated from the red calx of mercury has been obtained by removing phlogiston from the air. Their connection of what *we* see as two distinct modes of reference, that fix reference to oxygen and to nothing, respectively, rests on their acceptance of the following hypothesis:

H.   There is a substance which is emitted in combustion and which is normally present in the air. The result of removing this substance from the air is a gas which can also be produced by heating the red calx of mercury.

Phlogistonians all accept *H,* and, in consequence, they are all willing to use tokens of 'dephlogisticated air' whose references are fixed either by the description "the substance obtained by removing from the air the substance emitted in combustion" or by encounters with samples of the gas obtained by heating the red calx of mercury.

We can now give a precise sense to the claim that a term employed by a

---

15. Note that applying a pure causal theory of reference to Priestley and Cavendish misdescribes those occasions (the initial misidentification of oxygen, for example) on which they fail to refer. The pure descriptive theory misdescribes those occasions (when they are reporting on samples of the gas that they have prepared) on which they succeed in referring to oxygen. So, I claim, the appeal to heterogeneous reference potentials is needed to account for the mixed record of success and failure.

community is 'theory-laden.' Theory-laden terms have heterogeneous reference potentials; the theoretical hypotheses with which they are laden are claims that are, in conjunction, equivalent to the assertion that all the modes of reference fix reference to the same entity. Theory-ladenness does not simply stem from scientific irresponsibility. Scientists inevitably court ambiguity in using the same term from occasion to occasion (Hempel 1965 123–133, 1966 Chapter 7, Kitcher 1978 542–543).

Phlogistonians communicate easily for they share the same reference potentials. From the perspective of the new chemistry, however, uses of central parts of the language of the phlogiston theory are laden with false theory. As a result there are no expressions in Lavoisier's language (or in ours) whose reference potentials match those of 'phlogiston,' 'dephlogisticated air,' 'inflammable air,' and other expressions. Kuhn is quite correct (1983) to declare that a certain style of translation is impossible: there is no way to take Priestley's text and replace the expressions in it that do not belong to our language with expressions of our language that have the same reference potentials as those they replace. If translation has to preserve reference potential, we cannot translate Priestley—for there is no term of the language of post-Lavoisierian chemistry that has the same reference potential as 'dephlogisticated air.' Moreover, the mismatch among reference potentials captures the intuitive (but vague) idea that phlogistonians and modern chemists "carve up the world differently" (Kuhn 1983).

Even if we cannot translate Priestley's texts (according to the standards of translation of the last paragraph), there is no difficulty in recognizing how we (and Lavoisier) can understand Priestley's claims. For, as my brief analysis of his language shows, we are able to recognize the reference potentials of his technical terms and to specify the referents of the tokens he produces. We could even offer a *reference-preserving* translation of his texts—although, since this would replace tokens of the same type with tokens of different types, it would hide certain inferential connections unless we employed one of the standard devices of translators (parenthetical identification of the term replaced, footnotes, prefatory glosses, etc.). Not only can we comprehend what Priestley said, we can also see that some of the statements he advanced are true and we can explain how to improve his language. Lavoisier made a conceptual advance by revising reference potentials so as to avoid presupposing false hypotheses.

I claim that the foregoing account solves the puzzle about the languages of the protagonists in the chemical revolution. It also motivates a general thesis about conceptual change in science. Conceptual change is change in reference potential. Dramatic examples are those in which the community becomes disposed to use tokens of a term (possibly an old term, possibly a neologism) to pick out a new referent, and those in which its members acquire dispositions to fix the references of tokens of old terms through description, where no such referential specification had been possible before. The cases on my original list command our attention because, at least at first sight, they belong to one of these types.

To understand conceptual *progress* we should recall my account of the different kinds of intention that may dominate in the production of a token. Perhaps for most linguistic usages, the general intention to conform to the usage of others is far more important than any intention we may have to refer to whatever fits a description. When we examine scientific usage, the intention to conform is by no means the only one that has to be taken into account. Scientists usually also have the general intention of referring to natural kinds, picking out the real similarities in nature, and, in recognizing this intention, we sometimes construe the descriptions they offer as *mistaken* identifications of the referent rather than as successful identifications of a different referent: thus, when he talks of the antics of the mice in the vessel and when he describes the "lightness in his breast," we understand Priestley as talking about oxygen, even though he would describe the gas as "the substance obtained by removing phlogiston from the air." But there are also occasions on which scientists intend that the referent should be whatever satisfies a particular description. The *ideal* situation for a scientist would thus be to obey three maxims:

Conformity: Refer to that to which others refer.

Naturalism: Refer to natural kinds.

Clarity: Refer to that which you can specify.

There are many situations in which these maxims conflict. When they do, the scientist "chooses" among them—in the sense that there is a dominant intention to obey one rather than the others. If the choices are made in different ways on different occasions of utterance, different tokens of the same type can refer differently.[16]

Conceptual progress should be assessed in terms of proximity to the ideal state. One of the goals of science is the construction of a language in which the expressions refer to the genuine kinds and in which descriptive specifications of the referents of tokens can be given. For such a language the three maxims are in harmony.

We can now introduce a notion of *conceptual progressiveness:*

(CP)   A practice P_2 is conceptually progressive with respect to a practice P_1 just in case there is a set C_2 of expressions in the language of P_2 and a set C_1 of expressions in the language of P_1 such that

(a)   except for the expressions in these sets, all expressions that occur in either language occur in both languages with a common reference potential

16. This should be easily understandable in terms of the psychological picture developed in Section 2 of the last chapter. We can envisage scientists as subscribing to three different long-term goals. In situations of conflict, one of these goals is activated, and, in conjunction with other features of working memory it dictates the mode of reference of the term token produced.

(b) for any expression *e* in C_1, if there is a kind to which some token of *e* refers, then there is an expression *e** in C_2 which has tokens referring to that kind

(c) for any *e, e**, [as in (b)], the reference potential of *e** refines the reference potential of *e*, either by adding a description that picks out the pertinent kind or by abandoning a mode of reference determination belonging to the reference potential of *e* that failed to pick out the pertinent kind.

In this definition, clauses (a) and (b) are necessary to eliminate the possibility that refinements of reference potential might be achieved at the cost of expressive loss. We isolate those parts of the language that do not change [in (a)], and demand that any kinds that can be discussed in the old language should be specifiable in the new [in (b)]. Improvements come about by abandoning modes of reference that are not in accord with one of the maxims or by adding modes of reference that would be in accord with both *clarity* and *naturalism*.

Numerous examples from the history of science reveal conceptual progress in the sense captured in (CP). The shift from Priestley to Lavoisier shows retention of the ability to refer to oxygen, with replacement of the flawed reference potential of the term Priestley employed to refer to oxygen ('dephlogisticated air') with the refined reference potential of Lavoisier's expression 'oxygen.' Similarly, in the Copernican revolution, a term, 'planet,' some of whose tokens previously referred to a natural kind (the set of things we would count as planets of the solar system) and some of whose tokens referred to a set that is not a kind (the set whose members are Mercury, Venus, Mars, Jupiter, and Saturn), underwent refinement of its reference potential. After Copernicus, reference can be fixed descriptively to a kind (the planets of the solar system) and the flawed mode of reference to the partial subset is dropped. Later in the history of astronomy, 'planet' comes to have a reference potential that shows further refinement: a descriptive mode of reference is used to fix reference to the kind that includes the planets of all stars, and the subkind of planets of the solar system is picked out by explicit restriction. Similar accounts can be provided for the other examples on my original list.

5. Explanatory Progress

Explanatory progress consists in improving our view of the dependencies of phenomena. Scientists typically recognize some phenomena as prior, others as dependent. For example, ever since Dalton, chemists have regarded molecular arrangements and rearrangements as prior to the macroscopic phenomena of chemical reactions, and, since the 1960s, geologists have viewed interactions among plates as prior to facts about mountain building and earthquakes. General ideas about these dependencies can be shared while specific theses about the details of dependence are debated. Commitment to Daltonian atomism in the nineteenth century persists through any number of claims about the formulas of chemical compounds.

Judgments about dependency can be understood as concerned with the forms of ideal explanatory texts that the scientists in question envisage. One component of practice is a collection of patterns or explanatory schemata. Fields of science make explanatory progress when later practices introduce explanatory schemata that are better than those adopted by earlier practices.

What does 'better' mean here? One answer, the response of robust realism, is to declare that there is an objective order of dependency in nature.[17] The character of chemical reactions is objectively dependent on underlying molecular properties, facts about mountain building and earthquake zones are objectively dependent on the facts of plate tectonics, the characteristics of contemporary organisms are objectively dependent on the evolutionary histories of those organisms, and so forth. Recognizing these dependencies—and how to extend them further—is an important progressive step.

The robust realism outlined here is not the only way to say what is meant by improving the set of explanatory schemata. An alternative is to tie advances to enhanced ability to understand the phenomena, for example by proposing that a particular collection of schemata offers a more unified vision of the world. I shall postpone comparison of these two metaphysical options until the next chapter (see Section 8, where I also take up the question of how to explicate the discussion of kinds that I took for granted in the previous section). For now it will suffice to see the general shape of the view. Explanatory progress consists in improving our account of the structure of nature, an account embodied in the schemata of our practices. Improvement consists either in matching our schemata to the mind-independent ordering of phenomena (the robust realist version) or in producing schemata that are better able to meet some criterion of organization (for example, greater unification).

Let us consider some examples from the history of science. Almost everything that Dalton maintained about atoms was wrong. Nonetheless, we recognize his formulation of atomic chemistry as an important progressive step because it introduced a new *correct* explanatory schema. Dalton proposed to explain facts about the course of chemical reactions (and about the weights of reactants and products) by appealing to premises about atoms, premises that specify the "fixed proportions" in which atoms of different elements combine. Chemists ever since have endorsed the claim that this is a correct picture of the objective dependencies, and they have further articulated Dalton's schema.

17. This conception is as old as the Aristotelian idea of an order of being. For robust realists, the aim of science is taken to be that of delineating the fundamental mechanisms at work in nature. If the achieving of adequate concepts is understood in terms of an anatomical description, the display of objective dependencies corresponds to physiology— showing how things work. For a modern attempt to give literal significance to these metaphors, see Wesley Salmon's defense of the "ontic conception" of explanation in his (1984). I discuss some of the epistemological problems associated with Salmon's approach—as well as some of its considerable merits—in my (1989). See also Section 8 of the next chapter.

One leading question of post-Daltonian chemistry concerned the ratios of the weights of substances that form compounds together. Within nineteenth century chemistry, we can discern the following simple pattern.

Question: Why does one of the compounds between X and Y always contain X and Y in the weight ratio $m : n$?

Answer:

(1) There is a compound Z between X and Y that has the atomic formula X_pY_q.

(2) The atomic weight of X is x; the atomic weight of Y is y.

(3) The weight ratio of X to Y in Z is $px : qy$ ($= m : n$).

Filling Instructions: X, Y, Z are replaced by names of chemical substances; p, q are replaced by natural numerals; x, y are replaced by names of real numbers.

Classification: (1) and (2) are premises; (3) is derived from (1) and (2).

DALTON is elementary—although it was, of course, instantiated in many different ways during the early years of the nineteenth century by chemists who had very different ideas about the formulas of common compounds! What makes it important for our purposes is the way in which DALTON was preserved and extended by subsequent work.

An important nineteenth-century step in the extension was the introduction of the concept of valence and rules for assigning valences that enabled chemists to derive conclusions about which formulas characterized possible compounds between substances.

Question: Why does one of the compounds between X and Y always contain X and Y in the weight ratio $m : n$?

Answer:

(1) Molecules of X contain j atoms, molecules of Y contain k atoms.

(2) The valences ("affinities") of X and Y are . . . .

(3) The possible equations governing chemical reactions between X and Y are . . . .

(4) There is a compound Z between X and Y that has the atomic formula X_pY_q.

(5) The atomic weight of X is x; the atomic weight of Y is y.

(6) The weight ratio of X to Y in Z is $px : qy$ ($= m : n$).

The filling instructions extend those of DALTON by requiring that j and k be natural numbers, that the valence relations should be specified in accordance with a particular scheme, and that (3) be completed by presenting chemical equations in a particular way. The classification demands that (3) be obtained from (1) and (2) in accordance with the principles of balancing equations, and that (4) be derived from (3) by showing that the compound X_pY_q be obtainable on the right-hand side of some equation in (3). As in the case of DALTON (6) is to be derived from (4) and (5).

At this first stage, the attributions of valence are unexplained and there is no understanding of why the constraints hold. However, the original explanations of weight relationships in compounds are deepened, by showing regularities in the formulas underlying compounds.

The next stage consists in the introduction of a shell model of the atom to explain the hitherto mysterious results about valences. From premises attributing shell structure to atoms, together with principles about ionic and covalent bonding it is now possible to provide derivations of instances of (2). These derivations provide a deeper understanding of the conclusions than was given by the simple invocation of the concept of valence because they show us *in a unified way* how the apparently arbitrary valence rules are generated. Moreover, the appeal to the model of the atom enables us to derive instances of (3) from premises that characterize the composition of atoms in terms of protons, neutrons, and electrons.

SHELLFILLING

Question: Why does one of the compounds between X and Y always contain X and Y in the weight ratio $m : n$?

Answer:

(1)   Molecules of X contain j atoms; molecules of Y contain k atoms.

(2)   (a)   An uncharged atom of X contains g electrons; an uncharged atom of Y contains h electrons.

(b)   The first incompletely filled shell level around an uncharged atom of X contains g_1 electrons and g_2 free places, the first incompletely filled shell-level around an uncharged atom of Y contains h_1 electrons and h_2 free places.

(c)   The combinations of atoms of X and Y that will achieve shell filling either through electron transfer (ionic bonding) or through electron sharing (covalent bonding) are . . . .

(d)   The valences ("affinities") of X and Y are . . . .

(3)–(6) As for MOLECULAR AFFINITIES.

The filling instructions extend those of molecular affinities in obvious ways: g, h, g_1, g_2, h_1, h_2 are required to be natural numbers and to satisfy relations imposed by the general theory of atomic structure. The classification demands that 2(b) be obtained from 2(a) in accordance with the principles of the shell model of atoms, and that 2(c) be derived from 2(b) using the principles of covalent and ionic bonding. 2(d) is simply a rewriting of 2(c) in what now is viewed as old-fashioned language. (3) can either be obtained from 2(d), as in MOLECULAR AFFINITIES or directly from 2(c).

Finally, the derivations from premises about shell filling can be embedded within quantum mechanical descriptions of atoms and the shell structures and possibilities of bond formation revealed as consequences of the stability of quantum-mechanical systems. Although this is only mathematically tractable in the simplest examples, it does reveal the ideal possibility of a further extension of our explanatory derivations.

What this very simple, preliminary account of the development of atomic chemistry shows is how we can formulate precisely the intuitive idea that Dalton outlined a picture that was filled in in much more detail by his successors. Of course, they improved on his chemistry in two distinct ways. One trend (with which I am not concerned at this stage) consisted in the replacement of his erroneous ideas about chemical formulae, and, in consequence, the production of true instantiations of his simple schema. The notion of explanatory progress involves the improvement of the schema itself. This line of development is seen in the move from DALTON to MOLECULAR AFFINITIES to SHELL FILLING and beyond to the account in terms of quantum chemistry (where I have only indicated the schema).

The example of Darwinian evolutionary biology, considered in Chapter 2, shows a similar progression. *Some* of Darwin's schemata were correct; others—such as the schema underlying those of his explanations in terms of disuse that require Lamarckian mechanisms of inheritance—have been discarded. Particularly interesting is the fate of the selection schema, SIMPLE INDIVIDUAL SELECTION As we saw in Section 7 of Chapter 2, this is absorbed into NEO-DARWINIAN SELECTION a schema that offers a more articulated version of Darwin's picture of the workings of selection. That schema has been further completed in some of the work mentioned in Section 8 of Chapter 2 (in Hamilton's introduction of inclusive fitness and Maynard Smith's development of evolutionary game theory, for example). As in the Daltonian example, tracing the details of the relations among the schemata employed at successive stages enables us to make precise the idea that there is a cumulative process of extending, correcting, and articulating a basic picture of the order of some natural phenomena.[18]

Four distinct kinds of processes are at work in these examples of explanatory progress. First, we have the introduction of correct schemata, illustrated by the work of Dalton in recognizing the dependence of facts about the course

18. In the example from the history of chemistry I have been concerned to draw the line of development without introducing the historical details. The Darwinian example of Chapter 2 is meant to balance this by providing much more of the scientific background and linking the schemata to specific points in the history of biology. With the discussion of this section in mind, it should be possible for the reader to return to the more discursive treatment of Chapter 2, identifying the same structural relations and progressive picture that I display here. Another example, treated in more detail in (Kitcher 1989), focuses on the development of explanations of hereditary distributions from Mendel to Watson and Crick.

of chemical reactions (specifically about the weights of reactants and products) on the facts about atomic combination and by Darwin's insight that distributions and relationships among contemporary organisms are dependent on the course of descent with modification. Second, we have the elimination of incorrect schemata, such as Darwin's appeals to the inheritance of acquired characteristics. Third, we find the generalization of schemata, rendering them able to deal correctly with a broader class of instances: evolutionary theorists who appeal to classical individual selection alone are correct in identifying a certain type of dependence, but their proposals are less general—and therefore less complete—than those which allow for drift, migration, meiotic drive, inclusive fitness effects, developmental constraints, and so forth. Finally, there is explanatory extension, when the picture of dependencies is embedded within some larger scheme. The incorporation of Darwin's selectionist patterns within NEO-DARWINIAN SELECTION and the embedding of atomic chemistry in quantum physics show this process at work.

At the end of Chapter 2 I suggested that the history of evolutionary biology does not show an indecisive alternation but *something like* cumulative progress. The goal of this section and its predecessor has been to make that judgment more precise. There do seem to be cumulative processes of specifiable kinds in my examples, as schemata are introduced, refined, generalized, and extended.

But how typical are these examples? Am I cheating by simply drawing on periods that Kuhn would classify as belonging to "normal science," and that other students of scientific change would rank in analogous categories?[19] My reply consists of a historical reminder and a promise. The historical reminder is to note that the periods over which I have traced the development of schemata not only are long (of the order of 100 to 200 years at the stage of most rapid change in the histories of the relevant sciences) but also span "normal" and "revolutionary" episodes. In each case, even if all schemata after the first can be assigned to a single "normal scientific" tradition, the introduction of the first marks a progressive step over the accounts previously available: the schemata introduced by Dalton, Mendel, and Darwin provide answers to the questions they address that identify correct dependencies unappreciated by their predecessors. Moreover, if one wants to extend the notion of "normal science" (or some equivalent) to cover the entire periods through which I trace the refinement, generalization, and extension of schemata, then it will be necessary to downplay the significance of "revolutions" in science. These periods are so large, and the processes I identify so prevalent, that "normal science" will be everywhere once a field attains maturity.

But I do not want to claim too much for these examples. They serve to remind us of what scientists are sometimes brave enough to say, that domination of philosophical discussion by the vicissitudes of a few concepts from

19. These questions were posed forcefully to me by Larry Laudan and Rachel Laudan. I am grateful to both of them for much helpful discussion of this point, although I suspect that neither will be convinced by my response.

theoretical physics needs to be balanced by consideration of cases in which progress seems much more assured.[20] Instead of dogmatically claiming that my view will generalize across all cases, let me offer a counterquestion: How frequently do periods that do *not* exemplify the patterns I have discussed occur in the history of the (modern) sciences? This question leads me to my promise. At the end of the next chapter I shall explicitly discuss what seem to be the hard cases, the examples that critics of cumulative progress have turned to in an endeavor to expose the idea of explanatory losses.

I conclude by offering an explicit account of explanatory progress. Continuing to waive the large metaphysical questions that I postponed earlier, let us suppose that a schema is correct if it identifies a class of dependent phenomena and specifies some of the entities and properties on which those phenomena depend. (For the robust realist it formulates part of the objective order of dependencies in nature; for someone who does not believe in a mind-independent order of nature, it presents part of an ideal system for organizing the phenomena.) One schema is more complete than another just in case the former identifies a more inclusive set of relevant entities and properties or the former is correct for a more inclusive class of dependent phenomena. One schema extends another if and only if a schematic premise of the latter is derived from the former. The extension is correct if the properties attributed to the entities in instances of the conclusion depend on the entities and properties referred to in the corresponding instances of the premises. The historical examples I have discussed involve introduction of correct schemata, elimination of incorrect schemata, replacement of less complete with more complete schemata, and correct explanatory extension.

The following definition brings together the facets of explanatory progress in a straightforward fashion:

(EP) P_2 is explanatorily progressive with respect to P_1 just in case the explanatory schemata of P_2 agree with the explanatory schemata of P_1 except in one or more cases of one or more of the following kinds.

(a) P_2 contains a correct schema that does not occur in P_1.

(b) P_1 contains an incorrect schema that does not occur in P_2.

(c) P_2 contains a more complete version of a schema that occurs in P_1.

(d) P_2 contains a schema that correctly extends a schema of P_1.

Making explanatory progress in this sense advances our goal of recognizing the structure of natural phenomena (or, if you like, the best way of organizing

20. Thus Mayr protests the influence of a few theories from physics on the development of an allegedly general picture of science and scientific progress (1982 857). In a valuable review of Mayr's book, John Maynard Smith is admirably forthright: "Unfashionable as it may be to say so, we really do have a better grasp of biology today than any generation before us, and if further progress is to be made it will have to start from where we now stand" (see Maynard Smith 1989 11). I hope that my account explains clearly why Mayr and Maynard Smith are right.

various areas of our experience).[21] A suggestive (but not entirely adequate analogy) is to think of the work of children engaged on a large and complex jigsaw puzzle. Subregions of the puzzle correspond to the structure of dependencies among a particular class of phenomena. Identifying correct schemata is analogous to fitting a few pieces together, the correction of schemata corresponds to scrapping faulty efforts at fitting pieces, the completion and extension of schemata consist in putting the pieces already fitted into larger chunks of the puzzle. The ultimate aim, of course, is to complete the picture. (Here, perhaps, the analogy breaks down, for there may be no complete— or completable—picture.)

6. Derivative Notions

Significant questions arise against the background of the ordering of the phenomena captured in our explanatory schemata. There are two main ways in which significance can accrue to a question, generating *application* questions and *presuppositional* questions, respectively. Application questions are generated from projects of finding particular instantiations of the available schemata. Presuppositional questions investigate the conditions that must obtain if available schemata are to be instantiated.

In the early stages after a new schema has been introduced into consensus practice, almost all instantiations of it are important. Thus, in the early nineteenth century, chemists committed to Dalton's atomic theory regarded any problem of understanding the weight relations among compounds as significant. Because solutions to these problems required them to make judgments about the chemical formulae of compounds, the question of how to assign chemical formulae and to discriminate among proposed formulae obtained derivative significance. Similarly, in the years after the publication of the *Origin*, naturalists set to work to find convincing instantiations of Darwin's

21. My view of explanatory progress plainly has some affinity with the ideas of the logical empiricists about the unity of science. However, there are important differences— stemming from my rejection of the demand that there be accumulation *at the level of details*. For more exact discussion of the relationship, see (Kitcher 1984, 1989), which explore the question in the context of the growth of genetics.

The writings of Lakatos, Laudan, and Toulmin, all of whom hope to specify broader units (research programs, research traditions) behind individual theories are sensitive to the problems that beset the demand for accumulation at the level of details. But the emphasis on empirical prediction and the shying away from analysis of the explanations offered within a field of science at a time seem to me to interfere with the appreciation of the possibility of accumulative explanatory structure. Toulmin's conception of "ideals of explanatory order" comes closest to recognizing what I regard as the salient point: *the need to break away from concentration on accepted statements* (a feature of logical empiricism that survives in Lakatos and Laudan) *and to focus on the ways in which statements are used in answering questions.*

schemata, choosing questions for their apparent tractability and illustrative power.

But once a field has established a set of paradigm answers to application questions, further instantiations of its schemata are no longer on a par. Many questions to which an available schema could be directed are not regarded as significant because the record of success in instantiating the schema gives everyone confidence that they could (with time and effort) be answered, and the task of grinding out the details looks like hack work. The credentials of the schema are well attested by the roster of paradigm answers, and the questions that now appear significant are those that seem to involve special difficulties of producing instantiations. These questions raise the hope that when they are answered the community will obtain corrected, completed, or extended schemata. They are intrinsically significant. Other projects of application may be inspired by the needs of other fields. Questions arising in one field may be instrumentally significant because answers to them are needed for addressing some question of intrinsic significance for another field.

Intrinsic significance, I have suggested, often accrues to those questions that seem hard to answer—questions that challenge the ingenuity of a scientist (compare Kuhn 1962/1970 55; there are numerous points of contact between Kuhn's account of normal science and the proposals of this section). So, for example, the problem of "altruism" has loomed large in discussions of the evolution of behavior precisely because there were, until recently, good reasons for wondering how behavioral traits that appear to detract from the reproductive success of their bearers could be maintained in an animal population. When one of the explanatory schemata of a practice is intended to apply to phenomena, including some for which there are good reasons for believing that no instance of some premise of the schema is true, then there is a significant scientific question of showing how to instantiate the schema, how to complete it, or how to differentiate the problematic instances. Application questions can be inspired by the promise of explanatory extension as well as by the prospect of producing a more complete schema.

Presuppositional questions arise when instantiations of some accepted schema presuppose the truth of some controversial claim.[22] Darwin's selectionist schemata presupposed (as some of his critics emphasized) that the variation in natural populations could be maintained in reproduction. This lent considerable significance to the question of offering an account of variation and hereditary transmission that would show if (and why) the presupposition is true. That question was finally answered in the 1930s through the efforts of Fisher, Wright, Dobzhansky, and others.

Presuppositional questions have also inspired important research traditions in other areas of science. Consider the schema that extends MOLECULAR AFFINITIES, SHELL FILLING. That schema presupposes that it is possible for stable

---

22. There are filiations here not only to Kuhnian ideas but also to Larry Laudan's discussion of conceptual problems in his (1977).

atoms to satisfy the conditions of the Bohr model and generates the question of understanding how atoms meeting these conditions can avoid the kinds of collapse that would be expected from the principles of classical electromagnetic theory.

In these examples, the presuppositions of the schemata are problematic in that there are apparently cogent arguments from plausible premises, available to proponents of the practice, that seem to show that the presuppositions are false. What are required are a demonstration of the possibility of the problematic presuppositions and a diagnosis of the flaw in the apparently persuasive reasoning. Combining this idea with the earlier discussion of application questions, we can say that questions are intrinsically significant when (a) answers to them would exhibit the possibility of instantiating an accepted schema, or (b) would exhibit the possibility of instantiating an accepted schema in apparently problematic instances, or (c) would show the possibility of some problematic presupposition of an accepted schema. Questions are instrumentally significant when answers to them would answer some intrinsically significant question of some other field or answer some instrumentally significant question of some other field. (The last clause allows for the possibility of a chain of fields each of which turns for help to its neighbor; but, at the end of the chain, there must be some field with an intrinsically significant question.)

The account of significance I have offered is best seen as outlining the idea of an *apparently* significant question. Relative to a set of schemata certain questions should be regarded as significant because they demand instantiations for those schemata, or instantiations in apparently difficult cases, or demonstrations of the possibility of problematic presuppositions. We can say that a consensus practice is erotetically well grounded if the questions to which it assigns significance are indeed those that are significant relative to its schemata. *Genuinely* significant questions are those that are significant (in the sense I have indicated) relative to correct schemata. We make erotetic progress when we have an erotetically well-grounded consensus practice in which we pose genuinely significant questions that were not previously asked.

Scientists often describe some fields—especially those that are regarded as most exciting—by suggesting that they now know how to pose the right questions. A concept of erotetic progress should capture such descriptions. Sometimes erotetic progress can be a by-product of conceptual progress: Priestley's questions about the role of dephlogisticated air in various reactions are better formulated as questions about oxygen. On other occasions, as I have suggested in the last paragraph, incorporation of new explanatory schemata generates new genuinely significant questions: a striking example here is Darwin's initial introduction of his evolutionary schemata, which gave rise to a host of new application and presuppositional questions. However, there is a further facet of erotetic progress that has not yet been made explicit. We make progress by posing more tractable questions.

Scientific fields typically begin with big, vague questions. The introduction of explanatory schemata structures our large, cloudy wonderings, by suggesting new, more precise inquiries for us to undertake. Little of significance is achieved unless we have realistic hopes of answering these new questions. Thus, while atomists before Dalton had offered a correct schema, their explanatory pattern was so imprecise that it failed to furnish tractable questions of application. (Indeed, the history of eighteenth century efforts to develop Newton's program of dynamic corpuscularianism—outlined in the celebrated *Query* 31 to the *Opticks*—shows clearly how an explanatory advance can be too inspecific to allow for erotetic progress.[23])

Sometimes, we can make erotetic progress not only by adding significant new questions but by decomposing some of the significant questions of prior practice. Consider, for example, how the enterprise of finding instantiations of NEO-DARWINIAN SELECTION generates subsidiary questions within contemporary evolutionary biology. To complete an account of the maintenance of a trait under natural selection, one needs to be able to identify population structure, to assign selection coefficients, and to measure genetic variation. Thus, for example, questions of the form, What is the fitness of T for members of G in environment E?, become derivatively significant. Particular instances of such questions generate further investigations. If it seems that measuring the fitnesses of organisms of a particular species is tractable then scientists will bestow significance on questions about identification of certain types, assignment of organisms as offspring of others, measurements of fecundity, and so forth. Significance can ultimately accrue to quite technical, practical, and limited questions—How do you form a reliable estimate of the number of offspring produced by an ant colony?—precisely because those questions stand at the terminus of a chain of inquiries, each of which derives significance from its predecessor. In general, we can think of the significant questions of a field of science as hierarchically organized and represented by trees. At the vertex of each tree is some central question of the field, addressed by one of the most general schemata. Along each path, successive questions are generated because the provision of answers to them would aid in the resolution of a predecessor.[24]

At this stage, it is possible to broach an issue that may, quite reasonably,

23. For an illuminating discussion of the programs that descend from Newton, see (Schofield 1969). I have sketched an account of these programs in the terms of the present chapter in (Kitcher 1981 section 3). See also (Boscovich 1763/1966 and Kitcher 1986). In contemporary developmental biology, there is similar uncertainty about how to focus the big, vague question, How do organisms develop?

24. For an example of this hierarchy, see (Culp and Kitcher 1989). Obviously, there is a sense in which the accepted schemata of a practice play a similar role to the "hard core" of Lakatos or the "core assumptions" of Laudan. However, as I have already noted, one missing feature of the approaches of Lakatos and Laudan is their concentration on accepted statements rather than the ways in which accepted statements are used. A consequence of this seems to me to be that their views of science do not relate to much of the work in which scientists actually engage. They do not provide accounts that enable us to see how the very local and specific projects that occupy almost all scientists at almost all times come to be valued. Here I take Kuhn's analysis of normal science to be suggestive, and I have tried to articulate the suggestions.

appear bothersome. Close attention to my account of correctness of explanatory schemata and to the implications of (EP) will reveal that it is possible to have practices P_1, P_2 such that neither the transition from P_1 to P_2 nor the transition from P_2 to P_1 would count as explanatorily progressive. For, while P_1 may lack a correct explanatory schema present in P_2, it may also contain a correct explanatory schema absent in P_2. Nor is this merely a logical possibility. Devotees of the alleged phenomenon of "Kuhn loss" will insist that major shifts in science often involve the abandonment of explanatory insights that later developments in the field will reestablish. While Newton introduced the idea of deriving conclusions about accelerations from premises specifying the forces acting, he gave up the Aristotelian grounding of forces in the geometrical structure of space, an explanatory proposal that was later recovered by Einstein. Lavoisier's new chemistry sacrificed the possibility of accounting for the combustibility of the metals, which had been successfully undertaken by proponents of the phlogiston theory and which would be recaptured in the twentieth century accounts of chemical bonding. I shall not pursue these examples in detail here (see §9 of Chapter 5 and §9 of Chapter 7), but simply outline a solution to the problem of understanding the shifts from Aristotle to Newton and from Priestley to Lavoisier as progressive. Suppose we grant that Aristotle had a correct schema that was abandoned in Newtonian practice, and that Newton introduced a new correct schema that Aristotle had lacked (and similarly for Lavoisier and Priestley). From the perspective of (EP), then, Aristotle and Newton (Priestley and Lavoisier) cannot be ranked for progressiveness.[25] Nonetheless we may still score a considerable advance in erotetic terms by recognizing that the significant questions that are abandoned in the shift from Aristotle to Newton are intractable while that transition introduces a host of new significant questions. Indeed, we might even maintain that the possibility of decomposing the imprecise (but significant) Aristotelian questions requires the prior posing of the tractable (and also significant) Newtonian problems.[26]

The view I have sketched can easily be motivated by using the homely analogy that I introduced in dealing with explanatory progress. The task of arriving at a correct account of the structure of nature may require us sometimes to abandon primitive explanatory insights (which will much later be reinstated in articulated form), just as, in solving certain kinds of puzzles (particularly those of building three-dimensional models) pieces that actually fit together, and will eventually be rejoined, may have to be taken apart to permit the assembly of large configurations. *If* the sympathetic accounts of

25. Since I take progressiveness to be a relation between practices, and since practices are multidimensional, it is possible that rival practices could be incomparable on some dimensions but that one was superior to its rival along others.

26. Thus, if one views—as Kuhn (1970 206–207) seems to—the Aristotelian approach as containing an embryonic schema for addressing questions about forces, one that traces forces to underlying spatial structure, then it is quite pertinent to suggest that development of this insight only became possible through the detailed mathematical investigations begun by Galileo and Newton.

the achievements of Aristotle and Priestley are correct, then this approach seems to me to offer a satisfying diagnosis of what was progressive in the transitions to Newton and Lavoisier, respectively. The losses (if any) were vague insights that could not be articulated at that stage in the development of science; the gains, in both instances, were correct explanatory schemata that generated significant, *tractable*, questions, and the process of addressing those questions ultimately led to a recapturing of what was lost. Even if these transitions do not exhibit explanatory progress, they show erotetic progress.[27]

Once we have the concept of a significant question at our disposal, it is relatively simple to understand progressiveness with respect to other components of practice. Instruments and experimental techniques are valued because they enable us to answer significant questions.[28] One instrument (or technique) may do everything another does and more besides. If so, then we make *instrumental* (or *experimental*) progress by adopting a practice in which the former instrument (technique) replaces the latter. Making this conception of progress precise requires us to look more carefully at progress in the set of accepted statements: if we know what counts as improving the set of accepted statements, then we can characterize instrumental and experimental progress by recognizing the increased power of instruments and techniques to deliver improved statements.

We make progress with respect to the set of accepted statements in a number of ways. Sometimes, scientists eliminate falsehood in favor of truth, abandon the insignificant, add significant truths, or reconceptualize already accepted truths. Moreover, as I shall suggest in the next section, they often replace statements that are further from the truth with those that are closer to the truth. I shall start with the apparently naive idea that part of scientific progress consists in accepting statements that are both significant and true.

---

27. It is worth explicitly forestalling a confusion here. I claimed earlier that explanatory progressiveness was fundamental and erotetic progressiveness derivative. How then can one have erotetic progress without explanatory progress? The answer requires a distinction of levels. To make sense of the notion of erotetic progress, we need the concept of a significant question. The ultimate bestowers of significance are the correct schemata of a practice. Thus the *general notion* of erotetic progress is conceptually dependent on that of explanatory progress (more precisely on notions that figure in the analysis of explanatory progress). However, it is quite possible that, in specific cases, loss of a correct schema that cannot yet be instantiated should go hand in hand with the acquisition of the ability to pose, for the first time, questions that are significant and whose significance accrues from correct schemata (either those that are retained or some that are introduced in the new practice). Moreover, these questions may even pave the way for future refinement and instantiation of the now-discarded correct schemata. In either case, the transition will show erotetic progress without explanatory progress.

28. I owe to Gareth Matthews the observation that, while some of my relations of progressiveness concern the attaining of certain epistemic desiderata, others involve achieving the means for going further. Thus, if we are making conceptual or explanatory progress we have already gathered some good things (like the firm that is already making a profit). If we are making instrumental or erotetic progress then we have prepared ourselves well for gathering good things in the future (like the firm that has invested wisely in new ventures).

Philosophical reticence about the attainment of significant truth is the result of a failure of nerve, induced by thinking about the problem in a faulty way. Beguiled by the notion that scientific significance accrues only to systems of generalizations—theories, conceived as axiomatic deductive systems—the problem is framed first in terms of the truth of very general axioms. A pessimistic induction on the history of science (the product of reflection on the famous grand generalizations of our predecessors—notably classical physicists) instills a conviction that these axioms cannot be strictly true.[29] So the best we can hope to achieve is, apparently, scientific systems that are "close to the truth." Popperian moves to find measures of truth content become inviting at this juncture, and we are plunged into a technical morass of comparing the "sizes" of infinite sets of true and false consequences.

Tradition takes a misguided view of significance and so brings truth into the picture in far too ambitious a way, framing the crucial issues in terms of truth *for whole theories*. My approach circumvents these difficulties by offering a quite different view of scientific significance. A significant statement is a potential answer to a significant question. What we strive for, when we can get them, are *true* significant statements, that is, true answers to significant questions. Sometimes, and with differing frequencies in different sciences, those statements are universal generalizations. However, there are important fields of science in which exceptionless generalizations are not sought.

Consider some examples. "DNA molecules consist of two helical strands, wound around one another in opposite directions"—is this statement true? is it a universal generalization? what makes it significant? Molecular biologists (and many other scientists) would surely count the statement as among the most significant truths enunciated in the past half century (to be conservative about the time period). But if the statement is construed as a strictly universal generalization it is immediately clear that it is false. Not all DNAs are found in double helical form: contemporary molecular biology makes considerable use of single-stranded DNA. The existence of such molecules is not at all at odds with the intended interpretation of the original statement, which offered a *restricted* generalization about the DNA molecules typically found in cells in nature throughout most of the phases of the life of the cell. The restricting conditions are not formulated precisely, and, I suspect, nobody knows how to formulate them precisely. Nevertheless, the statement is both significant and true: significant because it answers the significant question, What is the structure of the genetic material?, true because counterinstances such as the one I have mentioned fall outside its intended scope.

Sometimes significant statements are even more obviously particular. Consider (1) "Part of Western California is at the junction of two plates that slide past one another," (2) "Gangs of male Florida scrubjays are sometimes able

---

29. I shall discuss the "pessimistic induction" in Sections 3 and 4 of Chapter 5. It is worth noting that Ian Hacking insightfully points out that our views about fluctuations in science would be rather different if we focused on *instruments* rather than on *theories* (see his 1983 55–57). See also the remarks by Mayr and Maynard Smith cited in footnote 20.

to expand the territories in which they reside." Such statements obtain their significance from the role they play in answering significant questions. (1) is an important part of an answer to the question, Why is there an earthquake zone in western California? (a question of cognitive significance for some, of practical significance for others, and of both kinds of significance for some of us!). Similarly, (2) plays a role in answering the question, Why do male Florida scrubjays often rèturn to the parental nest and assist in the feeding of siblings?—since, as detailed research on these birds makes clear, there are beneficial consequences for inclusive fitness not only in providing relief for overworked parents but also in increasing the possibility of gaining a territory by imperialist expansion (see Woolfenden and Fitzpatrick 1984).[30]

Is it reasonable to believe that all the particular statements that fill contemporary scientific journals will endure as parts of the consensus practices of future fields of science? No. But the main problem is not with truth but with significance. If you turn to the pages of *Nature* (or other scientific journals) fifty years ago, they will present conclusions that are sometimes, by our lights, oddly formulated but substantially correct—conclusions, however, that no longer seem to be significant. Many of them never became part of consensus practice (or at least part of the consensus practice of a significant subdisciplinary group). Others enjoyed a brief career in consensus before being discarded.

Why does this occur? Because the explanatory schemata of a practice generate significant *primary* questions, which spawn *derivative* questions: answering an appropriate sequence of the latter is seen as a way to address the former. Not all routes succeed. A community striving to answer Q formulates the strategy of doing so by tackling q_1, q_2, and q_3 (in that order). The initial answer to q_1 is hailed as significant, but then the endeavor becomes stuck. Ultimately another route to answering Q is found and successfully exploited, and the answer to q_1 disappears from consensus practice. Moreover, even attaining the answer to a primary significant question by solving some derivative problems may not endure as a permanent achievement. The same work may be done more decisively or more elegantly by later scientists. Consensus practice is economical. The exemplary studies of one generation can be displaced by better exemplars.[31]

---

30. Examples can be multiplied by looking at any issue of any scientific journal.

31. Thus, at a particular point in time, the hierarchy of significant question *forms* corresponds to numerous hierarchies of significant *particular* questions. Scientists often work on specific projects in the hope that the instances they address will prove the key to tackling the question forms. Although they may reap all kinds of successes, their own favored instances may vanish from the subsequent discussions of the field simply because some rival hierarchy of specific projects provides a clearer or more elegant treatment of the question form.

However, as Rob Cummins pointed out to me, it is possible that work done on one sequence of questions should fail to be adopted as the definitive treatment of the ultimate issue but should prove useful in some other area of research. So, for example, some research efforts channeled originally toward problems within physics are attributed enduring sig-

Part of the story of progress in the set of accepted statements is that it consists in eliminating falsehood in favor of truth, eliminating the merely apparently significant in terms of the genuinely significant, and using improved language to reformulate antecedently enunciated significant truths. Significant instruments and experimental techniques enable us to make progress in the statements we accept. We make progress in our methodological principles by formulating strategies that give us greater chances of making conceptual progress, explanatory progress, erotetic progress, or progress in the statements we accept.

However, this is only part of the story. It is time to fill an obvious gap.

7. Verisimilitude

The friends of verisimilitude rise up to protest: not all scientific statements are true. They are right. Moreover, we sometimes make progress by improving the statements we accept, even though we do not attain truth. For all my efforts to avoid it, are we not finally stuck with the old problem of verisimilitude?

That problem has two parts. The first part concerns the search for exceptionless generalizations. Some sciences do offer exceptionless generalizations, and truth for such statements is typically harder to come by. However, there are some circumstances under which we naturally rank later generalizations as closer to the truth than earlier ones. Consider, for example, the bare neo-Darwinian statement—popular in the early days of critical thinking about group selection—that alleles disposing their bearers to forms of behavior that would aid others at costs to themselves would be selected against. For present purposes we can focus on two complicating conditions: (i) if the allele is associated with a disposition to favor only (or primarily) kin, it may be maintained in the population through inclusive fitness effects; (ii) if there is a structured population in which groups found descendant groups at different rates, dependent on internal conditions that are fostered by the presence of altruists, then, provided that the group founding process goes forward quickly enough, the altruistic alleles can be maintained.[32] Now we can formulate three generalizations: "Altruistic alleles are always opposed by natural selection," "Altruistic alleles are always opposed by natural selection except when there is an offsetting inclusive fitness effect," "Altruistic alleles are always opposed by natural selection except when there is a sufficiently strong effect from population structure

---

nificance because of their import for resolving debates within biology and geology about the age of the Earth. Significance can sometimes be serendipitous.

32. The first type of exceptional case is, of course, treated in Hamilton's classic pair of papers on inclusive fitness (1964). The second emerges from Maynard Smith's criticisms of group selection (1964) and has been further developed by a number of writers (see Roughgarden 1979 for a survey, and D. S. Wilson 1980, 1983 for the most general treatment; Sober 1984 provides relatively nontechnical expositions of the main ideas).

and group founding." A natural evaluation of these generalizations is to declare that the first is relatively close to the truth, that the second and third are closer to the truth than the first, and that the second and third are incomparable. However, anyone who assessed both the second and third as being closer to the truth than the original generalization would be able to construct a fourth generalization—"Altruistic alleles are opposed by natural selection except when there is either an inclusive fitness effect or a population structure effect"—and this would be ranked as closer to the truth than either the second or the third.

One can generate these results by appealing to the classical (Popperian) notion of verisimilitude.[33] In general, if there is a class *A* whose members have property *B* except under rare conditions *C*, *D*, ..., then the generalization "All *A*'s are *B*" will be relatively close to the truth (exceptions will be infrequent). We will move closer to the truth by restricting the generalization to exclude one or more of the exceptional cases (*C*, *D*, etc.) and adding a true generalization about the properties of those *A*'s that are subject to the exceptional condition.[34] In cases where there are *partial* treatments of the exceptions (as with my second and third generalizations, which cope with inclusive fitness and group structure separately) it is possible to combine them to produce a more inclusive account of the exceptions and so to improve on either of the partially successful generalizations.

Some cases, then, are equally unproblematic, either on my account or on traditional approaches to verisimilitude. Other kinds of comparisons are also easily made: some generalizations are preferable to others because they embody conceptual advances or because they figure in improved explanatory schemata. The residual context in which talk of verisimilitude is attractive can be studied by confronting a problem that arises for *particular* statements.

As I have emphasized, success in achieving exceptionless generalizations is by no means a sine qua non for good science. There are successful sciences in which accepted explanatory schemata contain full sentences (degenerate cases of schematic sentences in which all the schematic letters are replaced). In the Newtonian schema for giving explanations of motions in terms of underlying forces, for example, Newton's second law occurs. There are also successful sciences in whose schemata there are no such full sentences. Darwinian evolutionary biology has served us as an example.

The difficulties of achieving truth do not only arise when we generalize, nor do they always beset our generalizations. Chemistry has arrived at enor-

---

33. The reason is quite straightforward. The conditions for Miller's argument are not met in this case because the later generalizations are restrictions of earlier ones that avoid some of the original false consequences.

34. The process of "mopping up the exceptions" in science is typically not just a matter of restricting the scope of the generalization. This is clear in my original example, in which one does not simply declare that cases in which there are inclusive fitness or group structure effects are off limits but presents an analysis of how these factors would affect the process of selection. Scope restriction is only tolerable when the abandoned instances are of no significance.

mous numbers of true generalizations about the molecular structures of particular chemical substances—starting with Cavendish's poorly formulated, but correct, account of the composition of water. Conversely, in our ascriptions of values to magnitudes, we know very well that we are likely to make mistakes. The concept of verisimilitude seems to find a natural application here.

Imagine that an early scientist refers to a physical object and assigns a real number r as the value of a magnitude of that object. Later, another scientist assigns the value r'. Let the actual value be r^*. Assume that r and r' are both different from r^*. What both scientists say is false. But we feel a strong temptation to say that the utterance of the later scientist is closer to the truth than that of the earlier scientist, a temptation grounded in the simple fact that

$$|r - r^*| > |r' - r^*|.$$

There is no need for any very complex notion of verisimilitude to do justice to cases like this. Even though we admit that both statements are false, there is an obvious respect in which the later statement is superior to the earlier one. Both statements assigned a value to a physical quantity, and one of the assigned values was closer to the actual value than was the other. Instead of trying to achieve a linguistic ordering of statements, we let the world do the work of ordering for us. The philosophical problems of understanding progress are resolved by appreciating the multidimensionality of scientific practice, and thus focusing on truth for individual significant statements. Once this is done, the artificial problems that have been at the focus of much logically ingenious work on verisimilitude can be bypassed.[35]

Philosophers concerned with the history of science may applaud my dismissal of artificial puzzles, but they are likely to offer a different type of protest. How does my account handle the really difficult cases in which rival theorists both seem to have different pieces of the truth? If I am allowed my approach to conceptual progress then the comparison of Newton and Einstein does indeed turn on a simple comparison of generalizations. But how do we compare rival approaches to the character of light? Was Fresnel's assertion that light is a wave closer to the truth than the corpuscularian claims of some of his contemporaries?[36]

My *total* account of scientific progress should be able to compare rival *practices* in optics. That does not entail that I am committed to the task of saying whether Fresnel's "Light is a wave" is closer to the truth than Brewster's "Light is a stream of particles." Recall the motivation for the project of this section. Practices have many dimensions. Along *one* dimension, we

35. I believe that the apparatus I have developed for understanding conceptual and explanatory progress enables us to tackle difficulties that emerge in nonmonotonic approaches to the truth, and to understand the approximate truth of universal generalizations. But I shall not explore the ramifications of these questions here.

36. Larry Laudan suggested to me that my approach will only handle the easy cases, and offered the example of the competing claims about light.

can ask whether the set of accepted statements of one practice (Fresnel's) is progressive with respect to the set of accepted statements of another practice (Brewster's): that inquiry might be undertaken by considering whether Fresnel replaced significant falsehoods or insignificant truths with significant truths; or, using the ideas presented in this section, we might consider whether Fresnel's magnitude ascriptions were closer to the actual values than Brewster's. The account of progress I am offering is thus keyed to investigating the *fine structure* of what Fresnel and Brewster wrote and said (and what they did—for we can compare the instruments and experimental techniques they employed) and to resisting the comparison of slogans. "Light is a wave" and "Light is a particle" are, at best, gross advertisements for the two practices, slogans to be traded in the opening moments of debate.

"Difficult cases" can easily be manufactured for an account of progress (such as mine) by pressing the concept of verisimilitude where it should not be applied, or, more generally, insisting that progressiveness be gauged along some particular dimension. My account of verimisilitude does not allow us to rank "Light is a wave" versus "Light is a particle." Nor should it, for there is no plausibility in the claim that one of these is closer to the truth than the other. But my *total* account of progress does allow us to assess the practices of Fresnel and Brewster, enabling us to see that Fresnel made conceptual and explanatory advances, that he was able to answer correctly significant questions that Brewster could not (questions about interference and diffraction, for example).[37] We can *loosely* sum up all these advances by suggesting that the wave theory propounded by Fresnel was closer to the truth than Brewster's corpuscular theory—but this is only shorthand for a complex of relations of the types I have been at pains to characterize throughout this chapter, not a claim about the restricted notion of verisimilitude that figures in *part* of the story.

In the light of this section we can refine the account of instrumental and experimental progress. Very frequently in the daily practice of a science, there is a significant question of the form, What is the value of x for S? (What is the age of that stratum? How many neurons project from this nucleus into the telencephalon?) There are instruments and techniques designed to answer classes of such questions. An instrument or technique improves on an earlier instrument or technique if the values that it generates are closer to the actual values than those given by the earlier experiment or technique. But is it not possible that an instrument or technique could yield better results in some instances but worse results in others? Of course. Geologists know very well that certain kinds of isotopes are better for dating some types of rocks, other isotopes more reliable in other cases. The experimental practice of radiometric dating uses a motley of techniques precisely because there is no all-purpose decay process that can be used on all rock samples. Geologists combine separate approaches to fashion a mixture of techniques that will apply across

---

37. For illuminating discussions of the work of Fresnel and the acceptance of his ideas about light, see (Worrall 1978, 1989).

the entire domain they hope to investigate. The claim that experimental practice is progressive rests on the idea that the mixture of instruments and techniques now employed yields in each case a value that is at least as close to the actual value as that previously generated—or, if there are exceptions, that these are either rare, insignificant, or both.

8. Two Refinements

The account offered so far is deficient in two main respects. First, as noted in Section 1, I have sidestepped problems posed by the splitting and merging of fields. Second, in the last two sections I have overemphasized the goal of attaining truth (or of improving false statements) and neglected the important role played by idealizing theories. The present section will attempt to set these defects right.

When fields of science split, merge, or hybridize, comparisons among practices cannot properly be made by considering only a single ancestor and a single descendant. Evidently, if we compare an earlier undifferentiated practice of chemistry with a later practice of inorganic chemistry, we shall expect there to be losses. The obvious remedy is to consider *total* practice at the earlier and the later times. We can construct a map of the state of science at a time, and, for each ancestral field, identify its successor fields at the later time. Then we assess progress within a field by seeing whether the combined practice of the successor fields is progressive with respect to the practice of the ancestral field. Alternatively, we can take an ancestor-descendant unit to be generated in the following way: Start with some ancestral field and identify all its successors. Now consider all those fields which are ancestral to some one of the successors already picked out. Next, take each of the ancestral fields already singled out in the unit, and include all successors. Iterate the process until no new fields are introduced. Progress is then assessed within ancestor-descendant units by comparing the combined practices of the successor fields in the unit with the combined practice of the ancestral fields.

But this does not yet address one of the most significant aspects of the splitting and merging of fields, to wit the fact that the large-scale organization of scientific activity itself reflects a conception of the order of nature and that changes in this conception may be progressive. Here (as so often) Kuhn is suggestive (1977 31–65), using the histories of the physical sciences to show how some changes are not so much modifications within existing fields as redrawings of the map of science.

We can accommodate the idea of *organizational* progress, conceived as improvement of the accepted relations among the sciences, by extending our view of consensus practice, of conceptual, explanatory, and erotetic progress. I shall suppose that the consensus practice of a field includes some claims about the relationship of the phenomena of the field to phenomena studied by other fields, and that, in the fashion discussed in Section 11 of the last chapter, we may combine these partial views to a *broad consensus* vision of the structure of nature. This broad consensus vision underlies the organization of scientific inquiry, recognizing certain problems and projects as connected with one another and granting authority to a subgroup of the community which tackles the problems and pursues the projects. Organizational changes may redraw boundaries, modify ideas about the dependencies of fields on others, or incorporate methods of one field into those of another. All these kinds of change can contribute to varieties of progress that have already been described. The redrawing of boundaries may constitute conceptual progress, revision of claims about dependency can be explanatorily progressive, and integration of concepts and methods from one discipline sometimes provides strikingly more tractable versions of old problems, thus counting as erotetic progress. Examples are ready to hand, in Maxwell's unification of the theories of electricity, magnetism, and light; in the incorporation of the theory of the chemical bond within atomic physics; and in the development of economic models in ecology.

Let us now turn to the second deficiency that I diagnosed earlier. One of the most obvious features of some sciences (notably parts of physics, but also subdisciplines of ecology and evolutionary biology) is their employment of idealizations. The generalizations of phenomenological thermodynamics, of the kinetic theory of gases, of statistical mechanics, are not, strictly speaking, true.[38] Nonetheless, it is entirely legitimate to hold that physics made progress by first achieving the phenomenological generalizations, then those of kinetic theory, then those of statistical mechanics. Using the ideas of the last section, we can recognize the progress of thermodynamics by suggesting that successive generalizations specify functions that are closer to those actually involved in the dependencies of various quantities (pressure, volume, and temperature, for example). There is nothing wrong with this—as far as it goes. However, it fails to recognize a certain aspect of the progress of thermodynamics: the goal of the enterprise is not to advance a set of generalizations that are *exactly* true.

Start with a distinction. The statements of phenomenological thermodynamics, of kinetic theory, and of statistical mechanics are rightly counted as false when we interpret the terms that occur in them as referring to actual magnitudes of actual gases or of actual molecules. However, if we conceive of the referents of those terms as fixed through stipulation—a molecule of an ideal gas is to be a Newtonian point particle that engages in perfectly elastic collisions, and so forth—then the theories in question can be seen as simply elaborating the consequences of these stipulations. To use an old (and disreputable) terminology, the statements are true by convention. As I have argued elsewhere, when Quinean morals are properly understood, there is room for truth by convention (see Quine 1966, Kitcher 1983 chapter 4). What

38. For a penetrating exploration of this claim, see (Cartwright 1983). My own discussion of idealization in what follows is brief and plainly needs supplementing by attending to the important distinctions that Cartwright makes. For present purposes, however, I am concerned to show how the elaboration of idealizations can be accommodated within my general approach to cognitive progress.

is important is that the conventions should be grounded in appreciation of aspects of reality.

In the present context, the grounding is readily specified. Thermodynamics made progress by recognizing first that there are important relationships among the temperature, volume, and pressure of actual samples of actual gases, that these relationships can be complicated in various ways, but that the behavior of actual gas samples can be predicted and explained by comparing them with entities in a story (the story of ideal gases). Later, physicists made progress by recognizing that the properties of actual gases are dependent upon the mechanical interactions among the molecules of which they are composed. They saw that the task of describing these molecular interactions would prove formidably complex if the sizes of the particles and the possibilities of inelastic collisions were taken into account. Once again, the actual behavior of gases could be understood by comparing actual gases with the characters in a story. The story preserves the new insight into explanatory dependence—the dependence of thermodynamic properties on mechanical properties of molecules—while providing a simple way of highlighting the most important features of the dependence. In the kinetic theory, we find an idealized version of part of the actual explanatory dependence; statistical mechanics provides a more complete story.

In general, I propose that we view idealizing theories as true in virtue of conventions, and that we regard the conventions (if they are valuable) as grounded in a double achievement. The first part of the achievement is to recognize a hitherto unappreciated explanatory dependence. The second part is to see that the form of explanatory dependence can be articulated in detail by forgetting about some entities or properties that complicate the actual situations. Fields in which we do not idealize are those in which we aim to develop the explanatory dependence by using statements that are strictly true. Idealization is an appropriate substitute when we appreciate that the search for exact truth would bury our insights about explanatory dependence in a mass of unmanageable complications.

Acknowledgments

Oppenheim, Paul, and Hilary Putnam. "Unity of Science as a Working Hypothesis."
In *Minnesota Studies in the Philosophy of Science*, Vol. 2, *Concepts, Theories, and
the Mind-Body Problem*. Edited by Herbert Feigel, Michael Scriven, and
Grover Maxwell (Minneapolis: University of Minnesota Press, 1958): 3–36.
Reprinted with the permission of the University of Minnesota Press.

Sklar, Lawrence. "Types of Inter-Theoretic Reduction." *British Journal for the Philosophy
of Science* 18 (1967): 109–24. Reprinted with the permission of Oxford
University Press.

Fodor, J.A. "Special Sciences (Or: The Disunity of Science as a Working Hypothesis)."
Synthese 28 (1974): 97–115. Reprinted with the permission of Kluwer
Academic Publishers.

Feyerabend, P.K. "Explanation, Reduction, and Empiricism." In *Minnesota Studies in
the Philosophy of Science*, Vol. 3, *Scientific Explanation, Space, and Time*. Edited
by Herbert Feigel and Grover Maxwell (Minneapolis: University of
Minnesota Press, 1962): 28–97. Reprinted with the permission of the
University of Minnesota Press.

Kuhn, Thomas S. "The Nature and Necessity of Scientific Revolutions." In *The
Structure of Scientific Revolutions*, 2nd ed. (Chicago: University of Chicago
Press, 1970): 92–110. Reprinted with the permission of the University of
Chicago Press.

Kuhn, Thomas S. "Revolutions as Changes of World View." In *The Structure of Scientific
Revolutions*, 2nd ed. (Chicago: University of Chicago Press, 1970): 111–135.
Reprinted with the permission of the University of Chicago Press.

Kuhn, Thomas S. "The Road Since Structure." *PSA 1990* 2 (1991): 3–13. Reprinted
with the permission of the Philosophy of Science Association.

Putnam, Hilary. "Explanation and Reference." In *Conceptual Change*, edited by Glenn
Pearce and Patrick Maynard (Dordrecht: Reidel, 1973): 199–221. Reprinted
with the permission of Kluwer Academic Publishers.

Laudan, Larry. "A Confutation of Convergent Realism." *Philosophy of Science* 48 (1981):
19–49. Reprinted with the permission of the University of Chicago Press.

Boyd, Richard. "Realism, Approximate Truth, and Philosophical Method." In *Midwest
Studies in the Philosophy of Science*, Vol. 14, edited by C.W. Savage

(Minneapolis: University of Minnesota Press, 1990): 355–91. Reprinted with the permission of the University of Minnesota Press.

Kitcher, Philip. "Varieties of Progress." In *The Advancement of Science: Science Without Legend, Objectivity Without Illusions* (New York: Oxford University Press, 1993): 90–126. Reprinted with the permission of Oxford University Press.